IMRE LAKATOS AND THEORIES OF SCIENTIFIC CHANGE

BOSTON STUDIES IN THE PHILOSOPHY OF SCIENCE

VOLUME 111

IMRE LAKATOS AND THEORIES OF SCIENTIFIC CHANGE

Edited by

KOSTAS GAVROGLU
National Technical University of Athens

YORGOS GOUDAROULIS
Aristotle University of Thessaloniki

PANTELIS NICOLACOPOULOS
National Technical University of Athens

KLUWER ACADEMIC PUBLISHERS
DORDRECHT/BOSTON/LONDON

Library of Congress Cataloging in Publication Data

Imre Lakatos and theories of scientific change.
(Boston studies in the philosophy of science;
v. lll)
Contributions from the International Conference in
Thessaloniki, Greece, in August, 1986.
Includes index.
1. Science—Philosophy—Congresses. 2. Science—
Methodology—Congresses. 3. Physics—Methodology—
Congresses. 4. Knowledge, Theory of—Congresses.
5. Lakatos, Imre. 6. Popper, Karl Raimund, Sir,
1902– . I. Nicolacopoulos, Pantelis, 1952–
II. Gavroglu, Kostas. III. Goudaroulis, Yorgos.
IV. International Conference in Epistemology (1986:
Thessalonikē, Greece) V. Series.
Q174.B67 vol. iii 001'.01 s 88-8953
[501]

ISBN 90-277-2766-X

Published by Kluwer Academic Publishers,
P. O. Box 17, 3300 AA Dordrecht, The Netherlands.

Kluwer Academic Publishers incorporates
the publishing programmes of
D. Reidel, Martinus Nijhoff, Dr W. Junk and MTP Press.

Sold and distributed in the U.S.A. and Canada
by Kluwer Academic Publishers,
101 Philip Drive, Norwell, MA 02061, U.S.A.

In all other countries, sold and distributed
by Kluwer Academic Publishers Group,
P.O. Box 322, 3300 AH Dordrecht, The Netherlands.

This book has been printed on acid-free paper.

Printed in the Netherlands

TABLE OF CONTENTS

PART VI

EDITORIAL PREFACE

How happy it is to recall Imre Lakatos. Now, fifteen years after his death, his intelligence, wit, generosity are vivid. In the Preface to the book of *Essays in Memory of Imre Lakatos* (*Boston Studies*, **39**, 1976), the editors wrote:

> . . .Lakatos was a man in search of rationality in all of its forms. He thought he had found it in the historical development of scientific knowledge, yet he also saw rationality endangered everywhere.
>
> To honor Lakatos is to honor his sharp and aggressive criticism as well as his humane warmth and his quick wit. He was a person to love and to struggle with.

The book before us carries old and new friends of that Lakatosian spirit further into the issues which he wanted to investigate. That the new friends include a dozen scientific, historical and philosophical scholars from Greece would have pleased Lakatos very much, and with an essay from China, he would have smiled all the more. But the key lies in the quality of these papers, and in the imaginative organization of the conference at Thessaloniki in summer 1986 which worked so well.

June 1988 ROBERT S. COHEN

INTRODUCTION

The stimulus for the papers in the present volume was an International Conference in Epistemology held in Thessaloniki, Greece in August 1986; its title was 'Criticism and the Growth of Knowledge: Twenty Years After', and it was dedicated to the memory of Imre Lakatos. It is already suggested in the title of the conference that among its aims was the critical appraisal of the developments in the philosophy of science in the twenty years that had passed since the very important 1965 London Conference which resulted, among other significant and far reaching accomplishments, in the volume *Criticism and the Growth of Knowledge*, edited by Imre Lakatos and Alan Musgrave. The evaluation of some of the programmatic statements articulated at the London Conference, and of Lakatos's particular contribution to the philosophy of science, were two of the focal points for examination and discussion in Thessaloniki. But in addition, a number of new theses and programmes indicative of the current progress in the philosophy and history of science as well as in the theory of knowledge (the various branches of the discipline that the term 'epistemology' may be taken to refer to) were presented, and quite a few fresh ideas were brought up and discussed.

The conference also signified the recently emerging situation of the philosophy and history of science in Greece. There is now a growing – even if only slowly growing – number of philosophers, scientists and historians who actively participate in the development of a complex discipline through work inspired by the modern, indeed the current, trends in the field. Given the opportunity of exchange and interaction with the contemporary problematique of the international community of the philosophy of science, this work would contribute to the understanding and the resolution of the issues involved. It has been argued that, slowly but surely, there is a realization on the part of philosophers in Greece that the heritage of the ancients cannot by itself sustain all that is significant and productive in philosophical thought in the current complex process of modernization of Greek society, the pace of which has been accelerated in the last fifteen years when, among other things, Greece has been faced with the task of becoming a full member of the EEC. So, as scholarship in many branches of knowledge has defined the key issues and expressed the dominant trends of the various disciplines, creating new specialities but also emphasizing interdisciplinary collaboration, the Greek university is currently establishing new units and branches, similar to the ones that were first formed and tested in some of the leading academic and scientific centers of the world in the sixties. It is hoped that the conference of Thessaloniki demonstrated the progress of epistemology in Greece and that it contributed to the further development of the field. Towards that purpose, a project to publish a volume of current studies in the philosophy and history of the sciences by Greek scholars is under way.

But back to the present volume. The papers included in it are diverse in topic and approach, reflecting the aims of the occasion that led to them, although, of course, most of them are re-worked versions of the original presentations. The diversity of topics provides the reader with a fairly thorough picture of the current interests and theories of the philosophy of science, as well as of the developments and achievements in the last two decades. The diversity of approaches, which indeed represent quite a few nations, academic centres and schools of thought spread over four continents, gives a good idea of the different evaluations of the aims, scope and progress of the discipline, but also shows that, despite the differences, there exists a community in this field. Philosophy, after all, grows through differences and arguments supporting opposite theses, and science does not stop with a discovery or a theory agreed upon.

We have grouped the papers of this volume on the basis of thematic interest and topic; thus, the volume is divided into six parts. However, we avoided giving titles to the parts lest we be unfair to some of the papers, since significant differences remain within each group. Part I consists of papers that, on the one hand, address some of the basic issues underlying this volume, especially with respect to the work of Lakatos, and that, on the other hand, reflect the different backgrounds and approaches of authors and theories stemming from a variety of countries and contexts. The first two papers, by John Watkins and Alan Musgrave, were the opening papers of the Thessaloniki Conference. Part II consists of case-studies that, to some extent, critically employ the work of Lakatos to deal with the philosophy and history of physics. Part III consists of papers aiming at a critical appraisal of some key notions of Lakatos's philosophy and at an evaluation of the role of philosophy of science. The papers in Part IV deal more specifically with the problems of the structure and incommensurability of scientific theories and of the growth and success of science. The papers in Part V have a thematic diversity, and they aim at the extension of the scope and application of the philosophy of science, or conversely at the enrichment of the philosophy of science with methods and ideas coming from other domains. Finally, the papers in Part VI deal more specifically with the work of Karl Popper.

This volume is dedicated to the memory of Imre Lakatos.

ACKNOWLEDGEMENTS

We would like to thank our colleagues at the Aristotle University of Thessaloniki and the National Technical University of Athens N. Avgelis, A. Baltas, T. Christides and A. Koutougos, with whom we collaborated in the organizing of the 1986 Thessaloniki Conference, for their cooperation through all the stages of this project. We also extend our thanks to our students Ch. Bala, M. Giamalidou, V. Kindi and C. Laurent, and to D. Koutougou and M. Mikou for their assistance. And we wish to acknowledge the generous support of the Latsis Foundation and especially thank Dr. Spiro J. Latsis.

KOSTAS GAVROGLU
YORGOS GOUDAROULIS
PANTELIS NICOLACOPOULOS

PART I

JOHN WATKINS

THE METHODOLOGY OF SCIENTIFIC RESEARCH PROGRAMMES: A RETROSPECT

By 'scientific theory' I shall mean a core of fundamental assumptions fleshed out by a suitable array of auxiliary assumptions. Let there be two or more competing scientific theories in a certain field, thrown up by rival research programmes in Lakatos's sense. Now consider these two questions:

(1) Which of these theories should I *accept*?
(2) Which of these theories should I *work on*?

They are surely very different questions. The personal pronoun 'I' could be dropped from question (1), which could be reformulated as 'Which of these theories should *anyone* accept as the *best* one, in the light of the presently available evidence?' In my (1984) I suggested that the best theory is the one that best fulfils what I depicted as the optimum aim for science; and this, I argued, will be the one that is best corroborated. Which means that we may turn question (1) into:

(1′) Which of these theories is *best corroborated*?

To answer this question one must, so to speak, press the 'Hold' button to freeze the theories and the evidence as they stand at the present time. It treats the theories in question as if they were finished products, even though some or all of them may be going to be modified later. One does not need to be a scientist working in this field in order to ask this question. It might be asked by any interested onlooker – a philosopher, say, or a historian of science (though the latter would presumably ask it of theories as they existed at some earlier time). One hopes that if this question were put to scientists working within the rival research programmes that have thrown up the theories in question, they would come up with the same answer, despite their different theoretical predilections. For to say that theory T' is better corroborated than T is to say that at least one test result is more favourable to T' than to T, while none is more favourable to T than to T'; and there should be no dispute over that, provided that the relevant test-statements are not in dispute. The French Academy judges

K. Gavroglu, Y. Goudaroulis, P. Nicolacopoulos (eds.), Imre Lakatos and Theories of Scientific Change. 3–13.

for the 1819 prize on diffraction were predominantly corpuscularians, but that did not stop them from awarding the prize to Fresnel for his wave theory, which had been corroborated in a most striking way.

Neither historians nor philosophers of science, nor interested spectators of the scientific scene, would ask themselves question (2). It would be asked only by a working scientist, and only in connection with theories within his own field. As I said, we could drop the personal pronoun from question (1), but this is not the case with question (2).

What will be the eventual aim of a scientist who decides to work on a particular theory? Unless his motto is that it is better to travel hopefully than to arrive, he will, presumably, hope to turn it into the *best* theory in its field. To put it another way, he will hope to turn this unfinished product into a finished product that will be better corroborated than its rivals. Which means that question (2) presupposes that question (1′) is answerable. But question (1′) does not presuppose that question (2) is answerable.

It may be said that, although the personal pronoun cannot be dropped from question (2), there is a related question which could be formulated in an impersonal way, namely:

(3) Which of these theories should scientists in this field *work on*?

There are many people whose main expectation of methodology is that it should answer this kind of question. For my part, I consider this an unrealistic and presumptuous expectation. Has anyone ever heard of a methodologist actually buttonholing a scientist and advising him to switch from what he is now working on to something different? Are methodologists supposed to be able to discern better than working scientists which are the more fruitful problems and lines of research in their field? But it would not be at all presumptuous of a methodologist, in a suitably clearcut situation, to say to a working scientist: 'You must agree that the theory you're working on is, at present, less well corroborated than its rival.'

But suppose, for the sake of argument, that it is the business of methodology to proffer the sort of advice called for by question (3). Doesn't this question presuppose that all scientists in a given field should work on the *same* programme, namely whichever of the currently competing programmes is the most promising. And wouldn't that be undesirable? Don't we want some theoretical competition? As Kuhn himself put it: 'If a decision must be made under circumstances in which even the most deliberate and considered judgement may be wrong, it may be vitally important that different individuals decide in different ways. How else could the group as a whole hedge its bets?' (1970, p. 186).

Alan Musgrave has put forward an interesting suggestion in this connection, namely that (the relevant sections of) the scientific community should devote *more*, or *most*, of its resources to the best programme (1975, pp. 479f.). But this suggestion faces the difficulty that scientists are not under central control. Suppose, by way of example, that methodologists agree that three times as many researchers should work on programme A as work on programme B. How would such a redistribution of labour be carried out, if the scientists in question do not work to order? Grant-awarding bodies may strongly influence the course of *experimental* work, especially where this involves expensive equipment; but we are discussing theoretical research interests. Suppose that I instruct an audience as follows: 'About three-quarters of you are to raise your right arm, the remainder your left arm, but no one is going to tell you who should do which.' And suppose, rather optimistically, that the audience is anxious to comply with this instruction. But how? So far as I can see, the only way would be for each of them to employ a randomising device. (For instance, they might spin a coin twice, and raise their right arm unless the coin came down heads both times.) But a scientist is unlikely to be willing to make a decision about which theory to work on in a random way.

I now turn to Imre Lakatos's views concerning the competence of his methodology of scientific research programmes (henceforth MSRP) to answer our different kinds of question. There are places where he stated as clearly as one could wish that its competence is restricted to something like our question (1), and does not extend to questions (2) or (3). For instance, he said that modern methodologies (and he was including his own) 'consist merely of a set of . . . rules for the *appraisal* of ready, articulated theories' (1978a, p. 103). Elsewhere he said that scientists are left free [by his methodology] to decide which programme to work on, but their 'products have to be judged. *Appraisal* does not imply *advice*' (1978b, p. 110). And there is the following longer passage, in a piece not included by the editors of his *Philosophical Papers*, which Musgrave (1975, pp. 474–75) quoted:

> my 'methodology' . . . only *appraises* fully articulated theories (or research programmes) but it presumes to give advice to the scientist neither about how to *arrive* at good theories nor even about which of two rival programmes he should work on.
>
> Whatever they [scientists] *have* done, I can judge: I can say whether they have made progress or not. But I cannot advise them – and do not wish to advise them – about . . . in which direction they should seek progress.

Although Paul Feyerabend and Alan Musgrave evaluated the view expressed in these quotations in opposite ways, they agreed about its nature. Feyerabend hailed it as an 'anarchism in disguise' (1975, p. 181), while Musgrave rather deplored the fact that Lakatos had 'gone a long way towards epistemological anarchism' (1975, p. 458). Musgrave added: 'Lakatos deprived his standards of practical force, and adopted a position of "anything goes"' (p. 478). My own view is that something did go seriously wrong in Lakatos's version of MSRP, but that Musgrave misdiagnosed what this was. One does not have to be issuing advice about what to work on for one's standards of appraisal to be demanding and to have practical force. Suppose that a prize is regularly awarded for the "best" product in a field. There is a well specified aim that the products are expected to fulfil, and the awarding body has clearcut criteria for judging, at least in a good many cases, which of them best fulfils this aim. (I say 'at least in a good many cases' to allow for ties and for the possibility that one product is both better in some ways, and worse in other ways, than another.) Then this body is equipped with demanding standards which have practical force; but it gives no advice to the producers of these products beyond the formal imperative: 'Produce something that satisfies our criteria better than any rival product'. How they go about this task it leaves to them.

I say that corroboration theory provides standards of this kind. I claim that the aim of science is to come up with theories that are deeper and more unified, and have more predictive-cum-explanatory power, than their predecessors, and which are possibly true, in the sense that there are no known inconsistencies either within the theory or between it and the available evidence. And at least in a good many cases there will be a theory, in a given field, that uniquely fulfils this aim better than any rival or predecessor. And this will be the one that is best corroborated.

Musgrave pointed out that Lakatos, in that passage where he was deprecating giving methodological advice to scientists, was reneging on earlier declarations. That passage came from a reply to criticisms of a paper in which there had occurred a passage (also quoted by Musgrave) with a very different tendency:

> ... I inject some hard Popperian elements into the appraisal of whether a programme progresses or degenerates or of whether one is overtaking another. That is, I give criteria of progress and stagnation within a programme and also rules for the 'elimination' of whole research programmes ... If a research programme progressively explains more than a rival, it 'supersedes' it, and the rival can be eliminated (or, if you wish, 'shelved') (1978a, p. 112).

But if a methodology rules that a research programme should be eliminated, it is surely advising scientists to cease working on it. This is a position from which Lakatos later backed down.

My own diagnosis of what went wrong runs as follows. The underlying mistake was Lakatos's insistence that research programmes must *replace* theories as the "unit" of appraisal: 'The basic unit of appraisal must not be an isolated theory or conjunction of theories but rather a *research programme*' (1978a, p. 110). The appraisal of such a unit leads to a decision to accept or reject it. What is it to accept or reject a research programme? Its rejection, Lakatos said, 'means *the decision to cease working on it*' (1978a, p. 70n.). I haven't found a place where he says that acceptance means the decision to start, or continue, working on it, but that is implied. So theory appraisal is replaced by programme appraisal, and programme appraisal means decisions about which programme to work on. But then, as we have seen, Lakatos got cold feet; was he, as a methodologist, entitled to advise scientists about what they should, and should not, work on? He came to believe, I think rightly, that he was not.

But MSRP would still be indicating to scientists what they should, and should not, work on *if* it could declare unequivocally that one programme has superseded another; for that would surely imply that scientists should cease working on the superseded programme. Having got cold feet about issuing such advice, Lakatos became, quite understandably, increasingly sceptical about the possibility of saying that one programme has superseded another. This change of heart has been much publicised, especially by Feyerabend, who never tired of pointing out that Lakatos's sanguine talk about his rules providing for the elimination of whole research programmes gave way to the admission that we can *never* say that a degenerating programme may not stage a comeback. He first replaced our question (1) by question (2) or (3), and subsequently shrank from answering the latter, thus ending up with no answer to any of our questions, delighting Feyerabend and saddening Musgrave.

After quoting the passage in which Lakatos said that he had no wish to advise scientists, Musgrave commented:

I find this latest position of Lakatos an extremely odd one. He develops an elaborate account of what is good science and what bad, but refuses . . . to give any advice to scientists. . . . He still claims that his rules of appraisal "explain the rationale of the acceptance of Einstein's theory over Newton's". But his rules of appraisal would equally well have explained (or rather, have failed to explain) the rationale of the acceptance of Newton's theory over Einstein's. If what I say gives a man no indication about how to act, then whatever he does is consistent with my (non-existent) advice and I can hardly claim that what I told him explains his behaviour (pp. 475–6).

I say that the question of what is good science or, more specifically, what is, at a given time, the best scientific theory in its field, should be determined by corroboration theory. This will indeed yield the appraisal that Einstein's theory is better corroborated than its rivals, including Newton's, and ought therefore to be accepted as the best theory in its field until it is superseded by an even better one. I also say that such a corroboration appraisal gives *no* indication as to what scientists in this field should be working on. That theory T' is better corroborated than T leaves it open whether they should work on T', or try to rehabilitate T, or seek a new theory T'' that will supersede both of them.

I think that Lakatos introduced an important innovation with the idea of appraising research programmes as progressing or degenerating. Popper is badly wrong when he suggests (1982, p. 32) that MSRP is only another name for what he called 'metaphysical research programmes', and that it can be dismissed as 'this latest fad' (p. 34). But I do oppose Lakatos's imperialist demand that research programmes *replace* theories as *the* units of appraisal. I will return to the question of the appraisal of research programmes shortly; but we have already seen that it is a tricky and uncertain business. If instead we seek to appraise two competing scientific theories, then I say *first* that we have two definite units before us; and *second* that the question as to which is now the better corroborated is a perfectly definite one.

Let me elaborate on the first of the above two claims, namely that a scientific theory is a definite unit. (I will touch on the definiteness of the question of corroboration later.) What constitutes a scientific theory? Given an arbitrary collection of factual propositions, could we, at least in principle, tell whether *any*, and if so, *how many*, scientific theories are contained in it? I say that we could. It is generally agreed that a genuine scientific theory is *unified* in a way that an arbitrary conjunction of propositions is not. But in what does this unification consist? In my (1984) I give rules for the *natural axiomatisation* of a scientific theory. And I say that a theory is unified if, when axiomatised in accordance with those rules, it satisfies what I call the *organic fertility requirement*; that is, for every partition of its axioms into two non-empty subsets, the union of the testable contents of the two subsets is less than the testable content of the whole axiom set. (Its testable content is understood as the totality of all the singular predictive implications, or negations of potential falsifiers, that are consequences of it.)

It might be the case that, although every proposition in our collection is falsifiable, none of them unites with any of the others into a scientific theory. But suppose that some of the propositions in our collection can be grouped together to form an axiom set that satisfies the organic fertility requirement; then we know that it contains at least *one* scientific theory. To discover whether it contains more than one, we must first check whether the addition of any of the remaining propositions to the present axiom set constitutes a more powerful system that again satisfies the organic fertility requirement. Having augmented it as far as we can without infringing that requirement, we remove from our collection any proposition entailed by this axiom set, thus augmented. We now repeat the exercise on the remaining propositions, carrying on until we find that no conjunction of the survivors, if any, satisfies this requirement. In this way we might find that our collection contains, say, three scientific theories, each of them a definite and well specified entity.

Let us now turn to the comparative appraisal of research programmes. As I see it, there are three ways in which one might proceed. (1) One might ask which of two rival programmes has performed best *so far*, excluding all consideration of their possible performance in the future. The obvious way to answer *this* question would be to take the latest theories which they have respectively thrown up and ask which is the better corroborated. *If* we exclude the question of future performance, then research programme appraisal reduces to theory appraisal. But should we altogether exclude consideration of their future performance? A *programme*, of whatever kind, is something to be worked on and, hopefully, to be *carried out*. If you are considering whether to work on it, won't you want to ask how promising it is?

(2) At the opposite extreme, one might try to guess which of these rival programmes is going to perform best in the future. But won't that involve crystal gazing? Even if programme A has been performing much worse than programme B, is that any guarantee that it will continue to do so? As we saw, Lakatos insisted that a stagnant or degenerate research programme may always 'stage a comeback' (1978a, p. 113).

(3) Peter Urbach (1978) has argued that one can assess the objective promise of competing research programmes, on the basis of presently known factors, especially the resourcefulness and efficacy of their positive heuristics, without engaging in crystal gazing; there is a middle way that

does not exclude all consideration of future performance. I accept that. But to assess two programmes in this way is to consider two arrays of branching possibilities; some of these possibilities may be very good, others not so good, and some pretty bad. To put objective probabilities on these possibilities would involve a mitigated kind of crystal-gazing. But to try to evaluate them in the absence of probabilities would be rather like trying to evaluate two lottery tickets that offer you *unspecified* chances of winning or losing various sums.

I conclude that to appraise scientific theories, as they stand at a given time, is to appraise objective structures that are actual and definite, whereas to apprase research programmes in a way that neither altogether excludes future performance nor involves crystal gazing is to appraise things that are neither actual nor definite.

Lakatos's equation of 'accept' with 'work on' had one very unfortunate consequence. Suppose that an inventor or defender of a scientific theory discovers inconsistencies in its foundations; it seems obvious that his answer to the question, 'Should I accept it as it now stands?' should be an emphatic No, and to the question 'Should I work on it?' an emphatic Yes. But if those two questions are conflated, he is obliged to answer them both in the same way: either that he should reject-it-*and-not-work-on-it*, or else that he should work-on-it-*and-accept-it*. Either he has such a horror of contradictions that he will not even work on them, or else he will work on them since he is not in the least squeamish about them. Lakatos imputed the first of these two attitudes to Popper: for Popper, 'working on [*an inconsistent system*] must be invariably regarded as irrational' (1978a, p. 147). (This means that Popper should have regarded Russell as irrational for working on the contradictions he had found in the foundations of mathematics. Popper actually held that contradictions are fertile for intellectual progress only so long as we are determined not to put up with them and *work* to remove them; 1963, pp. 316–17.) Lakatos himself was not squeamish about the contradictions; he applauded Bohr and Dirac for sailing ahead unconcernedly on inconsistent foundations. This allowed Feyerabend to congratulate him for holding that 'blatant internal inconsistencies' should not prevent us from retaining a theory that pleases us (1975, p. p183).

In both Lakatos and Feyerabend there is the curious suggestion that a proposition or theory that is rejected is thereby *eliminated* or *removed*; this suggests that a promising theory should not be rejected, even if flawed by internal inconsistencies, since it would then be lost to us. Lakatos claimed that a rule tolerating inconsistencies 'secured a sanctuary for the infinitesimal calculus hounded by Bishop Berkeley, and for naive set theory in the period of the first paradoxes' (1978a, p. 147). But there really was no danger that Berkeley's very cogent criticisms of the calculus would drive it right out of existence; it was still *there* afterwards, but now seen to be in urgent need of repair. The detection of contradictions, plus the feeling that contradictions are intolerable, led here to important progress.

But my main criticism of MSRP is that it annuls our question (1), or (1′), by first replacing it and then failing to answer its replacement of it. In the world of horse racing there are all sorts of interesting diachronic questions, to do with the horses' breeding, training, and form; and knowledgeable answers to these questions may give punters good indications as to which horses to back. But there is an independent question whose autonomy must be preserved, namely 'Which horse won?' In science, the race for corroborations has no finishing line; but we can always ask, with respect either to the present or to some earlier time, 'Which theory is in front?' This question, or to be more exact, the question 'Which theory is now (or at some specified time) the best corroborated?', is independent of all those interesting diachronic questions with which MSRP is concerned. That the autonomy of this question must be preserved is my main message.

A spokesman for MSRP may say that the question whose autonomy I wish to preserve is a vacuous one, since *no* scientific theory ever is positively corroborated. As Lakatos put it, all scientific theories are *born refuted and die refuted* (1978a, p. 5). That all theories *die* refuted is not very controversial from a Popperian point of view. To render it entirely uncontroversial, it should be worded a shade more cautiously as: in the past it has happened again and again that even the very best theories were eventually refuted, and we have no reason to suppose that this will not continue in the future. But it is also part of the Popperian view that a properly fleshed out version of a good scientific theory may enjoy a spell when it wins corroborations without yet running into refutations. The fleshed out versions of Newtonian theory that took Uranus to be the outermost planet was eventually refuted by that planet's "misbehaviour", and so in due course was its successor. But according to the first part of Lakatos's dictum, all these successive versions were already refuted when they were first put forward.

I have discussed this issue in my (1984, Section 8.6), focussing on a case cited by Lakatos and which is well documented, namely Kepler's First

Law, 'born' in 1605. It can be plausibly argued that this was not refuted until after 1684, the year in which Newton became interested in the implications of his theory for interplanetary perturbations, and asked Flamsteed to look into the case of Saturn and Jupiter when in conjunction. Lakatos suggested that it was first refuted as early as 1625. I could find no evidence for this, which in any case concedes that it was not *born* refuted.

The task of a historian of science who confined himself just to our question (1′) would be a demanding one. He would be primarily interested in what I call *strong corroborations*. To establish that an experimental result provides a strong corroboration for a theory T he would need to show the following: (1) the experiment was aimed at an empirical generalisation g that is entailed, not just by the auxiliary assumptions, but by these in conjunction with fundamental assumptions of T; and (2) g was either (a) *theoretically novel* or (b) *theoretically challenging*; and (3) g survived this test. I say that g is theoretically novel if no extant rival of T entails a counterpart of g; in that case T is breaking new ground with this g. I say that g is theoretically challenging if a counterpart g' is entailed by an extant rival, and g diverges, perhaps only very slightly, from g', and no test sufficiently stringent to be able to discriminate between g and g' had been carried out before the present test. In this latter case, for the experimental result to provide a strong corroboration for T, g' must fail the test. (I have gone into these details in order to bear out my earlier contention that the question as to which theory is best corroborated, though there may be no flip answer to it, is nevertheless a quite definite one.)

So our historian would need a very thorough understanding of the content and logical structure both of T and of its extant rivals, and of the nature and stringency both of the present experiment, and of the historical record of previous experiments in this field. If T is an important theory, and he carried out this demanding task satisfactorily, he would provide us with something of great value.

Yet something valuable would be missing from his account. If "history is about chaps", then he wouldn't give us much history; he would be dealing only with the objective results of people's theorising and experimenting, being concerned with finished products rather than with the human activities that went into their production. The creative process would get left out. It is here that MSRP, with its essentially diachronic character and its interest in the making and improving of theories, comes into its own. The story of a scientific theory *in the making* is full of human interest. I have claimed that MSRP, at least in Lakatos's version, failed in the end to answer any of our three questions. I think that this is connected with its tendency to see scientific theories as being forever in the making, and science as a venture where one prefers to travel hopefully than to arrive. But while MSRP fails to give us something we want, namely answers to question (1), it can, in its application to the historiography of science, give us something we do want and wouldn't get if historians of science confined themselves to question (1).

REFERENCES

Feyerabend, Paul, *Against Method*, London: NLB, 1975.

Kuhn, Thomas S. *The Structure of Scientific Revolutions*, Chicago: University Press, second edition, 1970.

Lakatos, Imre, *The Methodology of Scientific Research Programmes, Philosophical Papers*, vol. 1, ed. John Worrall and Gregory Currie, Cambridge: University Press, 1978a.

Lakatos, Imre, *Mathematics, Science and Epistemology, Philosophical Papers*, vol. 2, ed. John Worrall and Gregory Currie, Cambridge: University Press, 1978b.

Musgrave, Alan, 'Method or Madness?' in Cohen, Robert S., Feyerabend, Paul K., and Wartofsky, Marx W. (eds.), *Essays in Memory of Imre Lakatos*, Dordrecht: Reidel, 1975, pp. 457–491.

Popper, Karl R., *Conjectures and Refutations*, London: Routledge and Kegan Paul, 1963.

Popper, Karl R. *Quantum Theory and the Schism in Physics*, W. W. Bartley, III (ed.), London: Hutchinson, 1982.

Urbach, Peter, 'The Objective Promise of a Research Programme' in Radnitzky, Gerard and Andersson, Gunnar (eds.), *Progress and Rationality of Science*, Dordrecht: Reidel, 1978.

Watkins, John, *Science and Scepticism*, Princeton: University Press and London: Hutchinson, 1984.

The London School of Economics

ALAN MUSGRAVE

DEDUCTIVE HEURISTICS

Is there a logic of scientific discovery? An affirmative answer to this question is a thread which ran through all of Imre Lakatos's writings. In this Lakatos swam against the tide of logical positivist and Popperian orthodoxy. According to that orthodoxy, the 'context of discovery' is the province of empirical psychology; it is only the 'context of justification' which is the province of logic. Moreover, psychological facts about the way in which a theory was discovered have no bearing upon the logical or epistemological question of whether it is a justified theory. Now Lakatos and his followers were not, of course, the first to swim against the tide of orthodoxy. But they were, I shall argue, the first to make much headway against it.

1. LOGIC OF DISCOVERY: INDUCTIVE OR DEDUCTIVE?

Down the ages logic of discovery has had many friends. After all, science is a reasonable business. Scientists do not come up with new hypotheses at random or through mystical flashes of intuition or (as it is currently trendy to suppose) through the operation of social forces of which they are unaware. No, scientists *reason* or *argue* on new hypotheses. To think otherwise is to consign the most important scientific work to the limbo of the irrational.

Next, doubt sets in. Discovery is, by definition, coming up with something *new*. But the conclusion of a valid argument cannot be new, cannot contain any information not already contained in its premises. No new hypothesis could be reached as the conclusion of a valid argument. So there cannot be a logic of discovery.

But this casts doubt only on the possibility of *deductive* arguments to new hypotheses. Most friends of the logic of discovery actually accept the argument, and conclude that the logic of discovery must be a content-increasing or ampliative or non-deductive or inductive logic. Uncommitted bystanders accept the argument too, and conclude, more cautiously, that *if* there is a logic of discovery, *then* it must be a content-increasing or ampliative or non-deductive or inductive logic. (Henceforth I shall just use the term 'inductive' for all this, as is customary.)

K. Gavroglu, Y. Goudaroulis, P. Nicolacopoulos (eds.), Imre Lakatos and Theories of Scientific Change. 15–32. © 1989 by Kluwer Academic Publishers.

This conditional proposition has been a virtually unchallenged presupposition of the entire debate. Friends of the logic of discovery accept it, and become friends of inductive logic also. Enemies of inductive logic accept it, and become enemies of logic of discovery also. Lakatos and his followers (chiefly Elie Zahar and John Worrall) reject the conditional proposition. They think that when scientists argue to new hypotheses, as they typically do, their arguments are *deductive* ones. They seek to retrieve logic of discovery from the inductivists, and show that deductivists can have it too. (Deductivism is the view that deductive logic is the only logic that we have or need.) Hence my title: *deductive heuristics*.

Being a deductivist myself, my sympathies lie entirely with the Lakatosians here. But there is something odd about any confident statement that some argument simply is, or is not, a deductive one. Arguers seldom state all their premises. An argument which is an invalid argument as stated can be 'reconstructed' as a deductively valid argument with unstated or suppressed premises. If we are not fussy about what we will allow as a suppressed premise, logical tricks can make the point quite generally. Any invalid argument from premise P to conclusion C can be 'reconstructed' as a deductively valid argument by supplying the suppressed premise 'If P, then C'. Because of such ubiquitous tricks, deductivism risks triviality: it would be trivial to turn inductive arguments in the context of discovery into deductive ones by resorting to tricks such as this.

Inductivists too must beware of triviality. Any deductive argument can be 'reconstructed' (perhaps 'deconstructed' would be a better word) as a non-deductive argument by recasting some premise as a so-called *material rule of inductive inference*. Even a humble syllogism of the form "All A are B, x is an A, therefore x is a B" can be turned into an inductive argument by recasting its major premise as the material rule "From a premise of the form 'x is an A' infer a conclusion of the form 'x is a B'". Such a deconstruction might be thought more plausible if, as is likely, the syllogiser did not state that major premise.

The deductivist ploy regarding an invalid deduction he wishes to appropriate is to view it as an *enthymeme* and search for its missing premise. The inductivist ploy regarding a valid deduction he wishes to appropriate is to view some deductively necessary premise as a material rule of inference. And both ploys risk being applied trivially.

Does it matter which way we jump here? Does it matter whether we opt for the deductivist ploy and have only deductive arguments, or for the inductivist ploy and have inductive arguments too? Are these not equivalent and equally good reconstructions of our arguings?

Well, it matters to anyone who thinks there is a difference between matters of fact and matters of logic. "All ravens are black" states a matter of fact: whether or not it is true depends upon the way the world happens to be. If we make a material rule of inference out of it, then whether the inferences it 'licences' are valid will depend upon the way the world happens to be. Anyone who thinks that logic is not an empirical science will have no truck with material rules of inference.

That was just a bit of positivist orthodoxy. Yet strangely enough, another bit of positivist orthodoxy has been the chief inspiration for the inductivist ploy, material rules of inference, and inductive logic in general. I refer to the verifiability theory of meaning, according to which the only genuine statements of matters of fact are the *verifiable* ones. Given this theory, even humdrum empirical generalisations become problematic; scientific theories become more problematic; and metaphysical principles even more problematic still. But we can save all from the rubbish bin by perpetrating the inductivist ploy on a massive scale and recasting them all as material rules of inductive inference. This leads to a position I have called *Wittgensteinian Instrumentalism*, defended by Schlick, Wittgenstein, Ramsey, Ryle, Watson, Toulmin, Harré and Hanson (see my (1980)). Its ultimate provenance is, as I say, the verifiability theory of meaning, though some of its later defenders seem unaware of the fact. I shall not criticise it here, save to mention a family quarrel amongst its adherents. This concerns 'how far down' the theoretical hierarchy the inductivist ploy is appropriate. Some apply it to the most humdrum empirical generalisations – others reserve it for more theoretical 'laws of nature' as well as metaphysical principles. The latter have a problem: they have to draw a line below which they think the inductivist ploy is inappropriate and defend that line by philosophical argument. The Wittgensteinians have not seen the problem, let alone solved it.

Returning to logic of discovery, I think the situation with arguments to new hypotheses is this. Scientists (like everyone else) seldom explicitly state all the premises of such arguments. Deductivists and inductivists alike have to reconstruct those arguments from sometimes scanty materials presented to them. What can be shown is that such arguments are *as plausibly or more plausibly reconstructed as deductive ones*. To show this we must of necessity consider particular cases.

Such cases will reveal that among the premises of arguments to new hypotheses are *heuristic principles*. Some of these are general metaphysical principles characteristic of many different and competing research programmes in science. Others are more specific principles characteristic of particular research programmes. Those still in the grip of positivist orthodoxy will find the idea that scientists deduce hypotheses from (partly) metaphysical premises abhorrent. They go in for the inductivist ploy and disguise some of these principles as rules of inductive inference. But they have the same problem as the Wittgensteinians – and like them, they have not seen it let alone solved it.

2. EARLY ATTEMPTS, WHY THEY FAILED, AND POSITIVE ORTHODOXY

Before considering any example, let me briefly make good a claim I made earlier: that the Lakatosians, though not the first swimmers against the tide of positivist orthodoxy, were the first to make such headway. This is because earlier partisans of the logic of discovery succeeded only in producing patterns of argument which really belong in the context of justification.

For example, both Bacon and Descartes complained that the logic then available (Aristotelian syllogistic) did not "help us in finding out new sciences". But what did they put in its place? Bacon gave us so-called *eliminative induction*, which is actually a form of deduction, an extended disjunctive syllogism. Eliminative induction does not discover any new hypothesis: all the possible hypotheses on the matter in hand are listed in the major premise; what the argument does is show one of them to be true by eliminating all the others. Bacon's proposal really belongs in the context of justification.

As for Descartes, he gave us the handy hints for problem solvers of the *Regulae*, and a few secretive references to the method of *analysis and synthesis* of the Greek geometers. The latter too belongs in the context of justification, since it begins with a hypothesis, analyses it into 'parts' known to be true, and then synthesises it out of those parts thereby proving it.

Jumping to the nineteenth century, we find John Stuart Mill denying that deductions are inferences *because* they are not content-increasing. What does he give us instead? He gives us singular predictive inference, hardly a method of coming up with new hypotheses. And he give us what have come to be called *Mill's Methods*, which are also a means of finding out which of a number of previously-invented hypotheses is the true one.

The next abortive attempt was Peirce's *abduction*, which goes like this:

> The surprising fact, C, is observed. But if A were true, C would be a matter of course. Hence there is reason to suspect that A is true. (Peirce (1931–58), 5.189).

Hanson is regarded as the twentieth-century's most important champion of logic of discovery because he proposed various inelegant variations on this Peircean scheme. But neither the original scheme, nor Hanson's inelegant variations upon it, belong in the context of discovery. For the hypothesis A figures in the premises and nothing is said about how it was discovered. And the fact that the conclusion is couched in terms of having reason to suspect that A is true clearly locates the scheme in the context of justification.

The same goes for the chief intellectual descendant of Peirce's abduction, misleadingly called *inference TO the best explanation*. Here too the best explanatory hypothesis figures in the premises, and we are told nothing about how it was discovered.

(By the way, abduction and inference to the best explanation, as well as being wrongly located in the context of discovery, are also wrongly regarded as species of inductive reasoning. They are better regarded, I think, as *deductive* arguments in the context of justification. But I will not digress to argue the point.)

We seem to have drawn a complete blank! All these attempts to devise an ampliative logic of discovery issued, not in patterns of argument leading to new hypotheses, but in patterns of argument leading to the justification of hypotheses already proposed. Perhaps the logical positivists and Popper were right: perhaps there just *is* no logic of discovery.

Now when many acute (and some not so acute) thinkers fall into what looks like a blatant confusion, you wonder whether there is not some obvious explanation. In this case there is. 'Discover' is a success-word. No-one can discover that the moon is made of green cheese, since it isn't. To discover that P one must come up with the hypothesis that P and P must be true. So our authors might protest that they were contributing to the logic of discovery after all, but to a second stage of it in which some previously invented hypothesis is shown to be true. Several readers of Popper's *Logic of Scientific Discovery* have been astonished by his denial that there is any such thing. But what Popper says is that "The initial stage, the act of conceiving or inventing a theory, seems to me to neither call for logical analysis nor to be susceptible of it" (Popper (1959), p. 31). This is quite consistent with the idea that Popper's logic of empirical testing captures the second stage of scientific discovery.

Because 'discover' is a success-word, it is odd to speak of discovering *hypotheses*. It would be better to speak, not of the context of discovery, but of the *context of invention*. Then we will be able to separate the question of inventing a hypothesis from the question of justifying it. But the phrase 'context of justification' is not a happy phrase either. Some philosophers hold that while scientists rationally evaluate or appraise hypotheses, they can never really justify them. So as not to beg the question against such philosophers, it would be better to speak of the *context of appraisal*. These terminological suggestions, which I accept, are due to Robert McLaughlin (see his (1982), p. 71).

We can now restate positivist and Popperian orthodoxy. The context of invention is to be distinguished from the context of appraisal. The former is the province of empirical psychology, the latter the province of logic. There is a psychology but no logic of invention – and there is a logic but no psychology of appraisal. Psychological facts about how a hypothesis was invented are irrelevant to the logical task of appraising it. Is the orthodoxy correct?

Of course it isn't. For a start, there is a psychology as well as a logic of appraisal. People actually appraise hypotheses in all kinds of ways, and if we simply describe what they do we do not leave the realm of psychology (or perhaps social psychology). The logic of appraisal has a different task, to specify how hypotheses *ought* to be appraised, to say when, for example, a piece of evidence is *good* evidence for a hypothesis. Popper remarked, while defending the orthodoxy, that the confusion of psychological and logical problems "spells trouble not only for the logic of knowledge but for its psychology as well" (Popper (1959), p. 30). Quite so. But if we deny that there is a psychology as well as a logic of appraisal, how will we distinguish good appraisals from bad ones? Many people appraise scientific hypotheses by asking how well they conform to their literal interpretation of the scriptures. Is this part of the logic of appraisal? Or suppose we find that most people in the casinos of Las Vegas systematically commit the Gambler's Fallacy. Is this part of the logic of appraisal too – at least, part of the logic which reigns in Las Vegas casinos?

But our concern is the logic of invention. Is there a logic as well as a psychology of invention? The editor of a recent collection of articles devoted to the context of invention (discovery) says that "few or none of the friends of discovery are committed to the existence of a logic of discovery (in the strict sense of 'logic')" and thinks it "misleading to retain the term 'logic' of discovery except for special purposes" (Thomas Nickles (1980), p. 7). If the thought is that there is no *special* logic of invention, then I agree. If the thought is that the invention of hypotheses is never logical, then I do not. Scientists typically argue to new hypotheses. Such *inventive arguments* can be subjected to logical analysis like any others. Like the Lakatosians, I think that inventive arguments are best reconstructed and analysed as deductive arguments. There is a logic as well as a psychology of invention, and it is deductive logic. Let me substantiate these claims with a few examples.

3. EXAMPLES, AND THE SUPERIORITY OF THEIR DEDUCTIVIST RECONSTRUCTIONS

Actually, I have time to consider only two examples in any detail. Even worse, my first example is not a real one, but a schematic one, and a terribly simple one to boot. Suppose a scientist seeks a hypothesis about the relationship between two measurable quantities P and Q. He has the hunch, derived from general considerations of the simplicity of nature, that the relationship might be linear. The scientist does not start dreaming up linear hypotheses and putting them to the test. Instead, he makes measurements, to fix the values of the free parameters a and b in the equation $P = aQ + b$. Suppose he finds that when Q is 0, P is 3, so that b is 3 also – and that when Q is 1, P is 5, so that a is 2. He then *deduces* that the linear hypothesis he seeks is $P = 2Q + 3$. Such is the tacit reasoning employed by any scientist who draws a straight line through a couple of data points.

We can set out the deduction pedantically as follows:

Heuristic principle: Nature being simple, P and Q are linearly related: $P = aQ + b$ for some values of a and b.
Experimental fact 1: When Q is 0, P is 3.
Experimental fact 2: When Q is 1, P is 5.
Hypothesis: $P = 2Q + 3$.

The example is extremely simple. Yet it reveals the true nature of so-called 'deduction from the phenomena'. Ever since Newton scientists have been saying that they deduce theories from phenomena. They are only half right. They are right to say that they perform deductions. They are wrong to imply that their premises consist *solely* of phenomena. Among the premises are general heuristic principles of one kind or another. But because these are typically left unstated, or disguised as rules of inference, it may seem that a deduction from phenomena alone has occurred.

It was Newton who first talked of deductions from phenomena – and it was Newton who began the tradition of disguising metaphysical heuristic principles as rules of inference. He first called his metaphysical principles 'hypotheses'. But then, anxious to make it seem that there was nothing hypothetical in his work, he renamed them 'Rules of Reasoning in Philosophy'. The first two rules are (Newton (1934), vol. II, p. 398):

We are to admit no more causes of natural things than are both true and sufficient to explain their appearances.

To this purpose the philosophers say that Nature does nothing in vain, and more is vain when less will serve; for Nature is pleased with simplicity, and affects not the pomp of superfluous causes.

Therefore to the same natural effects we must, as far as possible, assign the same causes.

We shall encounter this metaphysic of the causal simplicity of Nature again.

My second example of an inventive argument is one of a number that have been reconstructed by Robert McLaughlin. McLaughlin reconstructs them as inductive arguments, specifically as *arguments by analogy*. I think they are better reconstructed as deductive arguments.

Rutherford noted from experiments that atoms were both 'dense and diffuse': most particles passed straight through them, but a few collided violently with them. In this respect atoms were analogous to the Solar System. Now the Solar System has a massive but relatively small central nucleus (the Sun), orbited by much smaller bodies (the planets). So Rutherford inferred that atoms have a similar structure.

Thus McLaughlin. He insists that the argument is *inductive* and sets out its general form as follows (McLaughlin (1982), p. 90):

(Q_1) Identifies or locates the problem phenomenon/situation, often by high-lighting some significant or striking feature of it.
(Q_2) Statement of a particular analogy between the striking feature of the problem phenomenon and some property of a model.
(Q_3) Description of some other interesting feature of the model.

(H) Ascription of the latter feature to the problem phenomenon also.

McLaughlin claims that arguments of this form are legitimated by a rule of inductive inference, the *rule of analogy* (McLaughlin (1982), p. 92):

(RA) From a similarity (analogy) between two phenomena in a significant respect, infer a similarity between them in some other respect pertinent to the problem phenomenon.

My objection to this account begins from the vague phrases which it contains: "some other *interesting* feature of the model", "in a *significant* respect", "some other respect *pertinent* to the problem phenomenon". Why these qualifications? What do they mean?

The reason for these vague qualifications is obvious enough. It is well-known that any analogy holds only in certain respects and not in others. The *unrestricted* use of analogical reasoning leads to absurd results, and the qualifying phrases are meant to block those results. For example, it is an interesting feature of the Solar System (arguably its *most* interesting feature) that it is inhabited by human beings. Why did Rutherford not draw attention to this interesting feature of the Solar System, and infer by analogy that atoms too are inhabited by human beings? Because, McLaughlin will say, this interesting feature of the Solar System was not "pertinent to Rutherford's problem phenomenon". But if we press further and ask why *this* is so, we will be led to a better reconstruction of Rutherford's argument. The structure of the Solar System was "pertinent", and its being inhabited by human beings not, because the former *explains* or *causes* its "dense and diffuse" behaviour with respect to bodies entering it. If Rutherford supposed (following Newton) that "to the same (or similar) natural effects we must assign the same (or similar) causes", then his argument can be reconstructed as follows:

(EC) The same (similar) effects have the same (similar) causes.
(O_1) Atoms and the Solar System behave in the same "dense and diffuse" way with respect to bodies entering them.
(O_2) The "dense and diffuse" behaviour of the Solar System is explained (caused) by its structure, a relatively small and massive nucleus (the Sun) orbited by much lighter bodies (the planets).

(H) Atoms are structurally similar to the Solar System, with a relatively small massive nucleus orbited by much lighter particles.

Notice that, reconstructed in this way, Rutherford's inventive argument is a straightforward *deduction*.

McLaughlin reconstructs several other historical examples as 'analogical inventive arguments'. Through lack of time I will not consider any of them in detail, but will merely say that similar considerations apply to them. They are all better reconstructed as deductions from principles of the *causal simplicity of nature*: the "Same effects, same causes" principle just mentioned, and its twin, a "Same causes, same effects" principle. He reconstructs several other cases as inductions which proceed according to a *rule of simplicity*: "Choose the simplest available admissable hypothesis". These are equally well reconstructed as deductions from the vague metaphysical principle "Nature is simple". In both cases what *counts* as 'simple' is heavily context-dependent.

McLaughlin (and others) point out that vague and general appeals to simplicity have often been turned into more precise and more specific metaphysical principles of theory-construction. Lakatos and his followers add that some of these principles are characteristic, not of the whole of science, but of specific research programmes in the history of science. Two of McLaughlin's examples are "There are no privileged directions in space" and "There are no privileged locations in space", both of which characterised the Newtonian programme and were rejected in the earlier physics of Aristotle.

Here a curious asymmetry enters McLaughlin's account. On the one hand he insists that inventive arguments simply *are* inductive ones. On the other hand he insists that particular simplicity and other heuristic principles simply *are* premises of inventive arguments rather than rules of inference governing them. So we have rules of inductive inference in some cases and metaphysical premises in others, and no well-motivated reason for the distinction. (The motivation cannot be in terms of *generality*, for one of the metaphysical *premises* is Ockham's Razor and you cannot get more general than that!)

So much for my two very simple examples. The literature already contains deductivist reconstructions of quite a number of much more interesting examples from the history of science. I especially have in mind those contained in several pioneering papers by Jon Dorling and by Elie Zahar (see Dorling (1971), (1973a) and (1973b), and Zahar (1973) and (1983)). Instead of considering those in any detail, I turn to consider some general objections to the whole approach.

4. SOME OBJECTIONS CONSIDERED

The first objection was given at the outset, when we had an argument to the effect that inventive arguments cannot be deductions because the conclusions of a (valid) deduction cannot contain anything *new*. My answer is that the conclusion of a valid deduction cannot be *logically* new (which is a fancy way of saying that it is 'contained in' the premises), but it may well be *psychologically* new. (When Wittgenstein said that in logic there are no surprises he was wrong: Thomas Hobbes was actually amazed that Pythagoras's Theorem could be deduced from Euclid's axioms!) More important, the conclusion of a valid deduction can be 'new' in the sense that it possesses interesting logical properties not possessed by any of the premises. Even in our trivial schematic case, the linear hypothesis in the conclusion is falsifiable by any further measurement of values of P and Q, while none of the premises has this property.

A second objection is that deductivist recontructions of inventive arguments, though possible, are not *necessary*. I concede that they are not necessary. But I claim, to repeat, that they are clearer, that they do not mix up matters of metaphysical fact with matters of logic, and that they do not land you with the problem of saying when you have metaphysical *premises* and when you resort to the inductivist ploy and disguise your metaphysics as 'rules of inductive inference'.

A third objection is that deductivist reconstructions take all the *inventiveness* out of inventive arguments and make theory-generation a matter of dull routine. Now this is an objection to any attempt to show that scientists *argue* to new hypotheses; it is not a specific objection to deductivist reconstructions of those arguments. Besides, it is mistaken. Inventiveness consists more in assembling the premises of the arguments than in deriving their conclusions – not that the latter is, in interesting cases, a trivial task.

A fourth objection is that the so-called general heuristic or metaphysical principles which figure among the premises are often *false*. It is false that similar effects always have similar causes, for example. Moreover, the falsity of these principles must have been known to the scientist in question. Is it plausible to suppose that scientists argue from premises which they know to be false?

Again, this objection does not specifically hit the deductivist. It does not help to disguise metaphysical principles known to be false as rules of inductive inference known not to be universally applicable. We could equally well ask whether it is plausible to suppose that scientists argue according to such ampliative rules. Second, the premises of inventive arguments are not *always* known to be false. Third, even when they are, this is not decisive because inventive arguments are not meant to *justify* their conclusions. If they were, the known falsity of a key premise would be fatal. But inventive arguments issue in *hypotheses*, to be subjected to subsequent tests and other forms of appraisal. There is nothing wrong with getting hypotheses from premises which are known not to be generally true but which have sometimes led to worthwhile results.

This response leads straight to a fifth objection, and to a restatement of the 'hard core' of positivist orthodoxy. To say that a valid inventive argument issues only in a hypothesis to be subjected to subsequent appraisal seems to concede that the positivists were *basically* right: the context of invention (wrongly described as the province of mere psychology, but no matter) is *irrelevant* to the context of appraisal (wrongly described as the province of logic, but no matter). Is this correct?

If it is correct, it marks a great retreat from earlier views. In the 17th and 18th centuries the dream was that a logic of discovery would simultaneously generate new theories *and* justify them. (This was the dream: the reality was that the proposed patterns of argument belonged entirely in the context of appraisal.) In the 19th and 20th centuries consensus emerged that to study the invention of theories contributed nothing to the solution of epistemological problems about their well-foundedness. Larry Laudan concludes his discussion of all this with a positivist challenge (Laudan (1980), p. 182):

> If this essay provides a partial answer to the question "Why was the logic of discovery abandoned?", it poses afresh the challenge: "Why should the logic of discovery be revived?"

Laudan's challenge can be met. That an inventive argument to a hypothesis *might* be relevant to its appraisal is really quite obvious. If all its premises are known truths, then so is its conclusion. Such was the utopian dream of early advocates of the logic of discovery. Again, if all its premises are rationally credible, then so is its conclusion. This too is a little utopian, in view of the fact that inventive arguments will sometimes contain premises that are known to be false. Finally, if the premises contain principles which have often led to successful conclusions in the past, but which are not generally true, then the conclusion might be described as 'plausible' or 'worthy of further investigation'. But to say this of a hypothesis is to make a *minimal appraisal* of it.

Perhaps such minimal appraisals are the most we can get out of many inventive arguments. Yet they are not insignificant ones. Philosophers of science tediously point out that there are always infinitely many possible hypotheses consistent with any finite body of data. Infinitely many curves can be drawn through any finite number of data-points. We observe some green emeralds and hypothesise that all emeralds are green – why not hypothesise that emeralds are grue or grack or grurple? Quite so. Scientists, however, are unimpressed with the philosopher's point, and would not even *consider* the gruesome hypotheses which philosophers produce in support of the point.

One philosophical question is: What *enables* scientists to narrow their intellectual horizons in this way? The answer, in a word, is *metaphysics*. The answer, in a sentence, is that general metaphysical principles which play a role in the context of invention narrow down the *kinds* of hypotheses which scientists propose and take seriously.

A second philosophical question is: Is it *reasonable* for scientists to narrow their intellectual horizons in this way? Positivists think not: metaphysics being the realm of whim and arbitrariness, any influence of metaphysics on science is a bad thing. But metaphysical principles which play a role in scientific theory-construction are not matters of arbitrary whim: they can be rationally assessed 'at first remove', so to speak. Roughly speaking, if theories constructed using one set of metaphysical principles are successful, while theories constructed using another set are not, then we have good reason to prefer the first set of metaphysical principles. The success of Newtonian theories and the failure of Aristotelian ones gives us good reason to think that Newton was right and Aristotle wrong about the metaphysical question of whether there are privileged directions and positions in space. Nobody now seeks Cartesian mechanical explanations of anything but the workings of (mechanical) clocks – and *rightly* so. I suppose that there is a possible world lacking natural kinds, where gruesome hypotheses would be the order of the day – but we have good reason to think that our world is not like that. We may also have good reason to think that Nature is simple, in some carefully specified sense or senses of the term 'simple'. Metaphysical principles used in theory-contruction can be rationally assessed, and therefore their influence upon science is not a bad thing, as positivists maintain. (Though having said that, we might give the positivists their due by adding that the trouble with metaphysical principles like "The Nothing nothings" or "History is the unfolding of the Absolute

Spirit" is that they do not impinge upon science, commonsense, or anything else.)

So my first answer to Laudan's positivist challenge is this. The way in which a hypothesis is invented can be relevant to its appraisal: an inventive argument can render a hypothesis plausible; and this minimal appraisal solves the problem of which hypotheses to take seriously out of an always infinite number of possible hypotheses.

My second answer to Laudan concerns the matter of *novel facts*. Many scientists and philosophers of science attach special epistemic significance, in the context of appraisal, to the prediction of novel facts. They say that the best sort of evidential support (some say the only sort) derives from such predictions. But what exactly *is* a novel fact for a theory?

This is not the place, and I have not the time, to canvas all the different answers that have been given to this question. The answer I do want to discuss is the so-called *heuristic view of novelty* defended by Elie Zahar and John Worrall. Some early formulations of this view were not happy ones: it was said that a fact is novel for a theory if the theory was not "*specifically designed to deal with*" it, if "*it did not belong to the problem-situation which governed the construction of the hypothesis*" (Zahar (1973), p. 103). I and others objected that this was too psychologistic a notion: it made the evidential support of a theory depend upon who designed that theory and what facts he was trying to deal with or explain (Musgrave (1974), p. 13). In response to this criticism the preferred locution changed: a fact is novel for a theory if it was not "*used in the construction of the theory*" (Worrall (1985), p. 318). It was argued that whether a fact is novel in this sense is an objective logical affair, and that a theorist may be trying to explain a fact but not use that fact to construct his theory.

But when is a fact "used in the construction of" a theory? The idea that inventive arguments are deductive arguments yields a simple answer to this question: the facts used to construct a theory are *those stated in the premises of the inventive argument leading to that theory*. And the basic intuition of the heuristic view becomes plain enough: if a theory has been deduced from some phenomena, then those phenomena cannot also support it. Reverting to our crude schematic example, the linear hypothesis $P = 2Q + 3$ certainly entails (predicts) that when Q has the value zero P has the value three. But this experimental fact does not support the theory, since it figured in the premises from which the linear hypothesis was deduced.

I share this basic intuition and happily subscribe to the heuristic view of novelty thus construed. But I wonder whether, thus construed, it can be the whole story about the evidential significance of novel facts. It is one thing to say that any fact used to construct a theory is not a novel fact and has no evidential value. It is quite another thing to say that any fact *not* used to construct a theory *is* a novel fact and has special evidential value. The latter gives no special significance to the kind of novel prediction which led to most of the fuss in the first place. These are predictions of effects which have not merely never been observed but which have also not been predicted by any extant competing theory. (Extant theories may predict that they will *not* occur, or be silent on the matter.) Now given our interest in getting theories that are predictively *better* than the ones we already have, crucial experiments to test such predictions do have special epistemic significance. The rhetoric of Lakatos and some of his followers say otherwise – and their heuristic view of novelty gives such predictions no special significance.

But to return to Laudan's positivist challenge. My second answer to it is that novel facts have special importance in the context of appraisal, and that the way in which a theory was invented can tell us that some facts are not novel for that theory. Hence the context of invention is relevant in this second way to the context of appraisal.

Let me now consider some further objections, both to the idea of deductive inventive arguments and to the heuristic view of novelty which depends upon it.

First, even though a scientist did deduce a theory from phenomena, we may not know *how* it was deduced and what phenomena figured in the premises – especially if the scientist has tried to suppress the heuristic route to the theory. Thus, on the heuristic view of novelty, we will not know which facts are novel facts for that theory. This objection does not seem to me to be fatal. We can always try to reconstruct the heuristic route to the theory. If a parameter in the theory is given some precise value for no theoretical reason, and if that value *might* have been read off from an experimental result, then it is natural to suppose that it *was* read off from an experimental result. That reconstructions arrived at by considerations such as these are hypothetical seems no more damaging than the fact that science itself is hypothetical.

Second, why assume that scientists always *do* argue to new hypotheses? It is certainly *possible* that a scientist should simply dream up a new hypothesis, conjure it out of nothing, perhaps have it come to him in a dream (as in the famous story about Kekulé). What, for example, of *chance discoveries*, examples of which are legion in the history of science? Indeed, does not the heuristic view of novelty put a premium upon generating hypotheses in this way? According to that view, it is better for scientists *not* to deduce theories from phenomena. Theoreticians are best advised to remain ignorant of the literature, lie back on their couches, and simply *dream up* hypotheses more or less at random. For every fact entailed by a theory arrived at in this way will be a novel fact!

To this I can only reply that *science is not like that*. I suppose that it is logically possible that I, ignorant of the relevant literature, should come up with a brilliant new hypothesis about the neurological kindling effect. Just as it is logically possible, I suppose, that a chimpanzee randomly bashing at a typewriter should type out a Shakespeare sonnet. Both are logically possible – neither is very likely. Theories conjured up out of nothing are more likely to be wildly implausible and to make no true novel predictions at all. Reverting again to our simple little example, it would be simply *crazy* for a scientist who suspects that the relationship between P and Q might be linear to start conjuring up specific linear hypotheses out of nothing by blind guesswork, when a couple of measurements would reveal what the relationship has to be *if* it is linear.

Chance discoveries are different. That they exist cannot be denied. But the best thing said about them was said by Pasteur: "Chance favours the prepared mind". The idea that hypotheses are deduced from phenomena explains what Pasteur meant. The 'prepared mind' has all the necessary premises of an inventive argument *except one* – and chance, some lucky accident, supplies the missing ingredient. That many discoveries contain a chance element is quite compatible with the views defended here.

The next objection arises from the phenomenon of *multiple discoveries*. What if two scientists independently discover the same hypothesis, but proceed by different heuristic routes? In this case, some fact might be novel for the hypothesis *as discovered by A*, but not novel for the same hypothesis *as discovered by B*. And the evidential support of the hypothesis will depend upon which scientists we take to have discovered it. But evidential support should not be person-relative in this way.

Multiple discoveries cannot be denied. What might be denied are real multiple discoveries through different heuristic route. Adams and Leverrier are both credited with the discovery of the planet Neptune, but they seem to have arrived at the discovery in precisely the same way. Indeed, the relevant facts and heuristic principles being common knowledge, it is hardly surprising that sometimes different scientists independently assemble the relevant premises, and argue to the same conclusion.

But even if real examples of it are hard to find, the *possibility* of multiple discovery via different heuristic routes remains. There are two things we might say about it. We might say that the *best* heuristic route (the one in which fewer phenomena are employed as premises) determines which facts are novel for the hypothesis. Or we might say that novelty depends upon three things (the fact, the hypothesis, and the heuristic route to the hypothesis), so that a fact can be novel for a hypothesis discovered one way, but not novel for a hypothesis discovered a different way. Or to put it another way, we might insist that strictly speaking facts are not novel for hypotheses, but only for hypotheses within particular research programmes and their associated heuristics. I do not think that it matters too much which of these two responses we make to what is, after all, more a theoretical possibility than a real one.

The next objection arises from the fact that inventive arguments, like any others, can be invalid as well as valid. Scientists are not logically infallible. What if a scientist invalidly deduces some hypothesis from phenomena? Is such a hypothesis plausible? And what facts are novel for it? It seems obvious to me that an invalid deduction of some hypothesis cannot confer any plausibility on that hypothesis (though it might be deemed plausible on other grounds than its having been invalidly deduced). It also seems obvious that if the phenomena from which a hypothesis was deduced do not really help to construct it (because the deduction was invalid), then these can be novel phenomena for that hypothesis. Lest this be taken as encouraging scientists into invalid reasoning, we can point out that invalid heuristic arguments are likely to result in hypotheses which are not merely implausible but quite likely false. After all, in the extreme case invalid reasoning could issue in a hypothesis which is actually *refuted* by the facts detailed in the premises!

A final objection arises from the fact that inventive arguments, like any others, can be valid yet contain *redundant premises*. What if a scientist deduces a theory from phenomena some of which are redundant premises in the deduction? What if a scientist actually uses a fact to construct a theory which he need not have used? Is such a redundant fact a novel fact for that theory? The answer to this last question is obviously 'Yes'. For the purpose of assessing novelty, and hence evidential support, heuristic arguments with redundant premises should be trimmed back to arguments with non-redundant premise sets.

It is time to sum up. Logical empiricist orthodoxy is wrong: there is a logic as well as a psychology of invention, and there is a psychology as well as a logic of appraisal. Moreover, the logic of invention is best regarded as deductive logic. Finally, logical empiricist orthodoxy is also wrong to say that the context of invention is irrelevant to the context of appraisal. Consideration of inventive arguments yields minimal plausibility appraisals, and it also can tell us which facts are novel for the theory being considered.

REFERENCES

Dorling, J., 'Einstein's Introduction of Photons: Argument by Analogy or Deduction from the Phenomena?', *British Journal for the Philosophy of Science*, **22** (1971) 1–8.

Dorling, J., 'Henry Cavendish's Deduction of the Electrostatic Inverse Square Law from the Result of a Single Experiment', *Studies in History and Philosophy of Science*, **4** (1973a) 327–348.

Dorling, J., 'Demonstrative Induction: Its Significant Role in the History of Physics', *Philosophy of Science*, **40** (1973b) 360–372.

Laudan, L., 'Why was the logic of discovery abandoned?', in *Scientific Discovery, Logic and Rationality*, T. Nickles (ed.), Reidel, Dordrecht, 1980, pp. 173–183.

McLaughlin, R., 'Invention and Appraisal', in *What? Where? When? Why?: Essays on Induction, Space and Time, and Explanation*, R. McLaughlin (ed.) Reidel, Dordrecht, 1982, pp. 69–100.

Musgrave, A. E., 'Logical versus Historical Theories of Confirmation', *British Journal for the Philosophy of Science*, **25** (1974) 1–23.

Musgrave, A. E., 'Wittgensteinian Instrumentalism', *Theoria*, **46** (1980) 65–105.

Newton, I., *Sir Isaac Newton's Mathematical Principles of Natural Philosophy and his System of the World*, Motte's translation revised by Cajori, University of California Press, Berkeley and Los Angeles, 1934.

Nickles, T., 'Introductory Essay', in *Scientific Discovery, Logic and Rationality*, T. Nickles (ed.) Reidel, Dordrecht, 1980, pp. 1–59.

Peirce, C. S., *The Collected Papers of Charles Sanders Peirce*, ed. C. Hartshorne and P. Weiss, Harvard University Press, Cambridge, Mass, 1931–58.

Popper, K. R., *The Logic of Scientific Discovery*, Hutchinson, London, 1959.

Worrall, J., 'Scientific Discovery and Theory-Confirmation', in *Change and Progress in Modern Science*, J. C. Pitt (ed.) Reidel, Dordrecht, 1985, pp. 301–331.

Zahar, E., 'Why did Einstein's Programme supersede Lorentz's?', *British Journal for the Philosophy of Science*, **24** (1973) 95–125 & 223–262.

Zahar, E., 'Logic of Discovery or Psychology of Invention?', *British Journal for the Philosophy of Science*, **34** (1983) 243–261.

University of Otago

HERBERT HÖRZ

DEVELOPMENT OF SCIENCE AS A CHANGE OF TYPES

1. FORMULATION OF THE PROBLEM

The development of science has been the subject of many historical and systematic investigations. They belong to three types of explanations. Firstly, attempts are made to understand scientific advance from the problem solving process of the sciences themselves. Thus, K. R. Popper thinks that we "have a certain innate knowledge" and that the cognitive advance "consists throughout of corrections and modifications of existing knowledge" (Popper, 1973). Despite many differences in detail, with regard to this attitude of Popper's, this type of explanation also includes the works by I. Lakatos on the research programmes, (Lakatos, 1982) those by G. Holton on the themes (Holton, 1973) and by Th. S. Kuhn on the change of paradigms (Kuhn, 1976). According to Kuhn, causes of scientific advance are scientific crises, i.e., theoretical difficulties in solving a programme problem within the framework of a given paradigm. However, the Ptolemaic system was not at all in crisis when the Copernican system emerged and supplanted it. Neither was the theory of relativity the result of a crisis in classical mechanics. The development of science must also have other causes than the scientific cognitive process itself. This process always provides a reservoir of ideas from which it is possible to take suggestions for progress. But why certain manners of thought impose themselves is not explained by this fact, however.

Secondly, it is indicated that science is embedded into the development of society (Böhme *et al*., 1978). It is important to highlight the role of external factors. But this does not mean that one has to give a Marxist explanation. This implies recognition of social existence as a determining factor for conscience by taking into account the dialectical relationship between productive forces and the relations of production, between the basis and the superstructure (Kröber and Laitko, 1975). If this vulgar Marxist theoretical approach explains the formation of theories directly from the mode of production, then its difficulties mostly become manifest in detail in case studies. However, they can be overcome by means of the

K. Gavroglu, Y. Goudaroulis, P. Nicolacopoulos (eds.), Imre Lakatos and Theories of Scientific Change. 33–46.

statistical conception of laws which recognize fields of possibilities and the role of chance as essential aspects of development (Hörz, 1980).

Thirdly, the role of the personality is emphasized. In histories of science of a wide variety of types, on the one hand, eminent scientists are appreciated in a justified manner for their achievements. On the other hand, this corresponds with a tendency to render the personality anonymous in the theoretical explanation of the development of science, and this development is understood only as problem-solving. However, the scientist's personality is a decisive condition for putting into practice such possibilities which emerge from the development of science. In the struggle against the thinking of élites and schematism, P. Feyerabend calls for a democratisation of science, and in this context, he exaggerates the role of the personality (Feyerabend, 1980). In fact, personality plays a decisive role in transforming social demands into scientific problems, in introducing cognitive results into social practice. The personality is the decisive connecting link between society and science, because the personality makes some possibilities which exist in the laws of the development of science come true.

In the attempt to understand the development of science as a change of types, the essential aspects of these three types of explanation are to be taken into account. A reduction to one group of factors contradicts the dialectic of the development of science. This dialectic embraces the interrelation between the development of science and the material process of life of a society, of social needs and human problems, of prerequisites for creative personal development and their utilisation by scientists in concrete historical conditions. The basis of society, a basis which is determined by the conditions of production, provides an ideological framework for the development of science to the material and to the social superstructure. This means that, in this manner, the scope of activities for the scientist is determined. Whether he uses this scope of the development of science, does not make full use of it, or opposes it, depends on his education; above all on his ideas of values, ideas marked by social values with regard to the meaning of scientific work, and on his abilities. These conditions of the development of science in concrete historical terms, which have emerged owing to the basis and superstructure, do not unambiguously determine the conduct of scientists' personalities, but admit an area of possibilities of development tendencies from which possibilities are realized by various scientific groups and personalities.

2. PHILOSOPHICAL THEORY OF DEVELOPMENT AND THE DEVELOPMENT OF SCIENCE

An understanding of the development of science as a change of types is based on the results of the philosophical theory of development according to which development is the tendency to the emergence of higher qualities, a tendency which imposes itself through stagnations and regressions and through the emergence of all elements of a phase in a cycle from the basic quality of an initial stage via the dialectical negation in a new quality until the dialectical negation of the negation in the higher quality as the final stage of the cycle (Hörz and Wessel, 1983). For our problem, this statement requires further specifications. The theory of development investigates development cycles, and in this context it is necessary to take into account the difference between the ideal cycle and the real cycle and the interrelation of major and minor cycles in a hypercycle. This leads to the fact that various levels of development are distinguished in the development of science. Thus, the development of science is to be understood as an entire social process from the emergence of science until the type of science of the scientific and technological revolution, then as a development cycle from an initial to a higher final quality if one understands as an initial quality for the development of science the unity of empiricism and of theory, a unity which is dominant in classless society, a unity in the cognition and in the application of knowledge and of skills in the satisfaction of human needs. Now this development of science has a comparable interconnection with an apparent return to the type of science of the scientific and technological revolution owing to the compulsion towards technology and owing to the unity of cognition and the design of the natural and social environment. The demands for a union between Man and Nature in the debate about ecology, for technologies as more humane means of Man's domination are pointing to the historical dimension of the conditions between Man and Nature. Man has ascended from Nature, and in dialectical negation, he overcame his subjection to Nature as a social being and he abandoned his relative dependence on Nature more and more by scientific cognition as the orientation of his activities.

In this process, there emerged a relative separation of science from the material process of life of society, and in this separation, the function of education for an élite was in the foreground. This élite acquired and used insights in the interests of the ruling class. Only owing to the type of science of the industrial revolution does science increasingly become an immediate productive force. But this type also belongs to the phase of separation between science and practice. This type demands the wasteful exploitation of Nature. This type allows an aggravated exploitation of the workers and in its autocratic orientation impedes a vision aimed at the interconnections between Man, Nature, technology, society and culture. There emerges a concept of Nature which only takes into account the technically available nature, which leads to romanticist counter-reactions. Owing to the type of science of the scientific and technological revolution, a type which serves to form the material and technological basis of socialism and communism, the dialectical negation of the negation is carried out on a higher level by the utilisation of scientific methods, by qualitatively new technologies and by the humane design of scientific and technical advance, the unity of cognition and of activity is brought about time and again as the unity of science and of practice.

Hence, the development of science as a change of types is, from the viewpoint of the philosophical theory of development, a development cycle going through the following stages: from the unity of intellectual and of physical work in the conscious creation of the environment on a low level of satisfaction of requirements, through social organization and the production of tools via the emergence of science owing to the separation of intellectual and physical labour as a dialectical negation which led to the emergence of subtle methods of cognition with an extensive material and technological basis, partly made cognition an aim in itself, made Nature an object of exploitation of Man and separated science from social practice. But at the same time this cycle provided deeper insights into processes of Nature and of society, and this served to prepare the dialectical negation of the negation, a process which is taking place with the type of science of the scientific and technological revolution. One can formulate in a provocative way that science has thus only received its full formation and will receive it, because now, without any socio-economic barriers, on a high theoretical and technological level, science can fulfil its function as a productive force in the effective organization of the metabolism with Nature, as a cultural force in the utilisation of all creative potential for cognitive advance, and as a human and social force in the human design of social relationships and of scientific and technical advance. This also makes understandable the statement of Karl Marx: "Only in the Republic of Labour can science play its real role" (Marx and Engels, 1982). Our insights into the development of types of science refer above all to scientific thought in the ancient world (Jüres *et al*., 1982), to the development of science until the industrial revolution and to the type of science of the industrial revolution (Bernal, 1961). The discussions about the new type of science in the scientific and technological revolution are still continuing. But in them it is already revealed that the investigation of the development of science brings forth advantages, when the character of the tendency of the emergence of higher qualities in development cycles, the existence of fields of possibilities and the probabilistic transition from one type to another are taken into account. Then the pre-scientific phase is a constituent part of the development cycle of science. The development of science is regarded in context with the process of material life and society, without being able to derive that direct correspondence of the mode of production and of the dynamism of theories. Its type formation is not bound to the subdivision into periods of political history, although it is always necessary to take into account the political and ideological interconnections between the type of science and the development in terms of policy and economics and of world-outlook and of culture. In this context, the explanation as a change of types is not any predetermined scheme, but a heuristic principle which needs further specifications by means of case studies.

The emphasis on the development cycle from the pre-scientific phase via the separation between science and practice until the true role of science in classless society includes other levels of development. However, first and foremost it must be indicated that the development is not terminated with a development cycle, but the higher quality is the starting point for a new cycle.

Although we still know very little about this, today it is already plain that the scientific and technological revolution can in the future go through various phases, which leads to a widening of the potential of explanation and the human potential in the type of science of the scientific and technological revolution. Thus, owing to the revolution of the instruments for thinking by artificial intelligence there can occur a wide application of scientific methods in everyday life. By this fact the expert élite still existing today will become superfluous, because the access to knowledge is facilitated. Space exploration allows us to conquer new, vital space, and genetic engineering yields human genetic assistance in gene deficiencies. It is worth discussing the development cycle of the scientific and technological revolution further, which, however, would by far exceed the framework of this paper which explains the conceptional foundations of the development of science as a change of types.

This also leads to the fact that other levels of development are discarded. They are to be mentioned, because they must always be also imagined in conceptional terms. Thus the development of science is differentiated in the type of the emergence of science as a specific practical and theoretical acquisition of reality, so as to obtain insights into regularities and laws of Nature, of society and of one's own activities, and these insights serve to satisfy material and cultural requirements of the ruling classes. The maturation period of this type is characterized by the synthesis of cognition in the writings of Aristotle. The type of science of corporate artisanship and of self-sufficient agriculture is determined especially by the separation of science and of the mode of production. However, in this type, important theoretical prerequisites for the further development of science emerge. Whether there exists a type of science of the manufactory has still not been elucidated. In Europe, the manufactory is the transition from corporate artisanship and self-sufficient agriculture to the industrial revolution. In other areas, however, the manufactory can play a greater role and bring about a type of science of its own. Then for the development of science the types of industrial and of scientific and technological revolution are essential.

In their turn, these various types of science consist of development phases. Thus, each type has its preliminary, it maturing and completing phase. In the necessary empirical and theoretical achievements, the preliminary phase coincides with the completing phase of the preceding type of science. The preliminary phase of the type of development of science is the zero type of science in the unity between physical and intellectual work, such a specific manner of rational cognitive advance which is not embodied by persons and institutions. If the preliminary phase of the type of science of the industrial revolution is already assumed in the 16th century, then by this are meant the preliminary achievements of the Renaissance engineers, natural philosophers, mathematicians and other thinkers. The maturing phase is directly connected with the appropriate mode of production. This does not mean that the type of science of corporate artisanship and of self-sufficient agriculture brought forth above all scientific achievements for the benefit of this mode of production. At that time, the education function of science for an educated élite, was prevailing that served the ruling class. It is in the completing phase of this type that it is shown that the law of inertia of the development of science applies: Science remains in a state of rest (cultivation of existing fundamental theories) and of uniform motion (expansion of this theory), if no social force (requirements from practice, personalities, school) calls forth an acceleration. The completion of theories can lead to contradictions in the theory and between theory and practice, and the solution of these contradictions is opposed to the prevailing completing phase of the type of science of corporate artisanship and of self-sufficient agriculture obtained its maturity in the Middle Ages, but during the Renaissance, this type is marked by the persistence of scholastics and by the theories opposed to it which are preliminary achievements of the type of science of the industrial revolution.

This internal dialectical contradictoriness in the development of types of science in connection with modes of production is also shown in the possible deformation of demands put forth by a type of science. Thus, it is impossible to make a full use of the human potentials arising from the type of sciences of the scientific and technological revolution under imperialist social conditions, and this is revealed by the socio-economic implications of the scientific and technological revolution. There are also other obstacles to development. Thus, on political and ideological grounds, such as anticommunism and vengeful policy, the necessary widening of the competence of scientists may take place, although this is important for the safeguarding of peace. Deformations are an expression of the contradictory dialectical formation and maturation of types of science, because the completing phase of the old type and the preliminary phase of the new type of science constitute a unity in the transition.

The development of science has the development of theories as a substantial constituent part, but cannot be reduced to it. Science is a manner of grasping, by cognition, the conditions and laws and of transforming certain discoveries into inventions by technologies. Technologies are means of Man's domination, in order to purposefully organize his environment and his own conduct in conformity with his aims. Hence, science is a search for truth and an evaluation of scientific insights, as well as their social utilisation. Apart from the development of theories, science also embraces the material and technological basis of scientific work for experiments, pilot studies, data processing, the institutions of education and of further training for scientists, the material and personal potential in institutions of research. In this sense, science cannot be understood outside the material and cultural life process. But also the development of theories has various phases which have an influence on the determining structure of science. Thus, the central process in the development of theories is the scientific revolution which leads to the emergence of new fundamental theories with repercussions on other sciences by new styles of thinking, on philosophy with its explained pictures of the world and on social practice by possible new principal solutions. The development of disciplinary theories takes place via the phase of the creative idea which is elaborated into a theory and then brings about a theoretically and experimentally feasible research programme towards the phase of the maturity of the theory and towards evaluation and social utilisation.

Now let us consider the role of the personality in the development of science. Then it is also necessary to take into account development stages of its influence, the process of de-subjectisation of a theory as a constituent part of social consciousness by the contributions of other scientists to its specification and utilisation and the role of scientific schools. Self-contained schools which only complete the doctrines of their teacher and are otherwise immune towards criticism can become impediments. However, open schools that take up suggestions and question what has already been time-tested, are accelerating factors in overcoming such an inertia which is opposed to modernism and which is supported by the monopolised formation of opinions.

The philosophical theory of development helps to justify the conception of the types of science if "type of science" is understood as a fundamental quality of scientific activities in the context of a mode of production.

3. WHAT IS A TYPE OF SCIENCE?

The conception of the development of science really imposes itself, if one wants to explain the new quality of science in connection with the scientific and technological revolution. A new type of science which is distinguished from that of the industrial revolution is maturing. Not infrequently, to characterize this type, we speak about classical and modern sciences, without analysing the theoretical foundations for this distinction. They consist in the different types of science exactly. To justify this conception, it is necessary to take into account the theoretical approach of Karl Marx. It reads like this, "Industry is the real historical relation of Nature and therefore, of natural science to Man: hence, if it is grasped as esoterical revelation of essential human forces, then the human essence of nature or the natural essence of Man is understood, therefore, natural science will lose its abstract material or rather idealist direction and will become the basis of human science, as it has already become the basis of real human life – although in an alienated form, and from the very outset, another basis for life, and another for science is a lie" (Marx and Engels, 1968).

If we take this idea seriously, then there arise from it implications for setting up types of development of science.

Firstly, Industry is the basis of human life and of science as a historical concrete expression of the relationship between Man and Nature. Thus, the development of science can only be understood in context with the concrete historical mode of production. Karl Marx has further developed his theoretical approach in the *Capital* and in the *Introduction to the Critique of Political Economy* by analyzing the dialectical relationships between the productive forces and the relations of production, between the basis and the superstructure and between science as a productive and human force.

Secondly, Marx demanded that the human beings "regulate their metabolism with Nature in a rational manner, bringing it under their joint control, instead of being ruled by it as if by a blind power: that they should carry it out with the least expenditure of force and under the conditions which are the most worthy of their human nature and the most adequate to it" (Marx and Engels, 1964). Thus Marx is pointing out the dialectics of effectiveness and of humanity. As a productive force, not only does science make its contribution to an effective production of material commodities, but it is only called upon to promote the regulation of the metabolism with Nature in a rational manner, owing to the discovery of new relationships and laws of Nature, of society and of consciousness, and owing to the transformation of discoveries into inventions. By this cultural achievement, science becomes a cultural force. As a human and social force, science founds the objectives of scientific research, obtains foundations for evaluating scientific insights and is conducive to the humane utilisation of the results brought about by scientific research. By calling for a social control of the rational metabolism with Nature, Marx indicates social conditions for mastering Nature. Science is concerned with an effective and human organisation of the mutal social relationships of human beings. In this context, the mastery of Nature is to take place in a humane manner. The functions of science are seen in the context with the concrete rule of Man over Nature.

Thirdly, a new quality of industry, of the mode of production, is a new quality in the relationship between Man and Nature with repercussions on the social relationships of human beings with one another, and in this context, the relations of production determine the manner and the objectives of the mastery of Nature. A qualitatively new mode of production is connected with a new quality of scientific work. This applies to science as a system of cognition, as social activity, as a social institution and as a setting forth of norms. These qualities are types of science.

Fourthly, let us include in our considerations certain results of the philosophical theory of development. According to this theory, in the development processes there are formed new qualities in development cycles. This also encompasses stagnation and regressions and the formation of all aspects of a development stage. In this context, dialectical contradictions are accelerating the transition from one quality to a new and higher quality. Then it is necessary to take into account various stages in the development of science and in the formation of types of science. Preconditions for a new type of science are possible theoretical preliminary achievements which also favour the development of a new mode of production. In this context, theory and industry are relatively independent of each other in their development. There ensues the maturity of the type of science which is connected with the completion of the mode of production. In this context, preliminary theoretical achievements can emerge which already characterize the transition to a new type of science.

A type of science is the concrete historical manner of recognizing new conditions and laws of Nature, of society and of consciousness, as well as of one's own conduct by Man and the transformation of discoveries into inventions by new technologies. The mode of production corresponding to the type of science and challenging this type is a dialectical unity between the development of productive forces and the completion of the relations of production. This mode of production is the basis of scientific work, because by the production of material goods, it creates the preconditions for the fact that human beings can occupy themselves with science. However, the mode of production as social practice is at the same time the starting point and aim of scientific cognition. The scientist personality transforms social demands and the requirements of production into scientific questions, so that they can be answered by an experimental and theoretical working out of research programmes.

Hence, a type of science is essentially determined by answers to questions about the goals of science, about the meaning of scientific work, about the object of scientific research, about the criteria of scientific rationality and about the philosophy which is adequate to the type of science.

4. DIFFERENCES IN THE TYPE OF SCIENCE OF THE INDUSTRIAL REVOLUTION AND OF THE SCIENTIFIC AND TECHNOLOGICAL REVOLUTION

The *aim* of the type of science of the industrial revolution consists in the contribution of the science to the development of the productive forces, a development which is determined by replacing the hands and muscles by machines. Since natural science had a great importance for industry, and therefore, not as in the Middle Ages, scientific cognition was not essentially an end in itself, science was less subjected to a compulsion for justification than before. Its cultural and educational function was recognized. Natural scientists withdrew from the major disputes on world-outlook, in order to engage in a specialized investigation of Nature. Therefore, it appeared as if the goal and justification of research in natural science consisted only in the curiosity of the researcher. The aim of the type of science of the scientific and technological revolution is founded above all in the essence of the scientific and technological revolution. Man becomes increasingly the creative designer and controller of his manner of work and of his way of life. Therefore, science must contribute to the human solution of global problems, to gain in liberty for the personality by social progress in peace.

Hence, the *meaning of scientific work* is directly connected with the goal of science and applies to the position of Man towards science. In the type of science of the industrial revolution, the salient point was to consolidate the pride of Man by his domination of Nature. By the scientific and technological revolution, science is faced with the task not only to make contributions to an increase in effectiveness, but to make the increase in effectiveness the basis for a widening of humanity. It is with the development of information technologies, hence with the revolution of instruments for thinking, that the question emerges, whether Man is capable of mastering these technologies of consciousness. By the multitude of information and by the spatio-temporal acceleration of information transfer, the separation between information and events is aggravated. Information becomes an entity in itself by which Man is orienting himself in his activities. If the world of information and the world of events do not coincide, then this can lead to disorientation. Therefore, it is always important to use the new means of effectiveness in the interests of Man, so as to prevent deformations of the personality by technocratic developments. The goal and meaning of scientific work in the type of science of the scientific and technological revolution are connected in expanding the functions of science as a productive force, a cultural force, as well as a human and a social force.

The object of scientific research at present comprises increasingly more complex tasks and global problems. Modern science provides a consistent explanation of Nature, of society and of the consciousness, of the genesis of the forms of motion in their interaction with the transition to the idea of structure, of process and of development since the mid-19th century until system analysis and the theories of self-organization. Complex tasks of scientific research which can be solved on a high level by interdisciplinary work are such complexes of needs as the rational utilisation of energy and of raw materials, the substitution of raw materials and the opening up of ecological cycles, the exploration of new and alternative sources of energy, the design of a natural environment which is conducive to Man, a healthy nutrition and the development of national health, as well as the development of the personality. A basis for these solutions is the development of interdisciplinary work. Hence the separation between natural and cultural sciences, their unilateral specialisation is overcome. Now we must remember that this separation had been characteristic for the type of science of the industrial revolution. At present interdisciplinarity proves to be an initial form of disciplinarity. Under present conditions, one-sided specialism can become a form of irresponsibility, if it prevents a widening of competence which is necessary, in order to solve scientific problems for the safeguarding of peace and for disarmament. Strategic programmes apply to constructive peace policy, they involve opportunities for social progress in the struggle against aggression and against oppression, they imply a gain of liberty for the personality by using the results of the scientific and technological revolution, and they include the design of scientific and technical advance along humane lines.

The criteria of scientific rationality have changed. In the type of science of the industrial revolution, exactness was connected with the example of classical mechanics. There came about a separation between truth (cognition by scientists because of curiosity) and value (evaluation by politicians according to the benefits) of scientific insights, of the exploration of Nature and of the mastery over Nature. The type of science of the scientific and technological revolution abolished the narrow concept of science which is only aimed at the search for truth. Criteria of scientific rationality are correctness, i.e., the examination of the consistency of theoretical statements, truth, i.e., the adequacy between scientific theory and the subject of cognition, simplicity, i.e. by a minimum of axioms, a maximum of explanatory power is obtained, as well as effectiveness, i.e., the possibility of practical utilisation in the social interest. Hence, science comprises the question for truth, the social evaluation of scientific insights and their social utilisation. Therefore, there is also a constraint towards technologicalisation applying to all sciences. Technologies are instruments for the domination of Man. These tools are formed by the transformation of insights into relationships and laws of Nature, of society and of consciousness and apply to the development of new equipment and to the functioning of existing equipment. Hence, the salient point is to transform discoveries into inventions, to elaborate scientifically founded strategies for a design of scientific and technical advance along human lines. The criteria of scientific rationality are connected not only with the dynamism of theories, but also with the design of the system of methods. In the type of science of the scientific and technological revolution, system analysis and the theory of self-organization play a major role. They suggest certain possibilities in order to specify dialectical principles. The system of methods which consists of crucial aspects in the experimental, in the historical, and in the logical mathematical method, is deployed in its unity. Substantial methods, such as modelling prove to be connecting links between the crucial aspects in the system of methods.

Mechanical materialism proved to be the *adequate philosophy* of the type of science of the industrial revolution. Mechanical materialism guaranteed the philosophical safeguarding of research programmes serving the goal of the industrial revolution of replacing Man-performed activities by machines. The type of science of the scientific and technological revolution requires philosophical materialism and materialist dialectics.

REFERENCES

Bernal, J. D., *Wissenschaft in der Geschichte*. Berlin, 1961.

Böhme, G. *et al.*, 'Die Gesellschaftliche Orientierung des wissenschaflichen Fortschritts. *Sternburger Studien* 1. Frankfurt am Main, 1978.

Feyerabend, P., *Erkenntnis für freie Menschen*. Frankfurt am Main, 1980.

Holton, G., *Thematic Origins of Scientific Thought*. Cambridge, Mass., 1973.

Hörz, H., *Zufall – Eine philosophische Untersuchung*. Berlin, 1980.

Hörz, H. and Wessel, K.-F., 'Philosophische Entwicklungstheorie'. *Weltanschaulische, erkenntnistheoretische und methodologische Probleme der Naturwissenschaften*. Berlin, 1983.

Jüres, F. *et al*., *Geschichte des wissenschaftliches Altertum*. Berlin, 1982.

Kröber, G. and Laitko, H. (eds.), *Wissenschaft, Stellung, Funktion und Organisationm in der entwickelten sozialistischen Gesellschaft*. Berlin, 1975, pp. 11 ff.

Kuhn, T. S., *Die Struktur wissenschaftlicher Revolutionen*. Frankfurt am Main, 1976.

Lakatos, I., 'Die Methodologie der wissenschaftlichen Forschungsprogramme'. *Philosophische Schriften*, Vol. 1. Braunschweig/Wiesbaden, 1982.

Marx, K. and Engels, F., *Werke*, Vol. 17, Berlin, 1982, p. 554.

Marx, K. and Engels, F., *Werke*, Ergänzungsband, Teil 1, Berlin, 1968, p. 543.

Marx, K. and Engels, F., *Werke*, Vol. 25, Berlin, 1964, p. 828.

Popper, K. R., *Objektive Erkenntnis. Eine evolutionärer Entwurf*. Hamburg, 1973, p. 286.

Akademie der Wissenschaften der D.D.R.

J. J. C. SMART

METHODOLOGY AND ONTOLOGY

In a conference in memory of Imre Lakatos it seems appropriate to try to relate my own ontological concerns to the methodological concerns of Lakatos, Popper, and other members of what might loosely be called 'the LSE group'. After all, the philosophy department at the London School of Economics is called 'the department of logic, methodology and philosophy of science'. Most of us have a good idea what 'logic' and 'philosophy of science' mean here, but 'methodology' is not so clear. I remember Gilbert Ryle many years ago remarking to me that he could not make out what the Popperians meant by 'methodology'. Since that time we could perhaps say that it is the sort of problem discussed by Lakatos in his book *The Methodology of Scientific Research Programmes* (Lakatos, 1978a). Unfortunately the word 'methodology' does not occur in the index to his book. It does occur in the index to Lakatos's *Mathematics, Science and Epistemology* (Lakatos, 1978b) but we are there (where it occurs by itself) referred simply to the fairly unhelpful entries under 'discovery' and 'heuristics'.

'Discovery' and 'heuristics' suggest something like the advice of a senior scientist to a junior one on design of experiments, ways of guessing at theories, or the sort of thing so beautifully done by G. Pólya in his two volumes on plausible reasoning in mathematics (Pólya, 1954). However these sorts of compendia of practical advice are not the kinds of things we find in the work of the LSE group. This is not to deny that Popper seems to have beneficially influenced eminent scientists, such as Eccles, Medawar and Bondi. In the case of Eccles Popper's philosophy seemed to be a vital morale booster. After being depressed at the falsification of an earlier theory Eccles was brought by Popper to see falsification as a scientific virtue, and he went on to propound another theory, on the basis of his experimental work, and so to win a Nobel prize.[1] In the case of Bondi it seems that it was at least partly the influence of Popper that led him to treat cosmological theories as testable by the hypothetico-deductive method, more continuous with normal physics and less metaphysical than they had hitherto seemed.[2]

I suspect that the word 'methodology' can best be understood as

K. Gavroglu, Y. Goudaroulis, P. Nicolacopoulos (eds.), Imre Lakatos and Theories of Scientific Change. 47–57.

meaning much the same as 'epistemology' or 'theory of knowledge'. Much philosophy of science is epistemology in this sense. However the word 'epistemology' itself can cover various very different types of enquiry. Traditionally it has tended to refer to a justificationist programme, an attempt to build knowledge on absolutely hard foundations, whether *a priori* or empirical. It is possibly in this sense that Gilbert Ryle (as I seem to remember) sometimes used the word 'epistemologist' as a term of abuse.

'Theory of knowledge' sounds like 'theory of electromagnetism' or 'theory of evolution'. So the words could naturally be taken to refer to a philosophically uncontentious type of scientific investigation. Animals (including of course human animals) know things. They can tell different shapes from one another, recognize prey, migrate at various seasons with wonderful navigational skills. Humans can do much more intellectual things. There clearly ought to be (and to some extent indeed are) scientific theories about how both human and non-human animals do these things. Also recent theories of evolutionary epistemology are essays in a related genre. Such theories argue plausibly that the cognitive abilities of animals must be generally successful because true beliefs (e.g. about the presence of prey) are conducive to survival so that abilities to form them arise by natural selection. These theories may become more questionable if the success of modern science is explained not by the survival of organisms but by the survival of theories in the face of disconfirming evidence.[3] This sort of selection is quite a different matter from natural selection and consideration of it leads us to the well known problem that there seems to be no way of getting from Popperian corroboration (survival of tests, genuine attempts to falsify) to confirmation. I shall go into this matter later in this paper, but at the moment and in anticipation I shall say that these considerations lead us on to questions of justification. If justification is thought of as Popperian corroboration, then there is the worry that it has only a minimal connection with the matter of truth.

There is another way in which the word 'methodology' may be understood. This is a sense in which methodology is not concerned with explanation (as in naturalized or evolutionary epistemology) but with *demarcation*. The Vienna Circle positivists distinguished metaphysics from science by arguing that the former was meaningless. Popper was not interested in questions of meaning, and anyway was quite willing to assert that traditional metaphysics was meaningful. Nevertheless he wanted to demarcate science from metaphysics and from pseudo-science (e.g. Freudian theory). This problem of demarcation was also prominent in

Lakatos's methodological writings, though what Lakatos sought to demarcate was not genuinely scientific from pseudo-scientific *theories* but progressive from regressive *scientific research programmes*.

Popper holds that the scientific character of theories lies in the possibility of refuting them. It does not lie in the probability that the theories are true. There is no inductive step from corroboration (unsuccessful though effortful attempts to refute) on the one hand to confirmation on the other hand. According to Popper even the best corroborated theory has a probability of zero. Similarly there is not even the smallest assurance that a progressive research programme will not suddenly begin to regress, and no assurance that progressive research programmes lead us towards truth. It is true that at one point Lakatos pleads with Popper for 'a whiff of inductivism' in the form of a conjectural synthetic inductive principle (Lakatos, 1978a, pp. 159–67). This shows that Lakatos is aware of the problem. The trouble is that if the synthetic principle is *merely* conjectural it does not really give us any of the assurance we need.

What is needed is a way to close the gap between rationality and the probability of truth or at least of verisimilitude. Consider the way in which, in a very ingenious and instructive book. J. W. N. Watkins has sought to defuse inductive scepticism by reference to an aim of science which he rightly says 'must involve the idea of truth' (Watkins, 1985, p. 124). However because of the impossibility of deducing confirmation from corroboration, the relevant clause in Watkins's definition of an optimum aim for science speaks only of a theory *possibly* being true, and possibly being true does not imply even the smallest positive probability of in fact being true, or even near to the truth. A little bit more precisely, Watkins holds as an aim for scientific rationality that a person should accept a system of hypotheses that (1) are possibly true for him or her, in that he or she has been unable to find refuting evidence for it despite the best efforts to do so, and that (2) is deeper, more unified, more predictively powerful and more exact than other systems of hypotheses that meet the first condition (Watkins, 1985, pp. 155–6).

Looking at the LSE group as a whole, despite their important individual differences, and despite Popper's own body of ontological conjectures, it would seem that their main interest in methodology arises from a desire to demarcate science from non-science and to clarify the nature of scientific reasoning (both theoretical and experimental) as so demarcated. Even though the notion of truth is brought into the discussion, this is of small help to the concerns of ontology. Interest in ontology is interest in the nature of the universe, and since apodeictic certainty is impossible, the interest must be in getting the most probable views about the world. Unfortunately the considerations of the LSE group suggest that there is no bridge from corroboration to confirmation, and so the dream of attaining probability is as incapable of satisfaction as is the older and discredited aim of obtaining apodeictic certainty in metaphysics. Somehow the ontologist, if he or she is to *be* an ontologist needs to make the connection between methodological rationality on the one hand and probability of truth or verisimilitude on the other hand.

As I conceive of metaphysics it is largely an activity of conceptual clarification in the service of attaining the most plausible view of the universe in the light of a synthesis of the various sciences. Plausibility in the light of total science is the ontologist's touchstone. On this view there is of course no sharp line between science and metaphysics. Metaphysics is the most conjectural and conceptually interesting end of total science. Many scientific theories give rise to highly conceptual discussions of the sort that we can recognize as typical of what is generally regarded as philosophy. Consider for example the well-known discussion between Bohr and Einstein (Bohr, 1959). The more the scientist challenges commonly held assumptions, assumptions that are so deep rooted they are rarely thought about but are taken for granted, the greater the similarity between the scientist and the traditional metaphysician.

Consider the assumption that the reality of external things is independent of our consciousness of them. This has been challenged by some physicists, who have given subjectivist interpretations of the collapse of the Schrödinger wave. According to the Copenhagen interpretation of quantum mechanics this must be understood in terms of a relation between the quantum mechanical system S^1 under consideration and a macroscopic measuring apparatus S^2 which is understood classically. Then to understand the total system S^1 and S^2 we have to relate it to a measuring apparatus S^3, also understood classically. Clearly it is unsatisfactory to be unable to dispense with classically understood systems. The problem becomes more acute still if we wonder about the Schrödinger wave of the whole universe. In this case there can be no relation to a measuring system outside. Some sort of Cartesian dualism (as seems to be favoured by E. P. Wigner)[4] is thus tempting, the collapse of the Schrödinger wave being brought about by consciousness outside the physical universe. Here the concerns of traditional metaphysics can hardly be separated from those of serious contemporary physics. Yet if metaphysical theories need to be judged in terms of plausibility in the light of total science, we should surely look for some non-subjectivist and non-dualist interpretation of quantum mechanics. If we look at the universe with the eye of a biologist, we think of the human mind as nothing over and above the brain, itself a physical system. From the point of view of physics, surely brains should be thought of as very small beer in the universe, and whether this be so or not, brains are not, as measuring devices, any the less physical systems than (say) photographic plates. A similar biological implausibility seems to belong to an ingenious theory recently put forward conjecturally by J. A. Wheeler,[5] according to which the cosmological big bang depends on subsequent knowledge of it — a theory of cyclic dependence which seems to put too much physical importance on those (admittedly marvellous) computer-like things that lie between our ears.

At this point, however, I should like to applaud the metaphysical imagination of the great theoretical physicists. (Wheeler's conjecture itself provides an example.) It used to be said, with some arrogance on the part of philosophers, that metaphysicians provided ideas that might usefully be taken over by scientists. It rather seems to me that the boot is on the other foot and that physicists have been conceptually innovative in a way that has rarely even been approached by those traditionally classed as philosophers. This judgement is of course complicated by the consideration that some of the great philosophers, e.g. Leibniz, were also great mathematicians and physicists. Moreover philosophers have seen problems that heve been lost on physicists of the time: consider Hume on the notion of physical necessity.

Metaphysics is the conjectural end of science. Its ontological claims must be tested by general scientific plausibility. Plausibility is largely a matter of maximal coherence of our beliefs in the light of often recalcitrant experience: in other words not only must theoretical beliefs cohere with one another but they must cohere with beliefs derived from observation and experiment. A coherence theory of knowledge (or warranted assertibility) should also include a requirement that we should be on the look out for new experiments and observations: coherence implies not only internal consistency and mutual support of our beliefs but also comprehensiveness, so far as this is possible. It should be stressed that I am advocating a coherence theory of knowledge, or more properly warranted assertibility, not of truth. The nineteenth-century and early twentieth-century idealists mixed up the two things. Thus F. H. Bradley was good on epistemology but bad on semantics and theory of truth. The same wrongheaded running together of epistemology and semantics seems to me to be the fundamental error of Michael Dummett and his sympathisers. The LSE group are clear about this distinction. See, for example, Lakatos's acknowledgement of the difference between truth and signs of truth (Lakatos, 1978b, p. 118).

I would argue that dualism in the philosophy of mind, libertarian theories of free will, phenomenalism about material objects, and many other metaphysical theories fail the test of coherence with well established scientific fact. Traditionally philosophers who have offered such theories have tried to produce *a priori* refutations. That is, they have tried to show that when such theories are clarified by conceptual analysis they lead to inconsistencies. However on examination it turns out that these inconsistencies are not directly derivable from the metaphysical theses themselves but are derivable only with the help of certain plausible background assumptions. It is always open to the defenders of such theses to challenge the background assumptions. The opponent of the theses will then have to try to deduce inconsistencies from the theses together with the new background assumptions, and this can be done only with the aid of further background assumptions. The idea is to chase the recalcitrant metaphysician into more and more implausible expedients.[6]

If one accepts science as the best touchstone of metaphysical truth one must be prepared to defend a realist account of the relevant scientific theories against instrumentalist or pragmatist accounts of such theories. Instrumentalism and pragmatism may be congenial to those who are metaphysically hostile to science, as well as to those, such as the logical positivists, who have endeavoured to emasculate theories of their ontological content in a mistaken attempt to do science a good turn. The defence of scientific realism depends on the contention that the literal metaphysical truth (or approximation to truth) of scientific theories provides the best explanation of the predictive success of these theories. The clause 'or approximation to truth' is important because it is likely that theories will come to be supplanted by better theories. So if we believe in electrons and nevertheless think that one day there may be a better theory in which electrons are not mentioned, we should take belief in electrons not quite to be belief in things of which the predicate 'is an electron' is true but to be belief in things of which 'is an electron' is *approximately* true. Many years ago Popper pointed out the need for a theory of verisimilitude in order to defend realism (Popper, 1963, p. 235; 1972, pp. 47–61, 101–3 and 331–5). His own definition was in terms of truth content, but this approach was shown to be unworkable by David Miller and Pavel Tichý (Miller, 1974; Tichý, 1974). More recently Ilkka Niiniluoto and Graham Oddie have explored a different approach, depending on the notion of distance from truth (Niiniluoto, 1984; Oddie, 1986). Whether or not these attempts are fully successful they are clearly of great importance.

Assuming that realism can indeed be defended successfully on these lines, a major difficulty for my position still arises. This is that a determined opponent can reject my assumption that science gives the best test of metaphysical truth. Many philosophers prefer common sense intuitions even when their objects lack scientific plausibility. Fitting in with common sense seems to them to be at least as important as is fitting in with well tested scientific hypotheses. Other philosophers reject scientific methods in favour of an alternative type of inquiry.[7] Still others may prefer theological to scientific plausibility.[8]

Consider also pseudo-science. It will be recalled that Popper ruled out Marxist political theory and Freudian psycho-analysis from the body of science because a proponent of such a theory is unable to say what would be taken as a refutation of it. Proponents of such theories can reply that respectable scientific theories are retained despite anomalies when no better theory that explains these anomalies is in the offing.

Lakatos therefore made the demarcation between science and pseudo-science by reference to sequences of theories (scientific research programmes). Pseudo-science, he held, arises by adherence to degenerating research programmes in which anomalies are explained in a blatantly *ad hoc* manner. Nevertheless degenerating programmes can turn into progressive ones, just as curves that turn downwards can eventually turn upwards. Watkins would criticise pseudo-scientific theories because of their flouting of certain definitions of rationality. Proponents of these theories might not accept the definitions. Methodology runs into the same sort of difficulty as traditional normative ethics. How do we get an 'ought' out of an 'is', whether the 'is' is a definition or whether it is some description, historical in nature, of established scientific practice?

It would be different if one could prove that science as characterised by some methodological theory is likely to lead us to truth. Ways of vindicating a scientific method (call it 'induction') argue that if any method works then this method works. The trouble is that there seems to be no way of arguing from outside science itself that *any* method works, and without this protasis of the vindicating hypothetical there is no assurance that the method in question leads to true theories. The sceptic about scientific method remains unrefuted. This is so even if scientific and philosophical method is subsumed (as I suggested earlier in this paper) under the general idea of a coherence theory of warranted assertibility. What fits into one philosopher's web of belief may not fit into another philosopher's web. The philosophers may have different assessments of plausibility. So it is still unclear how we have any good reason to think that science leads us towards truth. The coherence theory of warranted assertibility therefore does not absolve us from the traditional and intractable problem of induction. There is another reason for this too. Plausibility as a matter of fitting into a web of belief depends on probability relations between items in the web. Such probability relations are not independent of one another, so that plausibility considerations are holistic in nature. (Nevertheless this does not mean that confirmation theory is totally irrelevant.) Holistically or not, the problem is still how to get any positive probability for a scientific theory.

It is true that from *within* science it is often possible to show that alternative methodologies are mistaken. To take an extreme example, consider astrology. We know the nature of the planets, how their orbits are determined mechanically, and we know enough about neurophysiology and embryology to know that the minute gravitational forces due to the heavenly bodies can have no influence on human destiny. As a less extreme case, consider the philosophical tradition of phenomenology. We know enough to be sure that the sort of inner gazing at essences in which phenomenologists believe does not sit happily with human biology, neurophysiological psychology, and theories of inner and outer perception. We know why vision is reliable because we know about optics, the eye and neural transmission, and there are beginning to be plausible hypotheses about the sorts of information processing that go on in the brain when we see something. No such theory of phenomenological knowledge seems even possible.

The trouble is that if we are to defend a conception of ontology as based on considerations of scientific plausibility in such a say as to convince anti-scientific philosophers then we need an Archimedean point to enable us to do so from outside science itself. If not we are left with the sort of sad division we get in philosophy today between scientifically oriented and 'analytic' philosophers on the one hand and phenomenologists, existentialists and the like on the other hand. If scientific methods cannot be shown to have at least some probability of leading to truth, not just to be rational according to some definition of rationality (whether Popper's, Lakatos's, Watkins's, or some other) then the demarcation of science from non-science or pseudo-science will not do much to help the ontologist: it will give no outside point where the ontologist can get leverage, no way in which he or she can do better than by relying on a certain dogmatism in basing his or her metaphysics on accepted science. I suppose this is all right, but it does leave us with the sad division in the philosophical community memntioned above.

Interest in the problem of demarcation thus is quite different from interest in ontology. Popper has of course both interests. He interprets Tarski's theory as a correspondence theory. He supposes that what sentences correspond to are facts. However, there seems to be no need to allow facts into our ontology. Indeed the fundamental relation in Tarski's theory is not correspondence but satisfaction of a predicate by an object or sequence of objects. Nevertheless Tarski's theory is in the *spirit* of the correspondence theory (Davidson, 1969). There does not seem a great deal of relationship between this aspect of Popper's thought and his methodology which denies a connection between corroboration and confirmation. Lakatos therefore has (as I remarked earlier) my sympathy when he appeals to Popper for a 'whiff of inductivism', a conjectural synthetic inductive principle. The trouble is that if the principle is purely conjectural we are still open to a *tu quoque* from anti-scientific philosophers. Perhaps nothing less than a positive solution of Hume's problem of induction would do the trick and that seems like asking for an impossibility.[9]

I can take a sort of comfort in the reflection that if nothing less than the solution of the problem of induction would do the trick, we could also say that even such a solution itself would not do the trick. The non-scientific or anti-scientific philosophers that worry me are not those who, like 'creation scientists', outrightly contradict the findings of orthodox science. The worry is more that from my point of view the methods and theses of the philosophers in question do not fit in plausibly with total science. However these philosophers are not sceptics about induction: when they get into an aeroplane they are not apprehensive that the laws of aerodynamics may be false. They do not doubt that their tap water is mainly H_2O, or that they can speak on the telephone because of the transmission of electrical impulses. Thus both for us and them it is not merely, as Lakatos says it is, a conjectural metaphysical principle that saves us from scepticism. 'The whiff of inductivism' must go further than mere conjecture. We are not sceptics anyway. We may say that we are justified without knowing how we are justified. It would be nice (indeed breathtaking) to be able to say how and so close the gap between the LSE group's definition of rationality on the one hand and probability of truth on the other hand.

In this paper I have been concerned with one horn of a dilemma, the gap between rationality as defined by philosophers of the LSE group and probability of truth. As for the other horn, I hold that it is necessary to preserve a gap between rationality and truth itself. I want to be able to say that even Peircean ideal science could be false in important respects. Otherwise we get involved in something like turn of the century Oxford idealism (Smart, 1986). The issue is not only a difficult but a delicate one.

NOTES

[1] See Eccles, 1964.

[2] See Bondi, 1961, pp. 18–20.

[3] This shift in the notion of selection is made more tempting by the fact that the evolution of intelligence has almost entirely been a matter of pre-scientific millennia. Even if modern physics is conducive to survival (which is doubtful, if we think of the H-bomb) the heuristics for it could not have been programmed in by biological selection.

[4] In Wigner, 1964, he seems even to have gone as far as a Berkeleyan idealism.

[5] See Wheeler (1977); and Patton and Wheeler (1975) especially pp. 564–68.

[6] The strategy here can perhaps be illustrated by the progress of my controversy with C. A. Campbell, who was perhaps the most acute and clear headed defender of the libertarian position. See Smart (1984) Chapter 7, Campbell (1951) and (1963).

[7] See Watkins's remark on Husserl's phenomenology (Watkins, 1985, pp. 153ff.). According to Watkins, Husserl wished to replace positive science by 'a new kind of science' of a quite different sort.

[8] I fear that I have not done justice to many excellent metaphysicians who hold that scientific plausibility provides a good test of metaphysical truth, without holding that it is the *best* or *only* such test. I am grateful to Peter Forrest for this observation. Such philosophers are of course congenial to me: we may not be able to get agreement on all issues, but we can aim for agreement on many issues, in respect of which agreement with other sorts of philosophers could not be hoped for. I am grateful, also, to Peter Forrest for commenting on this paper while I was making last minute modifications on the paper as originally read at the conference.

[9] I am not sure of this. John Clendinnen has come pretty close to vindicating induction (Clendinnen, 1982). He argues that if any method leads to truth then scientific methods do. The trouble is that the protasis of this hypothetical can not be shown to have positive probability. Better still would be to validate induction. At the time of writing this I am not absolutely convinced that something on the lines of Donald Williams (1947), and the strengthened argument in Stove (1986) may not do what we want, despite the criticisms which have been made of Williams's book.

REFERENCES

Bohr, N., 'Discussion with Einstein' in P. A. Schilpp (ed.) *Albert Einstein: Philosopher–Scientist*, pp. 191–241. New York, Harper Torchbooks, 1959.

Bondi, H., *The Universe at Large*. London, Heinemann, 1961.

Campbell, C. A., 'Is Free-will a Pseudo-Problem?', *Mind* **60** (1951) 441–65.

Campbell, C. A., 'Professor Smart on Free-Will, Praise and Blame', *Mind* **70** (1963) 400–5.

Clendinnen, F. J., 'Rational Explanation and Simplicity', in R. McLaughlin (ed.), *What? Where? When? Why?*, Dordrecht, D. Reidel, 1982, pp. 1–25.

Davidson, D., 'True to the Facts'. *Journal of Philosophy* **66** (1969) 748–64. Reprinted in Davidson (1984).

Davidson, D., *Essays on Truth and Interpretation*. Oxford, Clarendon Press, 1984.

Eccles, J. C., 'The Neurophysiological Basis of Experience', in M. Bunge (ed.), *The Critical Approach to Science and Philosophy*, New York, Collier-Macmillan, 1964, pp 266–79.

Lakatos, I., *The Methodology of Scientific Research Programmes, Philosophical Papers Vol. 1*. Cambridge University Press, 1978a.

Lakatos, I., *Mathematics, Science and Epistemology, Philsophical Papers Vol. 2.* Cambridge University Press, 1978b.

Miller, D., 'Popper's Qualitative Theory of Verisimilitude', *British Journal for the Philosophy of Science* **25** (1974) 166–77.

Niiniluoto, I., *Is Science Progressive?* Dordrecht, D. Reidel, 1984.

Oddie, G., *Likeness to Truth*. Dordrecht, D. Reidel, 1986.

Patton C. M. and Wheeler, J. A., 'Is Physics Legislated by Cosmogony?', in C. J. Isham, R. Penrose and D. W. Sciama (eds.) *Quantum Gravity*, Oxford University Press, 1975, pp. 538–605.

Pólya, G., *Mathematics and Plausible Reasoning*. Princeton University Press, 1954.

Popper, K. R., *Conjectures and Refutations*. London, Routledge and Kegan Paul, 1963.

Popper, K. R., *Objective Knowledge*. Oxford, Clarendon Press, 1972.

Smart, J. J. C., *Ethics, Persuasion and Truth*. London, Routledge and Kegan Paul, 1984.

Smart, J. J. C., 'Realism v. Idealism', *Philosophy* **61** (1986) 295–312.

Stove, D. C., *the Rationality of Induction*. Oxford, Clarendon Press, 1986.

Tichy, P., 'On Popper's Definition of Verisimilitude', *British Journal for the Philosophy of Science*, **25** (1974) 155–60.

Watkins, J. W. N., *Science and Scepticism*. London, Hutchinson, 1985.

Wheeler, J. A., 'Genesis and Observership', in R. E. Butts and J. Hintikka (eds) *Foundational Problems in the Special Sciences*, Dordrecht, D. Reidel, 1977, pp. 3–33.

Wigner, E. P., 'Two Kinds of Reality'. *Monist* **48** (1964) 248–264.

Williams, D. C., *The Ground of Induction*. Cambridge, Harvard University Press, 1974.

Australian National University

FAN DAINIAN

IMRE LAKATOS IN CHINA

I feel very honoured to be able to attend this international conference entitled 'Criticism and the Growth of Knowledge: 20 Years After' dedicated to the memory of Imre Lakatos. I am grateful to Prof. Kostas Gavroglou, Secretary of the Organizing Committee, for his invitation. Today I want to use this chance to give a brief account of the dissemination of Lakatos's doctrines in China.

Although Lakatos was already famous in Western academic circles of philosophy of science as early as the 1960s because of his distinguished contributions to both the philosophy of mathematics and the philosophy of science, yet his doctrines began their dissemination in China only after the beginning of the 1980s.[1]

In 1979, when I worked in the Institute of Philosophy, Chinese Academy of Social Sciences (CASS), I started *Ziran Kexue Zhexue Wenti Congkan* (*Journal for Philosophical Problems in Natural Sciences*), and held the post of editor-in-chief. The second issue of 1980 (appearing on June 25, 1980) published the following papers by Lakatos.

(1) 'The Methodology of Scientific Research Programmes' (translated passages by Qui Renzong, read and revised by Fan Dainian from *Philosophical Papers*, Vol. 1, Cambridge University Press, 1978, pp. 47–52, 68–72).
(2) 'A Renaissance of Empiricism in the Recent Philosophy of Mathematics?' (Translated by Lin Xiashui, read and revised by Qiu Renzong from *British Journal for the Philosophy of Science*, **27**, pp. 201–23, 1976)

The second part of this paper was published in the third issue of the same journal (appearing on September 25, 1976). These are probably the first translations of Lakatos' papers to appear in China.

In the same issue, John Worrall's paper 'Imre Lakatos (1922–1974): Philosopher of Mathematics and Philosopher of Science' was also published (translated passages by Qui Renzong from *Zeitschrift für allgemeine Wissenschaftstheorie*, **5**(2), 211, 1974). This paper may be the

K. Gavroglu, Y. Goudaroulis, P. Nicolacopoulos (eds.), Imre Lakatos and Theories of Scientific Change. 59–67.

first report published in China on the life and academic contributions of Lakatos.

In *Congkan*, 1st issue, 1981,[2] Lakatos' paper 'Science and Pseudoscience' (i.e. the 'introduction' in *Philosophical Papers*, Vol. 1, pp. 1–7) was published (tr. by Lin Keji, read and revised by Qiu Renzong).

In *Congkan*, 3rd issue, 1982, the following papers by Lakatos were published:

'The Problem of Appraising Scientific Theories: Three Approaches' (tr. by Zhang Shirong from Imre Lakatos, *Philosophical Papers*, Vol. 2, *Mathematics, Science and Epistemology*, pp. 107–120).
'Infinite Regress and Foundations of Mathematics' (Lin Xiashui tr. passages from *Philosophical Papers*, Vol. 2, pp. 3–23.
'What Does a Mathematical Proof Prove?' (Lu Xudong tr. passages from *Philosophical Papers*, Vol. 2, pp. 61–69).

Meanwhile, J. J. Kockelmans' paper 'Remarks on Lakatos' Methodology of Scientific Research Programmes' was also published (Zhang Xingqiang and Chen Guang tr. passages from G. Radnitzky and G. Andersson, eds., *The Structure and Development of Science*, 1979). *Congkan* has published more articles by or about Lakatos than any other Chinese journal.

In November 1980, the Institute of Philosophy of CASS, and the Research Society for Dialectics of Nature affiliated with the Chinese Association of Science and Technology jointly sponsored by the first National Symposium on the Philosophy of Science. The main topic of the symposium was Karl Popper's philosophy of science. In December 1981, the Association for the Journal of Dialectics of Nature under the Chinese Academy of Sciences ,[3] the Institute of Philosophy of CASS and the Chinese Research Society for Dialectics of Nature jointly sponsored the Second National Symposium concentrating on the discussion of T. S. Kuhn's theory of paradigms. At this meeting, we stressed that we should strengthen research and discussion on Lakatos' philosophy of science.

In the fall of 1982, the *Journal of Dialectics of Nature* published two review articles on Lakatos. This journal is a comprehensive publication on the philosophy, history and sociology of the natural sciences. An article by Zhang Shirong entitled 'Scientific Thoughts Concerning the Evaluation of Scientific Theories: A Brief Introduction to Lakatos' Philosophy of Science' was published in the issue of August 10, 1982. This article introduced three methods proposed by Lakatos to evaluate scientific theories, i.e. scepticism, demarcationism and elitism. The article also commented on the origin and development of the issue of demarcation and on Lakatos' "scientific research programme" which the author called "a many layered theoretical structure". The author held that "this model accounts for both revolution and stability, continuity and discontinuity in scientific development. So this model is more flexible and has greater explanatory power, although it cannot explain every complexity. Generally speaking, such a model is primarily descriptive; it's logically weak and cannot adequately identify inherent causes. . . . This model also lacks the ability to explain the roles of scientific research and bold speculation".[4]

In the fifth issue of 1982 the *Journal of Dialectics of Nature* published an article by Qiu Renzong entitled 'Lakatos' Methodology of Scientific Research Programmes', which introduced sophisticated falsificationism, the methodology of scientific research programmes and the proposal that research on philosophy of science should be combined with research on history of science. Lakatos laid much emphasis on this combination. The author did not himself comment on Lakatos' ideas.

In 1982, the Commercial Press (Beijing) published a translation of A. F. Chalmers' book *What Is This Thing Called Science? – An assessment of the nature and status of science and its method* (Open Univ. Press, 1978) translated by Zha Ruqiang, Jiang Feng and Qiu Renzong. The Press of the Central China College of Technology published a translation of John Losee's book *A Historical Introduction to the Philosophy of Science* (2nd ed., Oxford Univ. Press, 1980) translated by Qiu Renzong, Jin Wulun, Lin Xiashui and others. Both books introduce Lakatos' methodology of scientific research programmes. These two books and the *Journal of Dialectics of Nature*, each of which has a circulation of nearly twenty thousand, promoted the dissemination of Lakatos' ideas in China, especially among university students and postgraduates.

In the summer of 1982, a graduate student Lan Zheng of the Department of Philosophy, Nanjing University, wrote a thesis on Lakatos under the guidance of Prof. Xia Jisong, entitled 'A Review of Lakatos' Methodology of Scientific Research Programmes: Criticism of Lakatos' Thought on Philosophy of Science'. This M.S. thesis was probably the most comprehensive introduction to Lakatos' philosophy of science written in China at that time. This paper included the following sections: (1) introduction, (2) the fallibilistic epistemology of Lakatos, (3) fallibilism versus naive falsificationism, (4) from naive falsificationism to sophisticated falsificationism, (5) the methodology of scientific research programmes, (6) testing the methodology of science through the history of science, (7) conclusions. The author concluded that Lakatos systematically criticized Popper's philosophy and developed Popper's methodology to its supreme form. Lakatos emphasized research on the history of science and also stressed the fallibility, relativity and wholeness of knowledge as well as the structure of knowledge. He also called attention to the subjective activity, creativity and imagination of the human mind. At the same time Lan Zheng argued that judged by the standards of dialectical materialism and historical materialism, the philosophical views of Lakatos are too numerous and jumbled. Lan Zheng opposed Lakatos' agnosticism in epistemology and his conventionalism in methodology. Lan believed that Lakatos overemphasized the fallibility of knowledge and overlooked the role of practice in the verification of truth. Lakatos denied the objective truthfulness of scientific theories, so he couldn't explain how one programme of scientific research supersedes another. But during a conversation with me this year, Lan noted that his criticism at that time lacked concrete analyses and arguments, and thus was rather arbitrary. After receiving his M.S. degree, he passed the examination and began to study for his Ph.D. in philosophy of science under the guidance of Prof. Jiang Tianji in Wuhan University.[5] While studying there, he wrote *Lakatos' Philosophy of Science* which was based on his Master's thesis, adding chapter entitled 'Problems of Demarcation and Induction' and expanding the section on philosophy of mathematics. He stressed the significance of Lakatos' fallibilism and anti-dogmatism. Lan agreed with Lakatos' emphasis on the autonomy of science. The manuscript of this book was sent to Hubei People's Publishing House in 1985, but unfortunately it has not yet been published.

In September 1983, the 3rd National Symposium on the Philosophy of Science was held under the joint sponsorship of the Association for the Journal of Dialectics of Nature, the Institute of Philosophy and the Chinese Research Society for Dialectics of Nature. This meeting focused on "Problems of Scientific Discovery". Related issues such as the objectives, content, methods, history, present status and current trends in the philosophy of science were also discussed.[6] Lakatos' philosophy of science and philosophy of mathematics and the London conference of 1965 were all subjects of discussion.

At this symposium, Zhou Jizhong, a young instructor in the Teaching and Research Section of Dialectics of Nature,[7] the Graduate School of the University of Science and Technology of China, read his paper entitled "Reflections on the Theory of Paradigms'.[8] Zhou summarized four representative international conferences on philosophy of science held since the publication of Thomas Kuhn's *The Structure of Scientific Revolutions* in 1962. Our present meeting commemorates the first conference held in July 1965, at Bedford College, London, entitled '1965 International Colloquium in the Philosophy of Science'. The second meeting held in Urbana, U.S., in March 1969, was entitled 'Symposium on the Structure of Scientific Theories'. The third, held in Pisa, Italy, in 1978, was the Second International Conference on History of Science and Philosophy of Science. The last of these four symposia was held at the University of Nevada, Reno, U.S., in October 1978 on the philosophy of science relating to scientific discoveries. In his description of the London Conference, Zhou briefly introduced the famous collection *Criticism and the Growth of Knowledge* with special attention to the three papers by Lakatos, S. E. Toulmin and M. Masterman. He held that "Lakatos' statement on scientific research programmes may be regarded as a development of the paradigm doctrine. They are also a distinctive product of this symposium." Zhou Jizhong has already translated *Criticism and the Growth of Knowledge* into Chinese, but unfortunately he has not yet found a press to publish it.

Wang Shunyi, of East China Normal University, delivered a paper entitled 'Mathematics is Quasi-empirical – commenting on Lakatos' philosophical thoughts on mathematics.'[9] After studying Lakatos' papers, including 'Infinite Regress and Foundations of Mathematics', 'A Renaissance of Empiricism in the Recent Philosophy of Mathematics', 'Proofs and Refutations', etc., Wang pointed out that Lakatos offered a new viewpoint on philosophy of mathematics after the three major schools of philosophy of modern mathematics had encountered serious difficulties. These three schools are logicism, formalism and intuitionism. Lakatos believed that "mathematics is neither purely rational nor purely empirical, but rather quasi-empirical". "Within the more general frame of epistemology and methodology, Lakatos explained why these three schools can get nowhere. His arguments attracted the attention of many mathematicians, including those who research basic theories'. Wang held that Lakatos' philosophy of mathematics needs to be further improved and developed. He wrote, "The Euclidean system is at least a subsystem within a larger mathematical system. Does Euclideanism exert any positive influence on epistemology and methodology in the development of mathematics? What is the relationship between the quasi-empirical system and the Euclidean system? Lakatos did not expound on these questions."

At the third National Symposium on Philosophy of Science, Ouyang Jiang and Fan Jiannian, both from Shanxi University, distributed their mimeographed translation of Lakatos' *The Methodology of Scientific Research Programmes* (*Phil. Papers*, Vol. 1). Later, the translation was distributed more widely in China. After correction by Wu Zhong of Nakai University and myself, it will be published by the Commercial Press (Beijing). Lan Zheng, who is teaching in the People's University of China (Beijing), has also completed a translation of the same book which will be published by the Shanghai Publishing House for Translations. Lin Xiashui of the Institute of Philosophy, CASS and others, are now translating Lakatos' *Mathematics, Science and Epistemology* (*Phil. Papers*, Vol. 2) and their translation will be published by the Commerical Press (Beijing). Kang Hongkui of the Central China College of Technology has translated Lakatos' *Proofs and Refutations: The Logic of Mathematical Discovery* (ed. by J. Worrall and E. G. Zahar, Cambridge Univ. Press, 1976) which will be published by Shanghai Publishing House for Translations. It is clear that, although some of Lakatos' major works and papers have already been translated into Chinese, problems still exist in printing and publication. Prof. Hong Qian, an early member of the Vienna Circle, took charge of editing *Selections of Western Papers on Philosophy of Science*. This book, which includes Lakatos' 'Falsification and the Methodology of Scientific Research Programmes' (translated by Huang Yaping, read and revised by Ji Shuili), will be published by the Education Press in Beijing.

In 1984, a book by Prof. Jiang Tianji of Wuhan University was published by the Chinese Social Science Press in Beijing with a distribution totalling 22,000. One chapter, one-tenth of the 286-page book, introduces Lakatos' *The Methodology of Scientific Research Programmes*. Prof. Jiang not only introduced Lakatos' papers 'From Naive Falsificationism to Sophisticated Falsificationism', 'Methodology of Scientific Research Programmes' and 'Meta-criteria to Evaluate Methodologies', but also gave an account of the criticism of Lakatos' historiographical methodology of scientific research programmes by N. Koertge, A. Musgrave, F. Feyerabend and H. Sarkar.

In the same year, Prof. Qiu Renzong of the Institute of Philosophy, CASS, compiled a book entitled *The Methodology and Dynamics of Science: An Introduction to Modern Philosophy of Science*. The book has been published by Shanghai Knowledge Press and the distribution totals 40,000. Three chapters totalling 41 pages and consisting of one fifth of the book are respectively entitled 'Sophisticated Falsification', 'Methodology of Scientific Research Programmes' and 'Philosophy of Science and History of Science'. These are introductions to Lakatos' philosophy of science. The rest of the book introduces Popper, Kuhn and Feyerabend. In his 'Epilogue', Qiu noted that "this book is primarily descriptive and not designed to shoulder the task of evaluation".[10] The two books by Prof. Jiang and Prof. Qiu have significantly contributed to the dissemination of Western philosophies of science, including that of Lakatos.

Zhexue Yanjiu (*Philosophical Research*, No. 11, 1985), a major philosophy journal in China, published a paper by Lin Xiashui entitled 'On Lakatos' Philosophy of Mathematics'. So far it is the only paper concerning Lakatos in this journal. Thus, it can be seen that the doctrines of Lakatos haven't exerted a very obvious influence on the Chinese philosophical community as a whole. In his paper Lin introduced Lakatos' quasi-empiricism in philosophy of mathematics, arguing that it 'has opened the way for a Euclidean type of closed deductive system and has struck a forceful blow against rationalistic conceptions in the philosophy of mathematics. It combines mathematical theories with empirical facts, and at the same time, acknowledges the deductive nature of theories. So his quasi-empiricisim is not a simple reversion to naive empiricism. His theory concerning potential falsifiers not only provides a specific form for the verification of formal mathematical theories, but also explains the complicated nature of the development of mathematical theories'.[11] But Lin asserted that the empirical elements of quasi-empiricism are not equivalent to the experiences of materialism, and "the problem of the truthfulness mathematical theories is still unresolved."[12] Moreover, Zhou Ying who is working in the Department of Social Sciences in the East China College of Chemical Engineering in Shanghai, wrote an article entitled 'Lakatos' Thought on the Philosophy of Mathematics'. This was based on his MS thesis, written for the Department of Philosophy, Zhongshan University. The article has been selected to appear in a collection of articles by Chinese scholars on the philosophy of science which is being compiled by Prof. Jiang Tianji and will be published in *Boston Studies in the Philosophy of Science*.

Since 1980, Lakatos' philosophy of science has not only had an impact on Chinese philosophical circles, but has also influenced historians of science, especially young researchers. For example, Wang Zuoyue, a graduate student in the Teaching and Research section of Dialectics of Nature, Graduate School, the University of Science and Technology of China, wrote a Master's thesis entitled 'The Discovery of the Equivalence of Wave Mechanics and Matrix Mechanics and the two Programmes of Quantum Theory'. The thesis, completed under the guidance of Prof. Xu Liangying, compared the two scientific research programmes in the history of quantum theory, i.e., the Einstein–de Broglie–Schrödinger research programme and the Bohr–Heisenberg programme. During May 20–25, 1986, 19 young Chinese historians of science held a 'Symposium on Western Theories of History of Science' in Qingdao. At that meeting someone introduced Lakatos' proposal to conduct research in history of science using the methodology of scientific research programmes, and heated discussions followed.[13]

The traditional Chinese models of thinking tend to be intuitive, ambiguous and dogmatic.[14] In the early half of this century, some Western doctrines were gradually introduced into China, including logic, scientific methodology, the New Realism of B. A. W. Russell, the logical empiricism and inductionism of the Vienna Circle and the convertionalism of H. Poincaré.[15] But these doctrines were not widely influential because at that time the Chinese people were engaged in the struggle for national independence and liberation. After the founding of New China, Marxist philosophy was widely popularized, including Engel's *Dialectics of Nature* and Lenin's *Materialism and Empirio-criticism*. In 1950's, Soviet research on philosophy of natural science was emphasized, while Western philosophies of science were all regarded as bourgeois and were dogmatically excluded. Since 1978, the Chinese Communist Party and the government have adopted the policies of opening to the outside world and "letting a hundred schools of thought contend". Western philosophies of science have been widely disseminated in China. These include Popper's falsificationism, Kuhn's theory of paradigms, Lakatos' methodology of scientific research programmes and philosophy of mathematics, etc. Although these doctrines have been rejected and opposed by some dogmatists, they have already had a positive influence on Chinese philosophers of science, especially the young scholars and students, encouraging them to smash the fetters of dogmatism and conduct many-sided research.

NOTES

[1] Owing to the lack of information, I only talk about what has happened in Mainland China and don't mention the situation in Taiwan.

[2] In September 1980, I left the post of editor-in-chief of this journal.

[3] I have held the posts of vice-editor-in-chief and executive director of the Association since September 1980.

[4] *Journal of Dialectics of Nature*, No. 4, 1982, p. 19.
[5] In the spring of 1986, I chaired the committee responsible for the examination of Lan's doctoral dissertation.
[6] *Journal of Dialectics of Nature*, No. 6, 1983, pp. 3–16, pp. 72–74.
[7] Since 1984, I have held the post of director of the section.
[8] *Journal of Dialectics of Nature*, No. 1, 1984, pp. 21–28.
[9] *Journal of Dialectics of Nature*, No. 6, 1983, pp. 17–24.
[10] Qiu Renzong, *The Methodology and Dynamics of Science: An Introduction to Modern Philosophy of Science*, Shanghai Knowledge Press, 1984, p. 194.
[11] *Philosophical Research*, No. 11, 1985, p. 53.
[12] Ibid., p. 52.
[13] *Journal of Dialectics of Nature*, No. 4, 1986.
[14] Bao Zunxin, 'On the Problem of the Nature of the Culture at the End of the Ming Dynasty and the Beginning of the Qing Dynasty', *Guang Ming Daily*, June 23, 1986.
[15] Fan Dainian, 'A Brief Review and Prospect on Researches in Philosophy of Science in China', *Studies in Dialectics of Nature*, No. 2, 1986.

Chinese Academy of Sciences

RISTO HILPINEN

ON THE CHARACTERIZATION OF COGNITIVE PROGRESS[1]

I

In this paper I shall discuss cognitive progress in terms of *belief systems*, and take "cognitive progress" to mean progressive changes in belief systems or the progressive development of beliefs. This discussion can easily be translated into talk about progress in science or about progressive theory changes; I prefer to speak about belief systems becuse of the great generality of this notion. By a "belief system" I understand a set of propositions (or beliefs) together with a system of evidential relationships among these propositions, in other words, a potential description of what some person might believe in some situation. I do not assume that belief systems have to be consistent or logically closed. Progressive changes are obviously changes which improve a belief system and make it more satisfactory than it was earlier. The evaluation of belief systems and judgments of cognitive progress depend of course on the assumed objectives of belief formation. In the history of methodology, we can find at least two "classical" theories (or types of theories) of cognitive progress, which may be termed *the accumulation theory* and *the convergence theory*.[2] According to both theories, belief formation has essentially one main goal which may be termed *the Truth*.

In the accumulation theory, the Truth is regarded as an aggregate or collection of truths (true propositions), and progress occurs when more and more of these truths are included in the belief system: progress is *growth*. This conception of cognitive progress has been ascribed to many philosophers and scientists of the 17th century: Bacon, Descartes, Locke, Newton, and others.[3] Its main drawback is that it presupposes an infallible method of arriving at new truths (or expanding one's belief system). The convergence theory assumes that it is meaningful to speak about the "distance" between a belief system and the Truth, and that cognitive progress takes place when the belief system comes closer to the Truth. This "approach towards the Truth" does not mean that more and more truths are included in the belief system (as it would according to the

K. Gavroglu, Y. Goudaroulis, P. Nicolacopoulos (eds.), Imre Lakatos and Theories of Scientific Change. 69–80.

accumulation theory), but rather that the (false) propositions in the system are replaced by propositions which are somehow "less false", i.e., closer to the corresponding truths. (But the "approach towards the Truth" may also involve the inclusion of new (true or false) propositions in a belief system.) The convergence theories are based on a numerical metaphor which seems difficult to apply to systems of propositions.[4] These theories include Peirce's theory of truth as the "limit of inquiry" and Popper's theory of verisimilitude and its contemporary descendants.[5] The convergence theories are more complex and in that respect perhaps more "realistic" than the simple accumulation theory. In view of the evident implausibility of the accumulation theory, many contemporary philosophers of science prefer some variety of the convergence view. Some authors have defended this approach on the ground that it is consistent and perhaps even required by scientific realism.[6]

However, the accumulation theory has one clear epistemological advantage over the convergence theories: it makes the progressive belief changes *manifest* or immediately identifiable. The cost of this advantage is the (unrealistic) assumption of an infallible method of belief acquisition. If a similar assumption is attached to convergence theories, viz. that a proposition which has been found to be false can always be replaced by one which is closer to the truth, the "approach towards the Truth" becomes manifest too, with a consequent loss in plausibility and realism. (This assumption might be termed the postulate of "limited infallibility". Laurens Laudan has called it the assumption that scientific method is "strongly self-correcting".[7])

In the present paper I shall outline a somewhat different conception of cognitive progress. I wish to argue that the judgments about cognitive progress are not necessarily dependent on any assumptions about the convergence of our belief towards a "true final opinion": it is possible to make meaningful judgments of progress even in areas where "the final truth" cannot be assumed to exist. Moreover, it seems to me that such judgments do not presuppose (scientific) realism in any strong sense.

The main features of the analysis presented below are suggested by Imre Lakatos's observation that[8]

> any scientific theory has to be appraised together with its auxiliary hypotheses, initial conditions, etc., and especially, together with its predecessors so that we may see by what sort of *change* it was brought about. Then, of course, what we appraise is a *series of theories* rather than isolated *theories*.

Formulated in terms of belief systems, this observation means that the judgments of progress always concern some *sequence* of belief systems B_1, $B_2, \ldots, B_m$ and in particular the *changes* between the successive members of such a sequence.

II

As was mentioned above, the evaluation of belief systems and the judgments of cognitive progress are relative to the assumed objectives of belief formation. I assume that the main function of a person's belief system is to provide satisfactory answers to the questions which are important for him and interest him. The content of this apparent platitude depends on what is meant by "satisfactory answer to a question". According to the traditional view of the pragmatics of questions, a genuine (or serious) question is an expression of ignorance and of the wish to know something. For example, according to Charles Peirce, serious interrogation or questioning involves[9]

> first, a sense that we do not know something; second, a desire to know it; and third, an effort – implying a willingness to labour – for the sake of seeing how the truth may really be.

Lennart Åqvist and Jaakko Hintikka have recently formalized this conception of questions by treating them as *epistemic imperatives* or *epistemic requests*: according to Åqvist and Hintikka, a simple alternative question[10]

(1) p or q?

can be analysed as

(2) Bring it about that I know that p or I know that q.

If the questioner is denoted by 'a' and the addressee (the person to whom the question is presented) by 'b', (2) can be presented formally as

(3) $ID_b(K_a p \vee K_a q)$,

where 'I' is an expression for the imperative mood, 'D_b' is an action operator which stands for "b brings it about that", and 'K_a' is the epistemic operator, read as "a knows that". Jaakko Hintikka calls the proposition in the scope of the action operator 'D_b' the *desideratum* of the question, and the proposition obtained from the desideratum by deleting (all occurrences of) the epistemic K-operator the *presupposition* of the question.[11] (Thus (3) has the disjunction '$p \vee q$' as its presupposition.) More generally speaking, a proposition is a presupposition of a question if the question has no correct answer unless the proposition is true. For example, a why-question, "Why p?" has the proposition p as its presupposition.

III

If we accept the view that a person knows something only if his opinion is justified and correct (i.e., some version of the justified true belief account of knowledge), the Åqvist–Hintikka analysis of questions entails that p is a satisfactory answer to a question Q (for a person a) only if

(i) p is true, and
(ii) p is justified within a's belief system B_a.

We might say that p is *prima facie* satisfactory with respect to Q if p is a complete answer to Q and a is justified in believing that p (p is justified within B_a).

In some earlier papers I have argued that if we take information and truth (or freedom from error) as the principal objectives of inquiry, as William James and many other philosophers have suggested, these objectives must be relativised to some set of questions or problems; thus the classical Jamesian view of inquiry turns out to be practically equivalent to the view that the purpose of inquiry is to provide satisfactory answers to certain questions.[12]

Let $B, B_1, B_2, \ldots$ be belief systems, and let QS be an arbitrary set of questions. Some questions in QS may be irrelevant to the evaluation of the systems B_i: only questions which *can arise* in the systems matter. A question can arise in a system only if its presuppositions are satisfied by that system: for example, the fact that a belief system does not give any answer to an explanation-seeking question "Why p?" is not a defect if p is false according to the system. Thus we may adopt the following conditions:

(D1) Q can arise in a belief system B only if its presuppositions hold in B.

(D2) B is satisfactory with respect to QS if and only if B gives a satisfactory answer to every question Q in QS which can arise in B.

The observations made earlier give us the following partial analysis of the notion of "satisfactory answer":

(D3) B gives a satisfactory answer to a question Q only if B contains some proposition p such that (i) p is a complete answer to Q, and (ii) p is justified within B.

(The concept of a *complete answer to a question* is here taken as given; I shall not try to analyse it here.) If a belief system B contains several conflicting (mutually incompatible) answers to a certain question Q, it contains incorrect information concerning Q, and we can say that such a system B is *not apparently error-free* with respect to Q. (A system which does not contain inconsistencies of this kind need not be free from error, but the error is not obvious or apparent.) Conflicting answers to a question "defeat" each other, and cannot be justified with B; thus we obtain the following condition:

(D4) A proposition p is justified within B only if B is apparently error-free with respect to every Q in QS to which p can serve as a potential answer.

IV

An *inquiry* is a process by which a belief system B_1 which is in some respect unsatisfactory is transformed into a new system B_2. If the inquiry is successful, the new system is in some respect more satisfactory than the original one.

Let us consider a single question Q. A belief system B may be related to Q in the following four (mutually exclusive) ways:

$R0$: Q cannot arise in B.
$R1$: Q can arise in B, but B contains no (complete) answer to Q.
$R2$: Q can arise in B and B contains an answer to Q, but the answer is not satisfactory (justified) because it is not supported by any evidence, or B is not apparently error-free with respect to Q.
$R3$: B contains a satisfactory answer to Q.

This classification suggests one way of making the much discussed distinction between the logic discovery and the logic of justification: we might regard the logic of discovery as an account of the conditions under which B is related to Q by $R2$ or $R3$, and the logic of justification as a theory of the conditions under which B is related to Q by $R3$.[13]

On the basis of $R0$–$R3$ we can define 16 possible belief change types Rij with respect to a question Q: Rij is a change from a belief system B_1

which is related to Q by Ri to a system B_2 which is related to Q by Rj. Some of these changes are obviously improvements of the system (given the two objectives of belief formation mentioned above, informativeness and error avoidance), whereas others seem cognitively undesirable. The following change types seem to be of the latter sort:

$R21$: loss of information, loss of an informative answer to Q,
$R31$: loss of justified (and thus presumably correct) information, and
$R32$: the replacement of a satisfactory answer by an unsatisfactory one.

Scientists seldom change their beliefs in the way described by $R21$ and $R31$; they are reluctant to give up an informative proposition h, that is, an informative answer to a question, unless it can be replaced by another, equally informative answer to the same question. (The latter proposition may be termed an "informational counterpart" of the former.) I have called this principle elsewhere "the principle of information retention".[14] This principle is taken for granted by Lakatos when he defines what he calls a "theoretically progressive problem shift" in terms of a sequence of theories: he assumes that each theory in the sequence has at least as much content as the unrefuted content of its predecessor.[15]

Change type $R32$ is also undesirable, but it cannot be avoided if an inquiry undertaken for some purpose unexpectedly casts doubt upon a proposition which was earlier regarded as a fully satisfactory answer to some question Q.

The reversals of the change types mentioned above can be regarded as improvements of a belief system, that is,

$R12$: the discovery (or invention) of an answer to a previously unanswered question (which however turns out to be unsatisfactory),
$R13$: the discovery of a satisfactory answer to a question, and
$R23$: the replacement of an unsatisfactory answer by a satisfactory answer.

Change types $R13$ and $R23$ represent fully satisfactory inquiries; they lead to what Peirce called "the settlement of opinion".[16]

Change types $R0j$ and $Ri0$ ($i, j = 0, \ldots, 3$) are in themselves neither desirable nor undesirable. Changes of the latter kind (with $0 < i$) simply make a question irrelevant to the evaluation and comparison of the two belief systems. According to the principle that a change type Rij is progressive if and only if its reversal *Rji* is a regressive change (this principle might be termed the Principle of *Symmetry*), changes of type $R0j$ $(0 < j)$ should also be regarded as neutral changes. Change types *Rii* are (in themselves) also neutral with regard to progress.

V

The analysis of belief change types presented above contains the following simplification: the distinction between $R1$ and $R2$ was described above in terms of the absence or presence of an answer to a certain question in a person's belief system; in other words, the distinction is based on the division of the members of B into non-answers and complete potential answers to Q. But it is also possible to make distinctions among the potential answers to Q on the basis of their informativeness; thus the distinction between $R1$ and $R2$ is a matter of degree. This holds for the distinction between $R2$ and $R3$ as well: we can also make comparative judgements regarding the degree of trustworthiness or reliability of an answer to a certain question. Thus the belief change types $R12$, $R13$, and $R23$ may be described as follows:

$R12$: increase in the informativeness of B_a with respect to Q,
$R13$: increase in the informativeness and trustworthiness of B_a with respect to Q, and
$R23$: increase in the trustworthiness of B_a with respect to Q.

I am assuming here that the "trustworthiness" of an answer depends on how firmly justified it is within the system. The reversals of these change types, $R21$, $R31$ and $R32$, can be defined in an analogous way in terms of the decrease in the informativeness or in the trustworthiness (or reliability) of the belief system with respect to Q. The account of belief change types given in Section IV is a simplification, but not a misrepresentation: the preceding observations can simply be taken to mean that the boundaries between the classes $R1$, $R2$, and $R3$ can be drawn in different ways, but this does not affect the classification of the basic belief change types given here.

Progressive, regressive, and "neutral" belief change types B_1–B_2 with respect to a question Q can be summarised in the form of the following table (Table I), where '-1' represents a regressive (or negative) change, '0' a neutral change, and '$+1$' a progressive (positive) change:

Table I. Evaluation of basic belief change types

		B_2			
		$R0$	$R1$	$R2$	$R3$
B_1	$R0$	0	0	0	0
	$R1$	0	0	+1	+1
	$R2$	0	−1	0	+1
	$R3$	0	−1	−1	0

VI

The view that $R23$ is always a progressive change and its converse a regressive one might be questioned. In the case of $R32$ we find out that a proposition which was regarded as a fully satisfactory answer to a question is actually something less, merely an unfounded and problematic assumption. The inquiry has taught us something; why is this not cognitive progress?

This objection is not completely unjustified, but it can be answered without giving up the basic conception of cognitive progress propounded above. Instead of the two belief systems B_1 and B_2, let us look at a longer sequence of systems, say one leading from B_1 to B_n. Suppose that the change from B_1 to B_n (with respect to a question Q) is of type $R33$, that is, in B_n a fully satisfactory answer to Q has been recovered. If B_n is *in other respects* (that is, with respect to some other questions) an improvement over B_1, we can say that the change from B_1 to B_n with respect to Q, and the change from B_1 to B_2 as well, has been *part of a progressive development* according to the criteria of cognitive progress outlined above. A similar observation can be made about changes of type $Ri0$ and $R0j$ in the context of longer sequences of belief changes. For example, consider a set of questions (Q_1, Q_2, Q_3). Let us assume that a person's belief system has changed in the way described in Table II (from B_1 to B_4) with respect to this set:

Table II. Belief changes in relation to Q_1–Q_3.

	B_1	B_2	B_3	B_4
Q_1	3	2	3	3
Q_2	2	3	2	3
Q_3	1	3	3	3

This table contains the indices of the R-relations which the questions Q_i bear to the belief systems B_j. It is clear that the change from B_1 to B_4 is a progressive belief change (according to the criteria of cognitive progress outlined above) if all the questions Q_i are taken into consideration, even though it is not a progressive belief change with respect to Q_1 alone.

According to the view of cognitive progress taken here, the judgments of progress are *defeasible*, subject to revision in the light of new evidence. In the example considered above, the replacement of B_1 by B_2 seems a progressive change with respect to Q_2, but the new evidence provided by B_3 makes it necessary to revise this judgment. On the other hand, if we consider only question Q_1, B_2 appears inferior to B_1, but taking account of questions Q_2 and Q_3 seems to reverse this judgment. It is important to note here that the revised judgments of progress are based on the same criteria of cognitive progress as the original judgments; no essentially new criteria are needed. It should also be observed that "a satisfactory answer to a question" means here an answer which satisfies an investigator's epistemic (or cognitive) requirements in a given knowledge situation, i.e., which should be judged to be satisfactory in a given situation.

I have assumed here that the judgments of cognitive progress are determined by two basic objectives (or *desiderata*) or belief formation, the acceptance of new (relevant) information and the avoidance of error. The analysis presented above might now be criticized as follows: In a change of type $R12$, a person accepts a certain answer to a question, and in so doing always risks introducing error into the system. A change of this kind improves the informativeness of the belief system with respect to a certain question, but the change need not unambiguously progressive, since it may conflict with the requirement of error avoidance.

But it is clear that in a situation in which an investigator has no answer to a question, he cannot avoid the risk of error if he is interested in finding an answer. We might say that in a situation of type $R1$, informational adequacy is the primary requirement, but in $R2$ type situations error avoidance is the investigator's principal concern. Different stages of inquiry are governed (or dominated) by different cognitive objectives; thus $R12$ can be regarded as a progressive change – with certain obvious qualifications. The proposition accepted must be a possible answer to the question (relative to the investigator's belief system); it should not be in an obvious conflict with the requirement of error avoidance, in other words, it should not introduce obvious or explicit inconsistency into the system. A change of type $R12$ represents what Charles Peirce called *abductive reasoning*, whereas $R23$ represents (or results from) *inductive* reasoning in Peirce's sense.[17] Progressive changes of the former kind are what Peirce might have called *plausible* or *reasonable* abductions.

VII

Above I have argued that the judgments of cognitive progress are dependent on the two main objectives of belief formation, viz., the acceptance of relevant information and the avoidance of error. These criteria of cognitive progress are *prima facie* criteria in the sense that the judgments based on them are subject to revision by new evidence, but they are the only criteria available to us: the revised judgments are based on the same general criteria of progress as the original ones. A belief change which is judged to be progressive does not necessarily bring the investigator "closer to the truth" (in an objective sense). Reasonable and justified judgments of progress are possible even in areas where no general "approach towards the truth" cannot be assumed to occur. The "approach towards the truth" depends on certain additional objective characteristics of belief formation, especially on its *effectiveness* and its *reliability*. The effectiveness of belief formation means that the investigator is likely to find a (potentially) satisfactory answer to a question in a given problem situation, and reliability means that an answer which has been found is not likely to be undermined by new evidence. If the methods of belief formation used in inquiry are both effective and reliable, progressive belief changes are also apt to lead to what Peirce called "the ultimate fixation of opinion".[18] But this "fixation of opinion" is obviously relative to the questions under consideration: when new questions and problems arise, new belief changes become necessary.

NOTES

[1] This paper is based on work supported by the U.S. National Science Foundation Grant No. IST-8310936 (Principal Investigators Jaakko Hintikka and C. B. J. Macmillan) and by the Finnish State Council for the Humanities Grant No. 09/053.

[2] Cf. Laurens Laudan, 'Peirce and the Trivialization of the Self-correcting Thesis', in L. Laudan, *Science and Hypothesis*, D. Reidel, Dordrecht 1981, pp. 226–51; Ilkka Niiniluoto, 'Scientific Progress', in I. Niiniluoto, *Is Science Progressive?*, D. Reidel, Dordrecht 1984, pp. 75–110; see especially Sections 2–4 (pp. 76–82).

[3] See Laurens Laudan, *op cit.*, p. 227.

[4] This is the basis of Quine's well-known criticism of Peirce's conception of truth as the limit or "the ideal result" of inquiry; cf. W. V. Quine, *Word and Object*, MIT Press, Cambridge, Mass. 1960, p. 23.

[5] See *The Collected Papers of Charles Sanders Peirce*, Vol. V, ed. by Charles Hartshorne and Paul Weiss, Harvard University Press, Cambridge, Mass. 1934, paragraph 5.407. For Popper's theory of verisimilitude and some of its contemporary varieties, see Graham Oddie, *Likeness to Truth*, D. Reidel, Dordrecht 1986, Chs. 1–3.

[6] Cf. Ilkka Niiniluoto, *Is Science Progressive?*, ch. 7.

[7] Laurens Laudan, 'Peirce and the Trivialization of the Self-Correcting Thesis', in *Science and Hypothesis*, p. 232.

[8] Imre Lakatos, 'Falsification and the Methodology of Scientific Research Programmes', in Imre Lakatos, *Philosophical Papers, Vol. I. The Methodology of Scientific Research Programmes*, Cambridge University Press, Cambridge, 1978, pp. 8–101; see p. 33.

[9] *The Collected Papers of Charles Sanders Peirce*, Vol. V, paragraph 5.584.

[10] See Lennart Åqvist, *A New Approach to the logical Theory of Interrogatives, Part I: Analysis*, Filosofiska Föreningen i Uppsala, Uppsala 1965, Introduction, section 2; Jaakko Hintikka, *The Semantics of Questions and the Questions of Semantics*, Acta Philosophica Fennica 28: 4, North-Holland, Amsterdam 1976; 'Answers to Questions', in *Questions*, ed. by Henry Hiż, D. Reidel, Dordrecht 1975. pp. 279–300; 'New Foundations for a Theory of Questions and Answers', in *Questions and Answers*, ed. by F. Kiefer, D. Reidel, Dordrecht 1983, pp. 159–190.

[11] Jaakko Hintikka, 'Answers to Questions', pp. 279–286; 'New Foundations for a Theory of Questions and Answers', pp. 174–175.

[12] I have discussed this issue in greater detail in my paper 'The Semantics of Questions and the Theory of Inquiry', forthcoming in *Logique of Analyse* (1987). The question theoretic (or interrogative) model of inquiry has recently been studied in an interesting way by Jaakko Hintikka; see his paper 'On the Logic of an Interrogative Model of Scientific Inquiry', *Synthese* 47 (1981) 69–83.

[13] This way of making the distinction between the logic of discovery and the logic of justification resembles the account of the logic of discovery given by Norwood Russell Hanson in 'Notes Toward a Logic of Discovery', in *Perspectives on Peirce*, ed. by Richard J. Bernstein, Greenwood Press, Westport, Conn. 1965, pp. 42–65.

[14] Risto Hilpinen, 'On the Principle of Information Retention', *Philosophia Naturalis* **21** (1984) 435–443.

[15] Imre Lakatos, 'Falsification and the Methodology of Scientific Research Programmes', in *Philosophical Papers, Vol. I: The Methodology of Scientific Research Programmes*, p. 33.

[16] See *The Collected Papers of Charles Sanders Peirce*, Vol. V, paragraph 5.375.
[17] For the distinction between abductive and inductive reasoning, see *The Collected Papers of Charles Sanders Peirce*, Vol. V, paragraphs 5.158–5.197, and Vol. VII, ed. by Arthur W. Burks, Harvard University Press, Cambridge, Mass., 1958, paragraphs 7.218–7.220.
[18] *The Collected Papers of Charles Sanders Peirce*, Vol. V, paragraph 5.430.

University of Turku

PART II

MARCELLO CINI

CONTINUITY AND DISCONTINUITY IN THE DEFINITION OF A DISCIPLINARY FIELD: THE CASE OF XXth CENTURY PHYSICS

1. THE DIVIDING LINE BETWEEN SCIENCE AND OTHER CLAIMS OF KNOWLEDGE

It is by now an almost trivial statement that the growth of scientific knowledge is not a cumulative linear process. I wish, however, to argue in favour of a less conventional claim: namely that a sharp line cannot be drawn dividing the rational reconstruction of reality allegedly performed by science by means of purely logical procedures firmly grounded on factual data, from other kinds of belief based on individual or collective experiences leaving more or less room for subjective factors.

To make the issue clear let me stress that I am not questioning that science represents actual features of the piece of reality which is being investigated. My point is that it does so only within an interpretative frame invented by the subject of this cognitive activity. The result therefore cannot be looked at as if it were a pure reflection of the object's properties, but always reproduces a relationship whose form leads back to the subject's active role, in spite of the accuracy of the object's portrait, no matter how abundant its details may be.

Traditional wisdom sometimes admits that this is the case, but adds immediately that it is possible, and indeed necessary, to erase all the tracks left by the subject, thereby attaining objective knowledge, provided definite universal rational rules are followed. My claim is, on the contrary, that these rules are neither universal nor given once and for all, but rather depend strongly on cultural traditions and social environment. This means therefore, that even when a new scientific contribution, after having been scrutinized in the light of the accepted rules, has been recognized as valid knowledge, we cannot assert that the subjective "impurities' introduced in the cognitive activity have been eliminated by the sieve of abstract rationality. The validity recognition simply means that the accepted contribution conforms to the standards established by the social subject entitled to perform this task.

K. Gavroglu, Y. Goudaroulis, P. Nicolacopoulos (eds.), Imre Lakatos and Theories of Scientific Change. 83–94.

The first thing to do, however, in order to explain what my claim implies, is to clarify the meaning of the word "subject". A careful distinction must be made between the individual scientist and the community to which be belongs.

This distinction corresponds to two different moments in the process of growth of scientific knowledge. The first, truly individual, moment is that of invention, in other words, the moment when a proposal is formulated for an innovation in the body of shared knowledge. The second truly social, moment refers to the appraisal of this proposal by the relevant disciplinary community and its final acceptance or rejection.

Innovation always causes a change in the rules of the game: by claiming that certain facts become irrelevant compared to other ones, by discovering analogies between groups of phenomena that had hitherto been considered unrelated, by inventing new concepts to explain the collection of already known empirical data, by eliminating unsolved problems as pseudo-problems, or by turning into questions statements which had been considered obvious up to that moment. The causes which lead the individual scientist to the formulation of this proposal of change are however numerous, different and hidden.

No reconstruction, however accurate, of the historical circumstances, the cultural traditions and the social environment in which the new idea was born, can therefore provide a really satisfactory explanation of its origin. These reconstructions are on the contrary fundamentally important for the understanding of change in the shared beliefs of a scientific community because they throw light on the main factors of the mechanism of acceptance and validation by this community of the individual new contribution to the construction of scientific knowledge.

The keepers of the rules of the game may accept or reject requests for their change. However, in order to reach a decision, they must base their judgement on other rules, of a different logical type, which are not given once for all, but are, nonetheless, not arbitrary.

2. THE HIERARCHY OF LEVELS OF KNOWLEDGE

It is one thing to judge whether a given contribution satisfies the validity conditions which stem from the ensemble of formalized rules that characterize a given discipline at a given moment in time. It is quite another one to judge whether a given proposal to change those rules is acceptable in the light of the metarules that fix the norms that should not be abandoned, at that given moment in time, by the practitioners of that particular discipline.

For those of you who are acquainted with the work of Gregory Bateson it will be clear that the distinction is completely analogous to the one made by this author between different levels of learning in his celebrated essay 'The logical categories of learning and communication'.[1] In particular, the distinction is analogous to that between two different types of change, introduced by Paul Watzlawick, a leading exponent of the Palo Alto school in his studies on the "pragmatics of human communication".[2]

But we are not dealing with simple analogy. In all cases, we are talking about separating the changes that come about within a given context from the changes of the context (and ultimately, in their turn, from the changes in the class of changes of the context). The growth of scientific knowledge, therefore, does not escape the general modalities of knowledge acquisition by humankind. These modalities are in fact based on the possibility of ordering the information contained in the messages that produce this acquisition in a hierarchy of classes, each of which is an element of the one immediately above it and which, at the same time, includes those at the next lower level. It is thus the identification of the information contained in each message, and the attribution of each piece of information to the level that is considered appropriate, which produces a growth of knowledge. Knowledge is not, therefore, any longer seen as the simple, undifferentiated accumulation of new contributions, but as a process of enrichment and reordering of this complex system of relationships between classes of propositions about the surrounding world, classes of propositions about the preceding propositions, and so on.

For the purpose of example, we may recall that the process of learning "by trial and error" is for Bateson only one given level, which must not be confused either with the lower one (learning level zero: a given response for each given stimulus) or with the higher one (learning level two: ability to change the ensemble of alternatives from which one makes the choice at the lower level).

This example appears as particularly significant for the purposes of our argument if we recall that Popperian evolutionary epistemology considers the growth of knowledge to be equivalent, in fact, to a process of learning by trial and error.[3]

In the light of what has been said above, the limit of this conception seems clear: it consists in the reduction of the complex hierarchy of the levels of control of the process of the development of knowledge to the sole level of the elimination of error in a context assumed given once and for all. It is not by chance that to this reduction there corresponds in Popper a unidimensional vision of reality which sees clouds at one extreme and clocks at the other.[4]

This way of looking at things also facilitates the discussion of the views of Popper's traditional opponent in the debate of 20 years ago, Thomas Kuhn. It is clear that his periodization of the history of science into phases of normal science, separated by brief periods of revolutionary change,[5] comes much closer to the process I have outlined, in the sense that the former clearly correspond to the acquisition of knowledge within a given context, while the second correspond to changes of the context (in Watzlawick's terms, $change_1$ and $change_2$). This dichotomy is, however, excessively schematic. Paradigmatic change is assuredly a change in the criteria that regulate the bounds of acceptability of the contributions that are regarded as valid for the development of a given discipline, but it too is also subject to *metarules* which are no less binding even if less formalized. What seems to be missing from Kuhn's conception is exactly an awareness of this hierarchy of levels of selection and control which has the function of ensuring the maintenance of the identity of a given discipline, at the same time as allowing the acceptance at the lowest possible level of the changes necessary for its survival.

This hierarchical structure explains, moreover, why the entire discussion between Kuhn and his opponents on the nature of the discontinuity implied by the concept of scientific revolution is a discussion that can go on *ad infinitum* if it is not understood that a change in paradigm consists simultaneously in a discontinuous change of the rules that define *how* to do, let us say, physics and the maintenance of the metarules that define *what physics is*. These latter, too, can in their turn change in a more or less discontinuous fashion, but even in this case it will always be possible to isolate *something* which ensures the continuity of the discipline.

This articulation of the concept of scientific revolution also resolves another problem that has been the subject of a long dialogue of the deaf: that of the circularity implied by the relationship between paradigm and community. If the community is defined as the ensemble of scientists who share a given paradigm, and this latter is nothing other then the rules which the scientists belonging to a given community have inherited and which they utilize for the development of their discipline, then neither the former nor the latter is defined.

If, however, the levels are clearly distinguished, everything becomes clearer: the community is composed of everyone who shares the *metarules* to be used for changing the rules of "normal" research activity by safeguarding what they maintain constitutes the indispensable identity of the discipline. This definition is not hollow nominalism. It is the only one that explains why an Einstein or a Schrödinger, who most certainly had every right to form part of the physics community, was in fact pushed out onto the sidelines after 1927. In the same way it also explains why a von Neumann, an outsider with respect to the physics community in that by profession he was a mathematician, became on the other hand one of its most authoritative members.

3. THE CRITERIA OF CHOICE OF THE DISCIPLINARY COMMUNITY

Before going in more detail into the identification of these *metarules* or criteria of choice, let me make a brief digression. I am not a professional philosopher. However, I believe that my point of view is not completely meaningless in connection with the problems we are discussing here. If anything, at least because, having been for many years the Editor of a well known Physics Journal, *Il Nuovo Cimento*, I have a good experience of the procedures adopted for accepting or rejecting new contributions to the body of my discipline's knowledge.

You probably will not be too surprised if I tell you that these procedures differ considerably from those recommended by professional methodologists. This is not because we are wicked our corrupted or emotional as some of you might think. Maybe we are, of course, but these are not the only drives behind our actions. The reason for having to adopt a variety of more flexible criteria is simply that logical consistency and empirical adequacy are not as precise and unambiguous terms as methodologists assume they are. To illustrate this point I will refer to the results of recent empirical investigations conducted by two sociologists of science, Gilbert and Mulkay, published under the title: 'Warranting Scientific Belief' in the journal *Social Studies in Science*.[6]

I am aware that sociologists do not enjoy a very high reputation in this audience, but I nonetheless insist that knowing how scientists behave is sometimes more fruitful than assuming that they do not exist. Let me quote therefore the conclusions of the work of these authors.

Scientists accounts of action and belief – the authors write – are systematically organized in ways which both explain and provide scientific justification for the speaker's actions, whilst explaining and condemning those of his opponents. Condemnatory accounts achieve their effect by linking actions and beliefs to contingent personal and social factors. In such accounts speakers "show" that specific actions and beliefs would have been otherwise if it had not been for the impact of scientifically irrelevant influences. On the other hand the justificatory account is organized to "show" that the speaker's actions or beliefs could not properly have been otherwise. This can be achieved on some occasions by presenting particular actions or judgements as following necessarily from a procedural rule which can be taken for granted. More frequently however our scientists formulate their justificatory accounts or theory choice in straightforward empiricist terms – that is present their actions and beliefs as following necessarily from what has been revealed experimentally to be the case in the natural world.

If things go this way the question immediately arises: who is right? The finding that competing theories and interpretations are credited with different epistemological status by their proponents and their opponents respectively, shows that additional criteria must be used, in order to chose between the two, if the scientific community has ultimately to resolve the controversy.

First of all we find criteria to evaluate the scientific status of the proposal, or, in other words, if the proposal is relevant for the discipline in question. Some of you probably think that this talk does not fall within the boundaries of philosophy of science.

As we shall see in the example I will discuss later, there are always some *a priori* features considered indispensable by a disciplinary community in order to make sure that the scientific image of the discipline is guaranteed. It is therefore in the light of this pattern of features that the proposals of enlargement of its boundaries are scrutinized.

Secondly, we find criteria for judging whether the issue at stand is an open problem. Once the question of scientific relevance has been settled, it has still to be decided whether a new proposition about a set at factual data needs to be proved or has to be taken as evident or perhaps should be assumed as a postulate. Theories have to stand on something. But the choice of this something, after the proposal by an individual scientist has been made, has to be endorsed by the community. The question whether the wave function collapse is an open problem is quite controversial in the physics community today. Another example is the choice made by the physicists at the beginning of this century between the opposite views of Lorentz and Einstein concerning the need of understanding why the electromagnetic phenomena appeared to satisfy the relativity principle.

Next we have different criteria to establish what is considered to be a valid explanation. In different situations communities are satisfied with claims of explanation possessing very different epistemological status. The most striking example is the difference between deterministic explanations and probabilistic ones. We all know that Einstein disliked the idea of "God playing dice". Today, however, an explanation which postulates the intrinsically random nature of a given phenomenon does not raise any objection.

Fourthly we recognize criteria of formal adequacy. Relevant examples are those pertaining to simplicity, elegance, beauty, internal coherence of a theory or a law. Schrödinger considered disgusting the idea of quantum jumps. Heisenberg, on the contrary, held "the physical content of Schrödinger's theory as hideous."

Finally we find criteria to judge about the empirical adequacy of a theory. We know from numerous examples that agreement or disagreement with a given experiment is considered more or less important according to the credit given by the community to the theory in question. Experimental results conflicting with the dominant paradigm have been disregarded until a new theory needed them as a corroboration. On the contrary other ones, which have been disproved later, were assumed immediately as important evidences in support of theories which were to be abandoned successively. Crucial experiments, as Lakatos has explicitly stressed, are recognized as such only in retrospective.

4. SCIENCES OF LAWS AND SCIENCES OF PROCESSES

Let us now abandon physics for a moment and have a look at the whole spectrum of natural sciences. It is common practice to divide them into two types: on the one hand the nomological sciences, characterized by the search for and the statement of necessary and universal laws of nature which are consequently capable of making rigorous predictions, and, on the other, the evolutionary sciences, considered incapable of acceding to the sphere of universality since they are given over to the investigation of irrepeatable processes, and thus are at last able to provide hypothetical reconstructions of successions of past events. To this distinction corresponds a substantial difference in epistemological status. In point of fact, it is the former (the sciences of laws, or of *why*) which are regarded as having the full right to be called sciences since they are verifiable or falsifiable (according to differing points of view) with respect to nature. The latter on the other hand (the sciences of processes or of *how*) are often maintained by epistemologists to be second class sciences.

A number of questions, then, immediately spring to mind. What is it that fixes the criteria of definition of the different disciplines and places them on one side or the other of the watershed? Up to what point do these criteria stem from the nature of the subject, that is to say from the collection of phenomena and facts that constitute the (approximately closed) ensemble to be interpreted and explained, and to what extent on the other hand do they depend on the choice made by the scientists who select the data and the experiment to construct their science, thus placing it on the one side or on the other? It is commonly maintained that it is the object investigated that determines the nature (the science of *why* or the science of *how*) of the discipline that studies it, the scientist being limited to taking note of what there is before him or her. But, in actual fact, this demarcation is much less objective than one might think. The case of the most recent developments in the field of biology provides a convincing demonstration of this.

Up to the middle of the seventies, this "science with a schizophrenic personality" – as M. Ageno defined it recently[7] – was present in two completely different guises. On the one hand functional biology, essentially a reductionist discipline, was given over to the analytical study of each organism, determining its structure and internal processes right down to the minutest details. It thereby tended to bring the explanation of all biological phenomena down to the events that take place at this latter level and therefore, in point of fact, to reduce biology to the physics and chemistry of the molecules involved. On the other hand evolutionary biology, instead, considered living organisms as indivisible entities whose particular characteristics come out only at the level of the totality and are only partially deducible from an analysis of the constituent subunits.

We are faced, then, with two biologies: the first is to be located in the group of the sciences of laws, or of *why*, while the second is that of the sciences of processes, or of *how*. But the revolution that has taken place over the last few years following on the discovery of unexpected properties of eukaryotic DNA has changed this picture quite radically. Let me quote Ageno:

> Far from being something invariant, which is in essence conserved within the framework of the dynamic of the organism, DNA now appears to be involved in a dynamic all of its own, one that is incessant and to a great extent dominated by random events. Face to face with the multiplicity of *a priori* equivalent solutions, the search for causes, the quest of the reasons,

shows itself to be surprisingly indecisive and unimportant, and molecular biologists are ever more led to ask themselves *how* each solution has come into being, through what chain of events and in what general environmenal conditions.

What has been happening in biology in fact indicates that the traditional demarcation, in this case at least, seems to have been created more by choice of the scientists themselves than by the properties of the objects under study. These events show, in point of fact, that the molecular biology community, aiming at retaining for its own discipline the normative science character that distinguished it from the start, has as far as is possible eliminated from this disciplinary field all those phenomena which would have compromised its image by the impossibility of fitting them into the framework of general and atemporal laws. Similarly, they show that evolutionary biology, by accepting this division of spheres of competence, continued – as long as this division did not enter into crisis – to consolidate its own original nature as a science of irreversible and non-repeatable processes.

At this point then, the doubt arises as to whether this mode of procedure, consisting in choosing the objectives of the investigation and the corresponding interpretive categories so as to ensure consistency between the development of the discipline and the epistemological status attributed to it by definition, is typical not only of biology but might be a common praxis adopted by the different scientific communities to define their own identities.

As we shall see, the history of twentieth-century physics acquires a new interpretive dimension if one sees it as the result of a consistent undertaking by the scientific community to keep for this discipline the characteristics of a science "of laws" by as far as possible excluding from its bounds all those phenomena, together with their interpretations, which could have introduced into it some characteristics of the sciences "of processes".

The formulation of quantum mechanics at the end of the twenties represents, if seen in this light, the success of this operation – albeit at the price of giving up determinism in the strict sense – by thrusting back and even outside its boundaries all that has to do with the unpredictability, irreversibility, randomness that characterize phenomena such as turbulence, dynamic instability, stochastic processes, the thermodynamics of irreversible processes, etc.[8]

This point of view also explains why physics has maintained and protected its image as science of the simple, excluding complexity as much as possible as one category that characterizes the reality selected as the subject of its investigation. Thus it is that we witness an unchallenged dominance of reductionist ideology, with priority being assigned to the search for the "elementary" constituents of particles, which still today constitutes the essence of the most prestigious branch of physics.

5. THE CONSEQUENCES OF THE CHOICE

My thesis is therefore that the scientific community, through its choice at the end of the twenties, wanted to restore to physics the character of being a science of general and immutable laws, a science founded on the Galilean conviction that "the great book of nature is written in mathematical language and its symbols are triangles, circles and other geometrical figures", a character that the crisis of the first decades of the century had thrown much doubt on. The essential element of this reassertion consisted in bringing the chance, probabilistic aspects of the new mechanics back within the rules of a logico-abstract algorithm through the elimination of their temporal character. From this point of view, the new mechanics emerges more as the legitimate heir of Newtonian mechanics than as its implacable opponent. For as we see, the equations of motion in both theories are deterministic and reversible, and the temporal evolution of the quantities that represent the state of a system is no other than the deployment of a succession of changes that contain nothing new, in as much as they are potentially included in any one of the past or future states of the succession itself, chosen arbitrarily.[9]

I now propose to illustrate this thesis by showing how the way out selected has implied the removal of a series of basic problems that the crisis of classical physics has raised by postponing their being taken into consideration for several decades. Here, I should like to make it clear that I do not wish to argue that the physics community ought to have tried to formulate an alternative theory to QM.

I should, rather, wish to underscore the ideological hostility that for a long time hindered any attempt to explore possibilities that existed in other directions, possibilities which are only after sixty years are currently being investigated with some success.

The first consequence of the choice made is the disappearance from physics of the concept of irreversibility from the end of the nineteenth century right up to our times.

I would recall only that Planck's programme, which then unexpectedly led to the black body law and to its interpretation in terms of quanta, set off with the intention of demonstrating that the electromagnetic field contained within a cavity would irreversibly reach the equilibrium state at a given temperature by virtue of no more than the equations of motion. Only when this objective proved unreachable did Planck abandon the description in terms of a temporal evolution and adopt the statistical one in terms of the probability of states that Boltzmann had already invented after having himself run into the same difficulties.[10] Time thus disappeared from physics in the sense that the evolution of a statistical ensemble towards equilibrium ceased to be a problem worthy of interest: statistical mechanics was reduced to the calculation of partition functions at equilibrium. And it is exactly this point on which was based the introduction of discontinuity, a discontinuity that was to become the seed from which QM germinated. One had to wait until 1954 for the Kolmogorov theorem to discover that the property of ergodicity postulated by Planck is far from obvious for complex systems. The problem, then, was opened up again, but history had by now run its course.

Another expulsion happened for another subject born within classical mechanics at the end of the nineteenth century with Poincaré's famous note on the three-body problem: that of dynamic instability. This was a theme which in its turn opened up an unexpected breach in Laplacian determinism, which seemed a necessary consequence of Newtonian mechanics. Developed in particular by the astronomers in the 1960s, and taken up independently in numerous other contiguous and allied disciplines, the subject is based on the fact that completely deterministic nonlinear dynamic systems can have a "wildly chaotic" behaviour.[11]

A third field of research, which was pushed out onto the sidelines by the physics community and which only recently has received a great impetus – coming back into physics by the window after having been pushed out through the door, as one might say – is that of stochastic processes. Even the foundation in the '30s, on the part of Kolmogorov and the Russian school, of such a fruitful new discipline as the classical theory of probability passed practically unobserved by physicists right up until the most recent times. Only in the '70s, the techniques of stochastic processes began, in point of fact, to become fashionable among theoretical physicists who dealt with statistical mechanics and field theory.

It is too soon to say whether this change, which for the moment seems to consist essentially in the adoption of more efficient and flexible techniques, is the prelude to a conceptual shift in respect of the attitude to be adopted towards problematics typical of the sciences of processes.

Personally, I am convinced that we do assist to a substantial change in the rules adopted by the physics community in the evaluation of acceptable contributions to the development of this discipline.

If this is the case, however, I doubt very much that what is going on may be interpreted as a consequence of a straightforward application of a set of universal methodological rules invented with the purpose of drawing a demarcation line between science and non-science.

NOTES

[1] Bateson, G., 'The logical categories of learning and communication' in: *Steps to an Ecology of Mind*. Chandler, 1972.
[2] Watzlawick, P., Beavin, J. and Jackson, D., 'The pragmatic of human communication'. W. W. Norton, New York, 1967.
[3] Popper, K., 'The Tree of Knowledge' in: *Objective Knowledge*, Oxford University Press, London, 1972.
[4] Popper, K., 'Clouds and Clocks', op. cit.
[5] Kuhn, T., *The Structure of Scientific Revolutions*, Univ. Chicago Press, Chicago, 1962.
[6] Gilbert, G. N., Mulkay, M., *Social Studies in Science* **12** (1982) 383.
[7] Ageno, M., *Le radici della biologia*, Feltrinelli, Milano, 1986.
[8] Heims, S., *John von Neumann and Norbert Wiener*, MIT Press, 1981.
[9] Jammer, M., *The Philosophy of Quantum Mechanics*, Wiley, New York, 1974.
[10] Kuhn, T., *Black Body Theory and the Quantum Discontinuity*, Oxford Univ. Press, London, 1978.
[11] Casati, G., and Ford, J. (eds.), *Stochastic Behaviour in Classical and Quantum Hamiltonian Systems*. Springer Verlag, Berlin, 1979.

Universitá 'La Sapienza', Rome

PETER CLARK

DETERMINISM, PROBABILITY AND RANDOMNESS IN CLASSICAL STATISTICAL PHYSICS

INTRODUCTION

In the early 1970s while still a young and ignorant graduate student at L.S.E., my supervisor Imre Lakatos suggested in his usual irresistable way that I should look into the historical evolution of thermodynamics and statistical mechanics. He was convinced that the introduction of statistical methods into physics was 'a second great scientific revolution' and that historians had 'got it wrong'. Anyway, he thought I was bound to learn something and shouldn't make too much of a mess of it. I like to think he was right on all counts. The result of my efforts was my doctoral dissertation, a case study in Imre's methodology of scientific research programmes entitled *Thermodynamics and the Kinetic Theory in the Late Nineteenth Century* submitted in 1977. My first paper in the area was also originally read in Greece in September 1974 in Napflion and was published as 'Atomism versus Thermodyamics' in Howson's *Method and Appraisal in the Physical Sciences*.[1] I am grateful to Imre for introducing me to this particular area of the history and philosophy of physics and for encouraging my research in it – for it is an area of immense historical and philosophical richness.[2] One of the major reasons for the philosophical richness of the area is that it is the scientific meeting point of two apparently irreconcilable metaphysical ideas, the first being determinism which arises naturally in the simple fact that the systems treated in classical statistical mechanics are (naturally enough) *dynamical* systems obeying Newtonian mechanics and they are deterministic in their time evolution. However, the second idea involved is that because we are dealing with an ensemble made up of a very large number of such dynamical systems or of a *single* system made up of a very large number of component systems (e.g. the molecules of a dilute gas) the behaviour of the system (the evolution of its macrostates) can be treated *probabilistically*, despite the determinism of the underlying mechanics. Indeed various stochastic postulates are employed in statistical mechanics, varying from very strong ones asserting the *randomness* of certain sequences, to weaker ones stating merely the

K. Gavroglu, Y. Goudaroulis, P. Nicolacopoulos (eds.), Imre Lakatos and Theories of Scientific Change. 95–110.

existence of a probability density. There is certainly something of a conceptual issue here: to what extent can *deterministic* systems, if any, satisfy stochastic postulates? It is essentially to this issue that this paper is devoted. I believe that for anyone interested in probability and randomness statistical mechanics has much to say that is surprising and interesting. I shall argue after setting up the problem in a little more detail that *deterministic* systems can most certainly satisfy non-trivial probabilistic constraints, when these are interpreted quite objectively and that that satisfaction has little or nothing to do with initial conditions. I shall further point out how on at least one account of randomness (the Von Mises–Church account) there are sequences which are simultaneously both deterministic and random (in a specified character). I take it then the relations between the three central notions involved are not, as they might *prima facie* appear, nor as they are often described, elementary. Let us however first try to sharpen up the initial problem from the rather vague formulation it has received so far.

1. STATISTICAL CONSTRAINTS ON DETERMINISTIC THEORIES: THE CONSISTENCY PROBLEM

Let us take statistical mechanics for these purposes to subsume three classes of theories, kinetic theories, equilibrium statistical mechanics, and non-equilibrium statistical mechanics. First kinetic theory: this theory seeks to explain by exact consideration of the molecular mechanics of a single system of moving molecules (in particular collisions given by a specific model of the atom or molecule – a paradigm example would be Maxwell's 1868 *Dynamical Theory of Gases*[3]) the macroscopic properties of the system. Of particular importance in the theory is the explanation of the transport properties of gases, that is their viscosity, thermal conductivity and thermal diffusion. Essentially such theories are based upon an integro-differential equation (referred to as the Boltzmann equation) an equation arrived at by adopting a specific *statistical assumption* concerning the lack of correlations obtaining among the positions and velocities of colliding molecules. In effect which particular specific statistical claim one makes depends upon exactly which dynamical correlations induced among the moving molecules by collisions one is prepared *to ignore*. Kinetic theories also seek to explain not merely the equilibrium characteristics of gases but their approach to equilibrium, in other words it seeks to explain irreversible phenomena, those subsumed under the

Second Law of Thermodynamics. A hierarchy of kinetic theories can be developed by systematically weakening the strength of the statistical postulate introduced. It is to be noted by way of an example (if perhaps somewhat extreme) that in the case of Boltzmann's deduction of 'the most probable distribution' (sometimes called the single system H-theorem[4]) combinatorial techniques from probability theory take over entirely from dynamical considerations. Here it is simply assumed that the points (in position-velocity space) representing the moving molecules are shuffled by collisions in a random way. The reshuffled points being then more likely to end up in a high probability distribution than in a low one. But here of course probabilistic postulates have taken over from mechanics, even though we know that the motions of the representative points are governed in complete exactness by the laws of mechanics.

Let us then turn to the second group of theories, statistical mechanics equilibrium and non-equilibrium, while recalling that our present purpose is essentially to raise an issue, the issue of dynamic determinism versus the satisfaction of stochastic postulates. Again we are dealing with a *class* of theories, each member of the class employing a different stochastic postulate to describe the macroscopic behaviour of the system. The most famous such theory is Gibbs' statistical mechanics of representative ensembles.[5] Here we deal not with a single system but with a very large collection of 'copies' of the single system all prepared in the same observational state (e.g. all having the same total energy). Each such system is represented by a point in phase space. The statistical assumption introduced is in that theory introduced (*a priori*) to the effect that all sets of dynamical states compatible with the observational state of the system, and occupying equal volumes in phase space, are equally likely. On this account the statistical expectation of a function of the dynamic variables (say those defining the thermodynamic properties of the system in equilibrium) is equal to the average value of that function over the region of phase space consistent with the macrostate state of the system. But again the behaviour of the system is fixed exactly in all its aspects by the dynamical evolution of the system. How is the *a priori* postulate to be justified?

The situation is of course compounded in the case of non-equilibrium statistical mechanics. The reason is that in theories of this kind we want to use deterministic *and time-symmetric theories* (once again classical dynamics) to obtain an explanation of the well-known phenomena that the behaviour of physical systems that are far from equilibrium is not symmetric under time reversal. That is, we want to obtain explanations of facts of the type subsumed by the Second Law of Thermodynamics. Now there is no possibility whatever of combining time symmetric deterministic dynamics to obtain non-time symmetric theories, some postulate (e.g. hypothesis of molecular chaos must hold for some but *only* some of the time in the approach to equilibrium) of a stochastic kind must replace a purely dynamical statement. But this reintroduces a *consistency* problem common to all three classes of theories viz: how can it be consistent to add statistical assumptions to treat of an aggregate motion, the component submotions of which, being governed by the laws of mechanics, are entirely deterministic. The dilemma which appears to haunt classical (the situation in quantum mechanics is dynamically identical *in this respect*) statistical physics then is this: to explain the macroscopic phenomena associated with thermodynamics we need to introduce stochastic postulates 'on top of' Newtonian mechanics, but the latter theory is deterministic (at least *prima facie*) *so* won't we by introducing such postulates have overdetermined the resulting motion, resulting in inconsistency. This is the *consistency* problem, and it is surely quite real.[6]

Before examining the consistency problem further I can't resist a historical note. The authors of kinetic theory and statistical mechanics were certainly sensitive to the existence and difficulty of the consistency problem. It was known by Maxwell to be non-trivial since he realised that the assumptions employed in his first theory (1860) to the effect that the components of velocity of a *single* molecule were statistically independent, was inconsistent with particle mechanics for a system of particles of fixed energy. Further of course Boltzmann's original proof of the Second Law involved the assumption that the hypothesis of molecular chaos was satisfied for all of the time (which induces an *analytic* increase in an entropy function defined purely in terms of dynamical variables), an assumption which is inconsistent with dynamic reversibility. Modern physicists too express concern over the consistency problem. For example, according to Grad, a contradiction between the statistical hypothesis and the underlying dynamics is always to be expected. In discussing his 'levels of description' account he remarks in respect of the consistency problem:

even when the statistical hypothesis is a very good approximation to the actual state of affairs, it will usually be logically inconsistent with the exact dynamical formulation. It requires judicious care to separate valid consequences from paradoxes.[7]

The situation then in Grad's view would appear to be analogous (if I do not stretch it too far) to that in naive set theory: it is perfectly alright to use naive theory provided that we restrict our references to those sets we consider are not too big, i.e. occur somewhere on the cumulative hierarchy. This model roughly informs us when our deductions are likely to be safe. So in statistical mechanics one can only really forget correlations at the thermodynamic limit, everywhere else they are strictly false, and our deductions are only safe at the limit, or at the thermodynamic level of description.

2. THREE RESPONSES TO THE CONSISTENCY PROBLEM

The three views I wish to consider in this section are what might be called the ultra-subjectivist theory, the relationist theory and the Popperian theory. By the ultra-subjectivist theory I have in mind a view often attributed to Laplace that in deterministic contexts the only appropriate interpretation of probability is the epistemic one.[8] On such a view assignments of probability distributions can *only* reflect incomplete knowledge, they would be as one distinguished physicist put it 'expressions of human ignorance: the probability of an event [being] merely a formal expression of our expectation that the event will or did occur, based upon whatever information is available'.[9] Now that probability ascriptions can be such reflections of partial knowledge I have no doubt, but to the question, 'Must they so be in statistical mechanics?' I equally have no doubt that the answer is no. Now the physicist Jaynes whom I quoted above is often labelled as expressing a subjectivist theory of probability in statistical mechanics, but when one actually examines his account it seems to me better described as a *relational theory*.[10] What seems to be going on in general is this. We interpret statistical mechanics as explaining not the behaviour of dynamical systems, but rather as explaining the *relations* obtaining between observers and the dynamical configuration of the system. As such there will now be no straightforward contradiction with the underlying determinism of the classical mechanic and the statistical postulates of statistical mechanics because such postulates will not be defined on the dynamical system or its macrostates *alone*. The theory so formulated will no longer make predictions about the evolution of the states of such systems, but rather it will make predictions about the time evolution of functions defined upon the 'coupled' system, i.e. of observer and ensemble. In other words, statistical mechanics is about composite systems comprising an observer together with a physical system, it is not any one of those subsystems alone. Of course, such a theory is definitely not just a reinterpretation of statistical mechanics as usually understood, it requires new postulates to be introduced describing the flow of information between dynamical system and observer.

Let us now look at one answer to the consistency problem which is very radical. It is Popper's and it asserts that the consistency problem can't be solved. He says:

> Today I can see why so many determinists, and even ex-determinists who believe in the deterministic character of classical physics, seriously believe in a subjectivist interpretation of probability: it is in a way, the *only reasonable possibility* which they can accept; for objective physical probabilities are incompatible with determinism; and if classical physics is deterministic, it must be incompatible with an objective interpretation of classical statistical mechanics.[11]

Clearly the reader is being invited to contrapose. The objective interpretation of statistical mechanics is the right one (with this I agree) so classical physics must be indeterministic. The argument then is as follows: since the statistical assertions of statistical mechanics are just as objective as the 'mechanical' assertions of that theory and since Popper argues those assertions cannot be treated as such if the underlying dynamical theory is deterministic, then the underlying dynamical theory must after all be indeterministic despite its *prima facie* deterministic character. Popper uses an example from non-equilibrium statistical mechanics thus: in order to explain irreversible approach to equilibrium, we have to assume that the measure of the set of possible initial states which produce 'pathological' (i.e. non-approach to equilibrium) behaviour is zero and the measure of the set of possible initial dynamical states which do eventually yield equilibrium is one. Now this probabilistic hypothesis, that the set of 'pathological' states has measure zero has to be physically interpreted as a claim about *propensities*, but propensities only obtain where exactly the same situation or 'state' may yield in the time evolution of a system, different subsequent states, but that is incompatible with *determinism*. That is, Popper says: 'propensities can be accepted as physical realities (analogous to forces) only when determinism has been given up'.[12] But the indeterminism of observational states is trivial here. Whatever propensities are present must be grounded in the *dynamics* (that which in complete exactness controls the behaviour of the system). So we must have to have dynamical propensities (of a non-trivial kind i.e. not always zero or one) and thus dynamical indeterminism. However, in the context of classical statistical mechanics there is no other way to appeal but to classical mechanics, so giving up determinism must be the same as saying, after all classical mechanics is indeterministic. Now it is an interesting claim and one worth pursuing, that classical mechanics in some contexts may be indeterministic. However, in the context discussed it is manifestly false. Paradigmatically as employed in standard formulations of statistical mechanics the underlying mechanics is deterministic. This is precisely because the equations of motion (assumed to hold by that theory) for the dynamical system, the canonical Hamiltonian equations (where the Hamiltonian is independent of the time), are such that in fact we have a system of first order differential equations whose solutions *always exist* and are *unique* given an arbitrary initial point $(q_1, \ldots, q_N, p_1, \ldots, p_N)$ at some instant t. Since existence and uniqueness of solution of the differential equations of motion obtains the dynamical states of the system are determined. Since I want to agree with Popper that the physical probabilities occurring in classical statistical mechanics are objective, but disagree completely with him that underlying mechanics in *this context* is indeterministic, I must, accepting the validity of his argument, deny that 'objective physical probabilities are incompatible with determinism'. And so I do. But this point makes it doubly important that we get clear what we have in mind by determinism, somewhat more exactly than has been given so far.

3. THE FORMULATION OF DETERMINISM – MONTAGUE'S ACCOUNT

Despite the long and embarrassingly lavish discussions of determinism in the literature it is difficult to find a formulation of the thesis which does not presuppose that we already grasp the claim of the thesis prior to its formulation. The difficulty is to steer a course between triviality, i.e. every system turns out to be deterministic on the one hand and an overstrong claim on the other. Thus famously Russell has established the charge of triviality against a formulation of the condition involving the existence merely of function dependencies among states of a *unique* Universe and the time.[13] On the other hand, some philosophers (following Laplace) have suggested that determinism be identified with a form of global *predictability* (according to the methods of science).[14] However it is clear that predictability could fail for all sorts of reasons, e.g. the failure of any functional dependencies to be effectively computable (who is to say *a priori* that every law of nature involves functions recursive in the data), or simply that they be beyond our capacities to survey or compute for practical reasons (as say in statistical mechanics). Rather the basic intuition seems to me to be this: if we consider an isolated physical system and think of the history of that system as the 'graph' of the values of its state variables in phase space through time, then the system is *deterministic* iff there is one and only one possible path consistent with the values of its state variables at any arbitrary time. Now I do not think there is available a completely adequate account of the deterministic thesis, however I do think that the Montague–Earman approach using the notion of a *deterministic theory* is by far the best account and is one certainly sufficient for our purposes. To cut a quite long story short, a *theory* T is said to be *deterministic in the state variables* (say $\delta_1, \ldots, \delta_n$) when any two standard models or *histories* as they are called, of the theory, (i.e. when any two relational structures which are models of the theory) $\langle A, \{\delta_1^A, \ldots, \delta_n^A\}\rangle$ $\langle B, \{\delta_1^B, \ldots, \delta_n^B\}\rangle$ agree at some given time (say t_o), i.e.

$$(\forall i)(\delta_i^A(t_o) = \delta_i^B(t_o))$$

then they agree at all other times, i.e. $\forall r \in R, \forall i (0 < i \leqslant n)$

$$\delta_i^A(t_o) = \delta_i^B(t_o) \rightarrow \delta_i^A(t_o + r) = \delta_i^B(t_o + r)$$

(For all t, for all i $(0 < i \leqslant n)$, $\delta_i(t)$ denotes the value of δ_i *at* t). In short the constraint entails that for a deterministic theory if two histories (or models) have identical states at one time ($t = t_o$) then they have identical states at all times. A physical system may be said to be deterministic in the state variables $(\delta_1, \ldots, \delta_n)$ when its history *realizes* (or satisfies) a *theory* deterministic in the state variables $(\delta_1, \ldots, \delta_n)$. This characterisation of a deterministic theory and deterministic system fits very well with the dynamical theory underlying statistical mechanics. A history or realisation of Hamiltonian dynamics is then just the unique trajectory in phase space Γ_S generated by the time evolution of a dynamical system (S) obeying Hamilton's equations. Now the important point about this is that, as was pointed out by Montague, it reduces the question of the determinism of classical mechanics to a more or less exact question in the theory of ordinary differential equations.[15] That is to say, the problem reduces to one of the existence and uniqueness of solutions in the theory of ordinary differential equations. Clearly Montague's characterisation entails uniqueness of solution. Since non-uniqueness requires two 'models' of the theory describing the system (i.e. two trajectories), agreeing at one instant, but subsequently disagreeing as to the values of the state variables, this contradicts the determinism condition.

So interestingly and quite generally Popper is surely right that one cannot say *globally* that classical mechanics is deterministic. It is in his phrase only *prima facie deterministic*, and this is true even when, as is the case here, a very much weaker characterisation of determinism is given than is used by Popper. However, will any of this help us with the sub-theory of classical mechanics, employed in classical statistical mechanics? The answer is *no*. This is a consequence of the fact that for conservative systems studied there, the differential equations of motion (the Hamiltonian equations) satisfy the existence and uniqueness of solution property. This point cannot be too strongly emphasised for it is precisely *because of this property* that the consistency problem has a positive solution at all.

4. NON-TRIVIAL PROBABILITY IN A DETERMINISTIC WORLD

To see that this is so let us note the well-known fact that because of the existence and uniqueness of a solution of the differential equations of motion the time evolution of the dynamical states of a dynamic system is given by a unique trajectory in phase space. The trajectory might simply be a stationary point, the trajectory might also be a periodic orbit (a closed curve), but what it cannot be is a curve which intersects itself, for at any such point of intersection the uniqueness of solution condition would automatically fail. Thus in any time interval Δt the point on the trajectory corresponding to the state of the system at t is transformed into the *unique* point on the trajectory corresponding to the state at $t + \Delta t$. But of course during the same time interval *every* other point in the phase space is transformed uniquely by the solution to the differential equations of motion into another point into the phase space. Essentially all this means is that as the dynamical system evolves through time, that evolution induces a *flow* of the points of phase space. This mapping has the very important property that it characterises a one parameter (the time) semi-group, which means that if we consider any point in phase space and look at where the flow takes the point in time t and then subsequently in time interval δ, the final position of the point is exactly that to which the flow takes the point in time interval $t + \delta$. This is the familiar *natural* motion of the phase space. (Since for classical mechanics the equations of motion are time symmetric for every transformation which is dynamically possible, there exists an inverse transformation of the natural motion of the phase space which has the structure of a one parameter *group*). In virtue of the well-known theorem of Liouville[16] this natural motion is *measure* preserving. Thus in the case of an ensemble of Hamiltonian systems all having the same energy, we have a realisation of the theory of probability, the sample space is the surface of constant energy (E), the set of relevant outcomes is the family of measurable subsets of (E), and the probability measure on this family (μ) is just that the existence of which is guaranteed by Liouville's theorem. It is precisely because of the properties of the dynamically induced (deterministic) phase flow that makes the fact that μ is a probability measure *significant*. Indeed for those systems which are ergodic in their motion important statistical properties can arise. If the surface of constant energy has the further property that it is metrically indecomposable then the famous theorem of Birkhoff[17] (now for an individual system) informs us of a very interesting statistical property that *except for a set of initial points* (of trajectories) of *measure zero* then for any integrable function f of the state variables (consequently including those whose values determine the observational states of the system) the *average* value of f along the dynamical trajectory of the system is equal (in the limit as $t \to \infty$) to the average value of f determined over E. This surely provides a striking justification of Gibbs' theory of the micro-canonical ensemble, and the introduction of the hypothesis of equal *a priori* probability. This justification is not entirely watertight but can I think be made so when the property of ergodicity is satisfied.[18] The philosophical point has been put with immense clarity by Reichenbach; he wrote:

> With the ergodic theorem, deterministic physics has reached its highest degree of perfection: the strict determinism of elementary processes is shown to lead to statistical laws for macroscopic occurrences, the probability metric of which is the reflection of the causal laws governing the path of the elementary principle.[19]

The only quarrel I have with this, is the claim (true of course in Reichenbach's time) that this is the highest achievement of deterministic physics since in fact abstract dynamics provides a hierarchy of systems exhibiting stronger and stronger stochastic postulates. The hierarchy is as follows, with the arrow as entailment:[20]

$$\text{Bernoulli} \Rightarrow \text{Mixing} \Rightarrow \text{Weakly mixing} \Rightarrow \text{Ergodic}$$

What in fact we have here is a classification of dynamical systems according to the probabilistic postulates which they obey.

A number of points can be made about this, the first is that the question of consistency is resolved for physically interesting systems between the underlying determinism and the satisfaction by observables of probabilistic postulates. In what must be a triumph of theoretical physics, Y. Sinai proved in 1963 that a system of hard spheres moving in a rectangular parallelepiped according to the laws of mechanics was ergodic, mixing and Bernoulli (i.e. the simplest model of the classical ideal gas).[21] But obviously none of this shows that the consistency problem for every version of statistical mechanics is solved. Nor is it directly relevant to non-equilibrium statistical mechanics and irreversible processes, for ergodicity is irrelevant to irreversibility while mixing is merely a necessary condition. Nevertheless there is no doubt that non-trivial objective probability measures exist which are naturally generated by deterministic systems.

5. DYNAMICAL DETERMINISM AND PROPENSITY

In the class of dynamical systems which we have been considering the interpretation of probability ascriptions is perfectly clear and their physical meaning entirely unambiguous and surely objective. Indeed, if the dynamical properties are objective and if this notion has any meaning they surely are, so must be probabilistic properties, which arise directly from them. There does not seem to be any further question as to what probability might mean here. The question which does arise however is to what extent their present interpretation is consistent with the propensity theory of probability. It seems to me that Popper is just wrong when he asserts that 'objective physical probabilities are incompatible with determinism' for the reasons given above. Is he right however to assert that 'propensities can be accepted as physical realities only when determinism has been given up'? I must say that I do not know enough about the notion of propensity to decide this issue, but I do not think that a propensity theorist should take much comfort in the results and methods so far discussed. (In this respect my view is different from that of Shimony and Von Plato[22].) The general point is this, it is not as if there is something left out of the account of the physical interpretation of the probability measures introduced in abstract dynamics. They are precisely measures on the phase space of a dynamical system, they are not anything else nor need they be interpreted as such. Of course, one could say that for example ergodic systems or mixing sytems have a *dispositional property* (barring a set of initial conditions of measure zero) to exhibit in their long-term behaviour an equality of time average and phase average of any integrable functions of their dynamical states. But this would hardly be very revealing, especially if we want the dispositional property in question to be grounded in the essential character of the systems which in this respect is given *entirely* by their *dynamics*. I do not suggest that the propensity theory is incompatible with the physical probabilities employed here, but just that it is unnecessary and unrevealing. However it is at least clear that on Popper's view of propensity, no propensity interpretation would be available, since if it is to be grounded in the dynamics that dynamics is deterministic.

6. RANDOMNESS AND DETERMINISM

Earlier in my paper I alluded to the existence of sequences which were *both* random and deterministic. In conclusion I should like to say something more on this. Now randomness or simply lack of order is a very difficult notion to capture, especially any reasonable sense of *physical* randomness, or randomness in Nature. What one might call the standard account of the randomness of countably infinite sequences of outcomes (say of zeros and ones) that due to Von Mises and elaborated by Church[23] has associated with it a number of difficulties, not least being the well-known one that *no* finite sequence can be random. Nevertheless it is one intimately connected with the existence of relative frequency and consequently one most often employed in physics. As is very well known the basic idea is this, that a sequence of outcomes of a statistically regular trial is random with respect to a given character of outcome (say, in respect of a countable sequence of 0s and 1s, the character 1) if there is no *effectively computable place selection function* subsequences selected according to which have a relative frequency for the occurrence of the character in question (1) different from that of the entire sequence. What is of importance here is the requirement (a natural one no doubt) of effective computability on the place selection function. Now suppose we have a *theory* which involves state variables where values are given by arithmetically definable but not effectively computable functions of some parameter. Now a model of such a theory will be a countable sequence whose members are the values taken on by the state variables (assume there is just one state variable δ) for successive values of the parameter. What is it for such a *theory* to be deterministic in δ? According to the characterisation due to Montague (which meshes so well with dynamics) it is simply that any two models (or histories) agreeing on the values of the state variables at some arbitrary time t, agree on the values of the state variables at *all* times. So the condition of determinism for the theory in question is automatically satisfied by the condition that δ be a *function* of t. But now look at the sequence of 'outcomes' of values of the state variable. Can we not find a theory of the above sort in which because of the failure of effective computability of the functions involved, the sequence of outcomes forms a *kollectiv*? Indeed the answer to this question, is yes. Montague conjectured that this was so in his paper in 1962 and Humphreys proved that there is a theory which is deterministic and which has as a model a sequence of outcomes which is random in the sense of Von Mises–Church.[24] It follows then that randomness does not require indeterminism. Further since it is immediate that indeterministic processes *may* generate non-random sequences (though with probability zero) it is clear that there is no immediate intimate connection between the two notions of randomness and indeterminism.

There are a number of important points to make concerning Humphrey's result. First and foremost there is clearly here no appeal at all to initial conditions. The randomness inherent in the sequence is not introduced as in the theory of unstable systems by variations in the initial conditions themselves forming a random sequence hence pushing the problem back. Nor are we dealing with a situation in which there is inherent lack of knowledge of exact initial conditions. The randomness of the sequence of the values of the state variable being an inherent property of the *kollectiv* characterised by the deterministic theory. Essentially the notion of determinism involved must not employ predictability because the trick of satisfying the two constraints is pulled precisely by employing a sequence which is arithmetically definable but not effectively computable. Now it might be suggested that what's really gone wrong here (my view is of course that nothing has 'gone wrong') is that we have now yet another example of the deficiencies of the Von Mises–Church characterisation of 'randomness'. There *may* be something in this but if we look at the so-called complexity measures of randomness which depend upon the algorithmic complexity of a sequence as characterised say in Kolmogorov's account,[25] then the connection between randomness and indeterminism becomes even more remote, just on the intuitive grounds that determinism is quite a distinct notion from level of complexity. Deterministic systems can generate *very* complex highly irregular outcomes as can clearly be seen from the study of simple iterative processes (e.g. the map $S(x) = 4x(1 - x)$ on the unit square and its iterations). There is no equation between determinism and simplicity or come to that regularity except it be with respect to *identical* initial conditions.

There is certainly some doubt as to whether the notion of *physical randomness*, of 'physical disorder', as opposed to mathematical randomness, has any definite meaning. In an article written a long time ago (1878) Maxwell remarked of the 'confused' motion we call heat, that 'confusion, like the correlative term order, is *not* a property of material things in themselves, but only in relation to the mind which perceives them'. I think he was wrong about entropy and the related concepts of thermodynamics but there is surely a lingering question mark over the notion of absolute randomness in physics.[26] However, my readers (if I still have any), will no doubt be relieved to learn that I do not propose to pursue these considerations here.

NOTES

[1] Howson (1976), pp. 41–105.

[2] The issue to be discussed here is only one of the conceptual problems typical of the relation between statistical mechanics and thermodynamics. A further issue is the problem of the theoretical *reduction* of one theory to another (on involving in this case transition to the thermodynamic *limit*). A third issue is the explanatory power of phenomenological theories, the realism–instrumentalism issue arises paradigmatically here, see my (1982). A fuller treatment of the consistency problem can be found in my (1987).

[3] Maxwell (1868).

[4] Boltzmann (1877).

[5] Gibbs (1902).

[6] The consistency problem for the various formulations of the stochastic postulates is very clearly discussed in Penrose (1970), pp. 39–44 and pp. 144–151.

[7] Grad (1967), p. 62.

[8] This view seems quite widespread, for example Hooker (1972), pp. 71–2; Gillies (1973), p. 133 and Popper (1982b), pp. 104–118.

[9] Jaynes (1957), p. 622.

[10] There is an excellent account of Jaynes' general approach to probability in statistical mechanics to be found in Lavis and Milligan (1985).

[11] Popper (1982b), p. 105.

[12] *Op. cit*. and cf. Popper (1982a), pp. 93–104.

[13] Russell (1917).

[14] Popper (1950). One major reason for not identifying determinism with predictability is that it would render unintelligible the study of systems which are unpredictable in their behaviour but which for very good reasons one would want to regard as intuitively deterministic. Their systematic study seems to have been begun by Ulam and Von Neumann (Ulam and Von Neumann (1947)).

[15] Montague (1962), p. 332, cf. also Earman (1971).

[16] For a proof see Khinchin (1949), pp. 15–19.

[17] Birkhoff (1931), and Sinai (1977).
[18] See particularly Malamet and Zabell (1980).
[19] Reichenbach (1956), pp. 74–81. The quotation is from p. 79.
[20] See especially Lasota and Mackey (1985).
[21] Cf. Sinai (1977), Lecture 8.
[22] Von Plato (1982) and Shimony (1977).
[23] Church (1940).
[24] Humphreys (1978), p. 100 and pp. 105–112.
[25] Fine (1973), Chapter 5.
[26] Quotation from Maxwell (1878), p. 646.

REFERENCES

Birkhoff, G. D., 'Proof of the Ergodic Theorem', *Proceedings of the National Academy of Sciences*, U.S.A., **17** (1931) 656–60.

Boltzmann, L., 'Uber die Beziehung zwischen dem zweiten Hauptsatze der mechanischen Wärmetheorie und der Wahrscheinlichkeitsrechnung' *Wissenschaftliche Abhandlungen*, **2** (1877) 164–223.

Church, A., 'On the Concept of a Random Sequence', *Bulletin of the American Mathematical Society*, **46** (1940) 130–5.

Clark, P., 'Atomism versus Thermodynamics', in C. Howson (ed.): *Method and Appraisal in the Physical Sciences*, 1976, pp. 41–105.

Clark, P., 'Matter, Motion and Irreversibility', *British Journal for the Philosophy of Science*, **33** (1982) 165–86.

Clark, P., 'Determinism and Probability in Physics', *Proceedings of the Aristotelian Society*, Supplementary Volume LXI (1987) pp. 185–210.

Earman, J., 'Laplacian Determinism, or Is This Any Way to Run a Universe?', *Journal of Philosophy*, **68** (1971) 729–744.

Fine, T. L., *Theories of Probability: An Examination of Foundations*, New York: Academic Press, 1973.

Gibbs, J. W., *Elementary Principles in Statistical Mechanics*, New York: Dover Publications, 1902.

Gillies, D. A., *An Objective Theory of Probability*, London: Methuen, 1973.

Grad, H., 'Levels of Description in Statistical Mechanics and Thermodynamics', in M. Bunge (ed.): *Delaware Seminar on the Foundations of Physics*, **1** (1967), 49–76.

Hooker, C. A., 'The Nature of Quantum Mechanical Reality', in R. G. Colodny (ed.): *Paradigm and Paradoxes*, 1972, pp. 67–302.

Howson, C., *Method and Appraisal in the Physical Sciences*, Cambridge: Cambridge University Press, 1976.

Humphreys, P. W., 'Is "Physical Randomness" Just Indeterminism in Disguise?', *PSA* 1978, Vol. 2, 1978, 98–113.

Jaynes, E. T., 'Information Theory and Statistical Mechanics, I,' *Physical Review*, **106** (1957) 620–9.

Khinchin, A. I., *Mathematical Foundations of Statistical Mechanics*, New York: Dover Publications, 1949.

Lavis, D. A. and Milligan, P. J., 'The Work of E. T. Jaynes', *British Journal for the Philosophy of Science*, **36** (1985) 193–210.

Lasota, A. J. and Mackey, M. C., *Probabilistic Properties of Deterministic Systems*, Cambridge: Cambridge University Press, 1985.

Mackey, G. W., *Mathematical Foundations of Quantum Mechanics*, New York: Benjamin, Inc., 1963.

Malamet, D. and Zabell, S., 'Why Gibbs Phase Space Averages Work – The Role of Ergodic Theory', *Philosophy of Science*, **47** (1980) 339–349.

Maxwell, J. C., 'Illustrations of the Dynamical Theory of Gases', *Scientific Papers*, **1** (1860) 377–409.

Maxwell, J. C., 'On the Dynamical Theory of Gases', *Scientific Papers*, **1** (1868) 26–78.

Maxwell, J. C., 'Diffusion', *Scientific Papers*, **2** (1878) 623–46.

Montague, R. C., *Formal Philosophy, Selected Papers* edited by R. H. Thomason, New Haven: Yale University Press, 1962.

Penrose, O., *Foundations of Statistical Mechanics: A Deductive Treatment*, Oxford: Pergamon Press, 1970.

Popper, K. R., 'Indeterminism in Quantum Physics and in Classical Physics', *British Journal for the Philosophy of Science*, **1** (1950) 117–33 and 173–95.

Popper, K. R., *The Open Universe: An Argument for Indeterminsim*, London: Hutchison, 1982a.

Popper, K. R., *Quantum Theory and the Schism in Physics*, London: Hutchison, 1982b.

Reichenbach, H., *The Direction of Time*, Berkeley: University of California Press, 1956.

Russell, B., 'On the Notion of Cause', *Mysticism and Logic*, 1917, pp. 132–51.

Shimony, A., Introduction to Carnap's *Two Essays on Entropy*, Berkeley: University of California Press, 1917.

Sinai, A. G., *Introduction to Ergodic Theory*, Mathematical Notes, Princeton: Princeton University Press, 1977.

Ulam, S. M. and Von Neumann, J., 'On Combination of Stochastic and Deterministic Processes', *Bulletin of the American Mathematical Society*, **53** (1947) 1120.

Van Fraassen, B. C., *The Scientific Image*, Oxford: Oxford University Press, 1980.

Von Plato, J., 'Probability and Determinism', *Philosophy of Science*, **49** (1982) 51–66.

University of St. Andrews

C. ULISES MOULINES

THE EMERGENCE OF A RESEARCH PROGRAMME IN CLASSICAL THERMODYNAMICS

The value of a metascientific model is to be judged not so much on a priori epistemological grounds but rather by trying to apply it to concrete pieces of its purported object, viz. science. And the more examples of application we get, the better. Here I intend to add one more case study to those already existing in the literature which make use of the idea of research programmes. The case I propose to investigate, and for which I can only provide a rough and brief picture today, belongs to a physical discipline which, in broad terms, we could call the "study of heat" or also "thermodynamics". For different reasons, none of these labels is really adequate, but I have no better proposal at hand. The discipline as such was born in the late 18th century with the systematic study of heat phenomena by Lavoisier, Black, and others, and in the course of its historical existence, it has given rise to several research programmes (R.P.s) in Lakatos' sense. That is, the discipline we now call "thermodynamics' consists of more than one identifiable R.P., both diachronically *and* synchronically. In different periods of time from the end of the 18th century on we find different R.P.s for thermodynamics. This is certainly not surprising, but what is perhaps a bit more interesting is that we may also detect different R.P.s overlapping in time. Sometimes these synchronically coexisting R.P.s are rivals but, interestingly enough, sometimes they are not: They may have mutual friendly relationships and even the same author may be associated to different coexisting R.P.s. In general, the main identity criteria for R.P.s are not of a sociological, but rather of a conceptual and methodological nature. Willard Gibbs, the hero of the story I wish to tell today, is one example of this: He worked within two programmes of thermodynamics – the phenomenological and the kinetic–statistical one.

Thus, disciplines typically consist of several, not necessarily incompatible R.P.s, and thermodynamics is no exception to this. But, moreover, disciplines not only consist of R.P.s. Especially in certain periods particularly confused from the conceptual point of view, they may consist of a kind of research that is not associated with a definite R.P. Many important

K. Gavroglu, Y. Goudaroulis, P. Nicolacopoulos (eds.), Imre Lakatos and Theories of Scientific Change. 111–121.

empirical discoveries and mathematical developments may be carried through during such periods of "fluctuation", but they do not add up to a R.P. because they lack conceptual coherence. Again, the evolution of thermodynamics is a case in point here. Without being able to expound my argument in detail today, I would contend that in the period lying between the agony of the caloric theory in the early 1830s and the clear formulation of the kinetic R.P. in the late 1850s no R.P. in a precise sense is identifiable, which does not mean that work done in this period was worthless. On the contrary, we find then some of the most remarkable discoveries which prepared the ground for the crystallization of two subsequent R.P.s: the kinetic–statistical one in the hands of Clausius, Maxwell, and Boltzmann, and the phenomenological one in the hands of Gibbs. With hindsight, of course, we may reinterpret those discoveries as a prologue to some definite R.P., but we should never forget that, strictly speaking, such an interpretation is an anachronism and obscures the true nature of scientific development.

Of course, all these distinctions and considerations only make sense if one has a reasonably clear conception of what an R.P. is. Unfortunately, Lakatos himself was never quite precise about the nature of the components of an R.P. and the way they are supposed to be interrelated. Some of his former disciples and collaborators have tried to refine the notion, with different outcomes.[1] Without embarking in the subtleties of a Lakatos-interpretation, I'll take from the Lakatosian model those components that seem to me to be both formalizable *and* useful for historiographical research. These are essentially the notion of a "hard core", its distinction from the "protective belt", and the idea of progressiveness of a R.P. The first component consists of the fundamental principles giving unity to the R.P. and having the virtue of being unfalsifiable; the second essentially consists in a series of particular hypotheses, obtained from the hard core not by a process of deduction but by a different formal process which I would like to call "specification": The general schematic form and the parameters of the fundamental laws are successively specified in order to construct testable hypotheses, which have, other than the hard core, a quite restricted range of application. Finally, the whole structure is said to be progressive if it can be shown that an increasing number of particular hypotheses and successful applications is added to the initial set as time passes. That these elements of the Lakatosian model are formalizable by model theoretic means in a reasonably plausible way has been shown elsewhere.[2] I cannot go into this here. I rather content myself with an informal use of these notions, which I think will suffice for the present purposes. A more technical discussion should be left for another occasion.

The degree of fruitfulness that we should assign to a metatheoretical model like the R.P.-conception essentially depends on the perspicuity with which it allows us to detect interesting structures in science of which we previously were not aware. It may also help in interpreting the historical data in a more cogent way than before. I think the search for concrete R.P.s in science has this value. This is at least what I propose to show for thermodynamics.

An R.P. as I understand it is a diachronic entity, i.e. a cultural process in historical time, which has a beginning, a distinct evolution, and a genidentic unity. So, when applying our model to thermodynamics we should be able to identify at least one entity of this sort: something with a clear beginning, a unity, and different stages of evolution. I claim that the emergence and evolution of Gibbsian equilibrium thermodynamics from the 1870s up to the 1930s is an example of this kind. The R.P. I want to consider is a process with three main stages, sociologically corresponding to three generations of scientists, all of them adhering to the same basic assumptions. These stages are: the "initial" or "founding" period due to Gibbs himself; then, the generation of the Dutch school of physical chemists together with some French and German physicists around the turn of the century, and finally the epoch of the Nernst school and the North American thermodynamicists.

Admittedly, this picture somehow runs counter the received view of the history of thermodynamics. According to the popular view we gather from textbooks, the emergence of thermodynamics can be summarized in this way: By the end of the 18th century, Rumford's canon-boring experiments definitely refuted the caloric theory; then, in the 1820s Sadi Carnot set the stage for the new discipline but was unduly ignored until the work of Joule, Helmholtz, and Kelvin around 1850 established thermodynamics on a clear and solid base. All of this story is nothing but a legend. Suffice to note that many years *after* Rumford's experiments, the leading scientists who studied the phenomena of heat, like Dalton and Laplace, still were convinced calorists; in fact, the caloric theory was the only recognizable R.P. in this field to attain the status of a well-developed theory of mathematical physics around 1820 (cf., e.g., Fox, 1971). Furthermore, when Carnot wrote his famous memoir he was still explicitly thinking in terms of caloric. As for the work of Joule, Helmholtz, Kelvin, and others around 1850, for all its significance in establishing the equivalence between heat and work, it was a far cry from being a conceptually clear and unified paradigm. I may just quote the judgement of Keenan and Shapiro on these authors:

> Their concepts were confused and their definitions vague . . . For example, no single definition seems to fit the various usages of the term *heat*, whereas other terms of frequent occurrence were often even more vague in meaning. (Keenan and Shapiro, 1947, p. 915).

We have therefore to abandon the textbook picture of the history of thermodynamics, which is alas!, still quite widespread even among philosophers and historians, and look for a more accurate interpretation. The R.P.-model may help us in putting things in the right perspective.

Whatever the significance and nature of the developments in the study of heat during the decades just before 1870 – whether they were the last remnants of the caloric R.P. (as in Carnot and the first Kelvin) or the first attempts at a particle-mechanical program (as in Joule, Helmholtz, and Clausius), or none of this (as in Regnault) – what I wish to claim here is that phenomenological equilibrium thermodynamics in the present-day sense started as a clearly identifiable R.P. only with Gibbs.

If it can be said that the "founding act" of the research program for classical particle mechanics was the work of one single man, viz. Newton, then, in a strikingly similar sense, it can also be said that phenomenological equilibrium thermodynamics (PET) was initiated by just one person, viz. Gibbs. And in the same way we nowadays speak of "Newtonian mechanics" as a "scientific style", we could speak of "Gibbsian thermodynamics" as another fundamental style of science. The parallelism between Newton and Gibbs was already noticed by a scientist who mastered *both* "styles" as no one else did in his time: Ludwig Boltzmann. Indeed, he characterized Gibbs as "the greatest synthetic philosopher since Newton" (quoted in Arveson, 1936). Now, of course one cannot say that *thermodynamics* as such was invented by Gibbs alone; what I do claim is that the first encompassing "hard core" synthesizing all previous results into one conceptual mould and giving rise to a quite definite tradition and "way of looking" at thermodynamic systems was Gibbs' feat. Modern mechanics was not invented by Newton either. Newton's synthesis presupposed the scattered discoveries, concepts, and laws of almost a century of previous mechanical investigation starting with Stevin and Kepler, and going through Galileo and many others up to the Royal Society in Hooke's times. In the same way, Gibbs' work must be seen as the "classical synthesis" after a complicated crystallization process which had started with Carnot half a century before. The only other possible candidate as a founder of a thermodynamic R.P. in our sense would be Clausius. Some historians seem to evaluate him in this sense. Ironically enough, Gibbs himself, with his characteristic modesty, essentially contributed to this evaluation (cf. his quite remarkable review of Clausius' work in Gibbs, 'Rudolf Julius Emanuel Clausius', p. 262). Of course, there is no question about Clausius' great merits in providing the first complete analysis of the Carnot cycle, both in reversible and irreversible processes, and in introducing the entropy concept; nor is it my intention to play Clausius down to favour Gibbs and get thereby involved into a sort of "personalities struggle". The interesting issue for us is to find out whether an R.P. in our sense may be detected within thermodynamics and when its origins are to be located. I claim that these origins are quite definitely identifiable with Gibbs' monograph *On the Equilibrium of Heterogeneous Substances* of 1876 (in the following abbreviated as *Equilibrium*) in quite the same way as classical particle mechanics as such started with Newton's *Principia*, and not before. Actually, to give a more flexible picture of the emergence of PETs R.P. we could include in this phase Gibbs' two initial articles of 1873 (see the bibliography), which in a significant way prepared the ground for the "great synthesis" of 1876. (They could be compared to the role Newton's manuscript series *De motu* played in preparing the ground for the *Principia* – cf. Newton's *Unpublished Scientific Papers*.) Neither the historical process leading from Galileo to Newton's *De motu* nor the one leading from Carnot to Gibbs' papers of 1873 seem to be definite R.P.s; they rather belong to the category of "crystallization" processes for which a different kind of formal scheme still has to be devised. The main reasons I have for identifying PETs birth with Gibbs' writings published between 1873 and 1876 and not with, say, Clausius' monograph *Mechanische Wärmetheorie* of 1865 are the following. Gibbs' work for the first time introduced or explicitly constructed:

(1) the idea of a conceptually independent phenomenological theory of thermodynamics, completely freed of the conceptual remnants of the caloric theory and also of any molecular notions. (Clausius and others made their conceptual frame still dependent on the molecular hypothesis.) Gibbs' approach was "pure thermodynamics" for the first time – though he by no means was opposed to the molecular hypothesis, as his later work demonstrated; he just neatly differentiated both approaches;

(2) the notion of a thermodynamic system in its most general sense, not restricted to cyclic processes;
(3) the concept of entropy in a clear and distinct way as a function of state on a par with inner energy, and independent of the concepts of heat, cyclic processs, etc. and also independent of any molecular hypothesis – contrary to Clausius (see Clausius, 'Über die Bestimmung der Disgregation eines Körpers und die wahre Wärmecapacität'; Clausius, *Mechanische Wärmetheorie*): Clausius was unable to introduce entropy as a fully independent concept, and "defined" it in terms of a body's "*disgregation*" plus the transformational content of its heat (see further Garber, 1966 and Klein, 1978);
(4) the fundamental law of PET as a functional relationship between entropy and the rest of extensive parameters, or equivalently between energy and the rest. (In his papers of 1873 Gibbs for the first time expressed his "fundamental equation" by putting energy as a function of entropy and volume solely, whereas in *Equilibrium* he generalized the idea to cover systems with material exchange (see Hornix, 1978); there, the fundamental law already appears in the same form as in present-day textbooks;
(5) the notion of a chemical potential as a brand-new theoretical concept (cf. Brønsted, 1933);
(6) the additivity principles for entropy and energy (see Gibbs, 'A Method of Geometrical Representation of the Thermodynamic Properties of Substances by Means of Surfaces', p. 36);
(7) the 'Maximin' Principle for entropy and energy in equilibrium states (see Gibbs, *Equilibrium*, p. 56; further Duhem, 1907);
(8) most important theorems directly derived from the basic core, like the so-called 'Gibbs–Duhem' and 'Gibbs–Helmholtz' equations (cf. Duhem, 1907 and Hiebert, 'Ostwald')';
(9) the phase rule (cf. Duhem, 1907 and Partington, 1972).

In sum, there is to my mind little reason to doubt that the PET R.P. emerged within a quite definite lapse of time between 1873 and 1876 and that it was the work of Gibbs alone. Moreover, this core – though almost completely ignored in its birth country – soon gave rise to a research tradition abroad, especially in Continental Europe. Physicists like Maxwell, Boltzmann, Duhem, and Nernst, and physical chemists like Pirie, the Dutch school (Van der Waals, Van't Hoff, Rozeboom, etc.), Berthelot, Le Châtelier, Ostwald, and many others, soon grasped the

R.P.'s basic core and made important contributions to it. In the maturity phase of this R.P., one of its most conspicuous progatonists, Le Châtelier, writing about Gibbs, felt compelled to say in a characteristic "normal-scientist" mood:

> This creation of his [=*Equilibrium*] was so complete and perfect as it came from his hands that the fifty years that have passed have been able to add little or nothing to it. The numerous scientists who have in the meantime concerned themselves with like questions have accomplished little more than a paraphrase of his work. They have perhaps completed some points in more detail; but more often they have only applied to particular cases laws formulated by Gibbs (quoted in Stevens, 1927, pp. 161–162).

Though Le Châtelier is obviously making the typical understatement of a "normal scientist" involved in a fruitful research tradition, the expression of his feeling about the state of equilibrium thermodynamics in the 1920s gives further support to the hypothesis that here we are confronted with a clear case of a research programme in our sense.

Let me add some historical data to make the picture more complete. I already suggested that, in a first overview, one may divide the evolution of the R.P. here considered into three phases. The *first period* starts with Gibbs' preparatory papers: 'Graphical Methods in the Thermodynamics of Fluids' and 'A Method of Geometrical Representation of the Thermodynamic Properties of Substances by Means of Surfaces'. Both were published in 1873. Their modest titles do not reflect the "revolutionary" significance of their content. They already introduce energy and entropy as "parallel" functions of state in the present-day sense and PET's fundamental law for the simple case where there is no change in the mole numbers of the substance(s) in the system. The additivity principles for U and S explicitly appear in the second paper. In 1876, Gibbs publishes the first part of his epoch-making monograph *Equilibrium*. The second part had to be published in 1878. The part of 1876 is the one devoted to the hard core of equilibrium thermodynamics. The 1878 extension introduces further concepts in order to deal mainly with strained solids, capillarity, and electrochemistry. In *Equilibrium* (1876), Gibbs introduces the fundamental law in its complete form (which includes consideration of changes of masses of the different substances involved and the novel notion of a chemical potential), the "Maximin" Principle for entropy and energy, and the definitions of the intensive parameters. In sum, the essential elements of the core are fully and explicitly constructed.

Gibbs was not very explicit about concrete applications of his formalism, but from the context it becomes clear that he thought it was obviously applicable to all usual kinds of thermodynamic processes known before him, especially gases at high to room temperatures. His new applications were composite systems of the same substance in different phases (like water in the gaseous, liquid, and solid phase), equilibrium between saturated solutions and the corresponding crystals, and some cases of chemical equilibrium *sensu stricto*.

The first well-known scientist to react very positively to Gibbs' work was Maxwell. He immediately saw that *Equilibrium* offered a new and most fruitful approach to thermodynamics (cf. Maxwell, 1876), and avowed in a letter to Tait that he himself had been confused about the basic notions of thermodynamics before (see Garber, 1966). Subsequently, he started divulging Gibbs' ideas as much as he could. He was quite instrumental in that some physical chemists, especially Pirie, studied *Equilibrium* and applied it to problems of chemical equilibrium already during this seminal period (see Pirie, 'Letters to Gibbs'; further Seeger, 1974).

The main advances in the R.P. during the *second period*, theoretically and also (even more) on the level of applications, were made by the Dutch group of physical chemists. Early in the 1880s, Van der Waals became acquainted with Gibbs' work and made Rozeboom aware of it. Rozeboom, Van't Hoff, and other Dutch scientists subsequently applied PET's formalism, especially the "Maximin" Principle and the phase rule to many problems of chemical equilibrium, e.g. the behaviour of much diluted solutions, mixtures, and alloys. Rozeboom popularized Gibbs' paradigm in his treatise, *Die heterogenen Gleichgewichte vom Standpunkt der Phasenlehre* (1901–1904), which may be seen as a sort of culmination of PET for this period (cf. Partington, 1972). Also, the behaviour of gases was studied more systematically and new laws were devised. Van der Waals had stated his law already in his Dutch dissertation independently of Gibbs' formalism but after its translation into German in the 1880s it became integrated into the new theoretical framework by Van der Waals' own collaborators (cf. Partington, 1972). A very important specialization of PET obtained in this period by the Dutch group was the virial expansion: Kamerlingh Onnes formulated this law in a version slightly different from the present one in 1901; in 1912 he and Kesson brought it into the present form (cf. Partington, 1972 and Rowlinson, 1958).

In the 1880s and 1890s several French scientists, especially the group around Berthelot, incorporated PET's ideas into their own researches. Within this context, Berthelot formulated his law for real gases in 1899.

German scientists also began to know and use Gibbsian thermodynamics in the early 1880s. Boltzmann was the most outstanding example (see Boltzmann, 'Zur Energetik').

Another great German follower of Gibbs was Ostwald. He translated *Equilibrium* into German (1892), thereby making it available to a wider public, since the American original paper was very difficult to obtain; but more importantly he used PET's formalism for a great variety of concrete applications in physical chemistry (cf. Hiebert, 'Ostwald'). He also speculated about using PET as the conceptual grounding stone for his "energetics" program.

The most characteristic feature of the *third period* is the introduction, modification, and ever-growing application of Nernst's so-called "Third Law". This is not to say that the other specialization lines of PET went into oblivion. Much work on and with the virial expansion was done before and after the First World War; quite useful new laws for real gases (like the law of Beattie–Bridgeman) were developed in the 1920s; and surprising applications of the phase rule continued to be found (see, e.g., Partington, 1972 and Rowlinson, 1958). I omit detailed consideration of these developments because they would add nothing to the essential argument. On the other hand, Nernst's law really meant a striking new development for equilibrium thermodynamics in general, both theoretically and in the domain of applications.

Strictly speaking, I should divide the third period into two subperiods: the first from around 1906 to the end of the First World War, the second comprising the period between the wars. During the first one, Nernst himself was very much involved in "his" law. He (and Planck) reformulated it several times and started a huge experimental program on processes at very low temperatures to confirm its validity. Afterwards, he became disinterested in active research on it, while his followers pursued research in this empirically elusive field with much ingenuity. The most remarkable among them was Franz Simon, who confirmed the applicability of the law beyond doubt in the late thirties, after strong criticisms had been raised against Nernst' claims (for all this, see Hiebert, 'Chemical Thermodynamics and the Third Law', and Hiebert, 'Nernst').

It is to be remarked that, from the very beginning, Nernst used Gibbs' basic core quite explicitly. To arrive at his "Third Law" he essentially made use of the Gibbs–Helmholtz equation, which is a central theorem of the core. He even preferred Gibbs' own original formulation and notation for this theorem (cf. Hiebert, 'Nernst').

To sum up, this evolution shows the essential features of the general R.P.-model: It has a clear origin in time with the construction of a clearly identifiable hard core which is unfalsifiable by its very general, schematic nature; to this hard core special laws for special cases are successively added; some of these laws were previously known and became integrated into the R.P. by a process of reinterpretation; some others were inspired by the "positive heuristics" (to use Lakatos' own terminology) of the programme, viz. to specify the concrete form the schematic fundamental equation $S = S(U, V, N)$ should take for concrete thermodynamic processes. Finally, we may add that, at least for the historical period considered, the R.P. was definitely progressive, since not only new special laws were constantly added but the range of successful applications increased steadily.

NOTES

[1] See e.g. E. Zahar, 'Logic of discovery or Psychology of Invention?' *British Journal of the Philosophy of Science*, **34** (1983) 243–261; T. Kulka, 'Some Problems Concerning Rational Reconstruction: Comments on Elkana and Lakatos'. *Brit. J. Phil. Sci.* **28** (1977) 325–344; A. Chalmers, 'Towards an Objectivist Account of Theory Change'. *Brit. J. Phil. Sci.* **30** (1979) 227–233.

[2] Cf. W. Stegmüller, *Structure and Dynamics of Theories* New York, 1976; C. U. Moulines, 'Theory-Nets and the Evolution of Theories: The Example of Newtonian Mechanics'. *Synthese* **41** (1979); W. Balzer, C. U. Moulines, J. D. Sneed, *An Architectonic for Science*, Dordrecht, 1987.

HISTORIOGRAPHICAL REFERENCES

Arveson, M. H., 'The Greatest Synthetic Philosopher Since Newton', *The Chemical Bulletin*, **23(5)** 1936.

Boltzmann, L., 'Zur Energetik'. In: L. Boltzmann, *Wissenschaftliche Abhandlungen*, III. Chelsea Publication Co., New York, 1968.

Brønsted, J. N., 'On the Definition of the Gibbs Potential'. *Mathematisk-fysiske Meddelelser*, **XII** (6) 1933.

Clausius, R., 'Mechanische Wärmetheorie'. *Poggendorffs Annalen*, **CXXV**, 1865.

Clausius, R. 'Über die Bestimmung der Disgregation eines Körpers und die wahre Wärmecapacität'. *Poggendorffs Annalen der Physik und Chemie*, **CXXVII**, 1866.

Duhem, P., 'Comptes rendus et analyses'. *Bulletin des sciences mathématiques*, **XXXI**, 1907.

Fox, W., *The Caloric Theory of Gases. From Lavoisier to Regnault*. Clarendon Press, Oxford, 1971.

Garber, E. A. W., *Maxwell, Clausius and Gibbs: Aspects of the Development of Kinetic Theory and Thermodynamics*. Ann Arbor, Michigan University Microfilms International, 1966.

Gibbs, J. W., 'A Method of Geometrical Representation of the Thermodynamic Properties of Substances by means of Surfaces' (1873). In: *The Scientific Papers of J. Willard Gibbs*, **I** (*Thermodynamics*). New York, 1961.
Gibbs, J. W., 'Graphical Methods in the Thermodynamics of Fluids' (1873). In: *op. cit.*
Gibbs, J. W., 'On the Equilibrium of Heterogeneous Substances' (1876/78). In: *op. cit.*
Gibbs, J. W., 'Rudolf Julius Emanuel Clausius'. In: *op. cit.*
Hiebert, E. N., 'Chemical Thermodynamics and the Third Law: 1884–1914'. In: *Human Implications of Scientific Advances* (ed. by E. G. Forbes). Edinburgh University Press, 1978.
Hibert, E. N., 'Nernst', *Dictionary of Scientific Biography*, **XIV**, 1978.
Hiebert, E. N., 'Ostwald', *Dictionary of Scientific Biography*, **XIV**, 1978.
Hornix, W. J., 'The Thermostatics of J. Willard Gibbs and 19th Century Physical Chemistry'. In: *Human Implications of Scientific Advances*, *op. cit.*
Keenan, I. H. and Shapiro, A. H., 'History and Exposition of the Laws of Thermodynamics'. *Mechanical Engineering*, **69** (1947) 915–921.
Klein, M. J., 'The Early Papers of J. Willard Gibbs: A Transformation of Thermodynamics'. In: *Human Implications of Scientific Advances*, *op. cit*.
Maxwell, J. C., 'On the Equilibrium of Heterogeneous Substances'. *Proceedings of the Cambridge Philosophical Society, II* (1876). In: *The Scientific Papers of James Clerk Maxwell* (ed. by W. D. Niven), vol. II. Dover, New York, 1965.
Newton, I., *Philosophiae Naturalis Principia Mathematica* (ed. by A. Koyré and I. B. Cohen). M.I.T. Press, Cambridge, Mass., 1972.
Newton, I., *Unpublished Scientific Papers* (ed. by A. R. Hall) Paperback ed., Cambridge University Press, 1978.
Partington, J. R., *A History of Chemistry* **4**. London, 1972.
Pirie, G., 'Letters to Gibbs (January and April 1878)'. In: *Scientific Correspondence of J. Willard Gibbs* (ed. by R. J. Seeger), Oxford, 1974.
Rowlinson, J. S., 'The Properties of Real Gases'. In: *Handbuch der Physik* (ed. by S. Flügge), vol. XII. Berlin, 1958.
Seeger, R. J., *J. Willard Gibbs – American Mathematical Physicist par excellence*. Pergamon Press, Oxford, 1974.
Stevens, F. W., 'Josiah Willard Gibbs and the Extension of the Principles of Thermodynamics'. *Science*, 1927.

Freie Universität Berlin

KOSTAS GAVROGLU

THE METHODOLOGY OF SCIENTIFIC RESEARCH PROGRAMMES AND SOME DEVELOPMENTS IN HIGH ENERGY PHYSICS

I

What follows is an attempt to critically appraise the basic claims of Imre Lakatos' epoch making paper in the proceedings of the 1965 London Conference by testing them against some of the developments in elementary particle physics during the twenty years following the Conference. During this period we witnessed spectacular successes in nearly every branch of physics, and it was a period when physicists came to grips with reinterpreted concepts, new theoretical frameworks, a multitude of research strategies and a subtly redefined role of experimentation.[1,2]

The study of the theory of relativity (both the special and the general), but more importantly, the study of the new physics fully established during the twenties, can be regarded as having provided the material for a convincing criticism of logical positivism as well as the background for a series of new proposals comprising the mainstream of contemporary philosophy of science. It is really not unreasonable to expect that the systematic study of the developments of the past twenty years in physics will eventually provide the basis for radical revisions of our present views about scientific change. Will the methodology of research programmes be able to bear the brunt of an exhaustive study of the past and the surprises to be met in the ensuing developments? The answer, on historical grounds at least, is almost certainly no. But what is remarkable is that twenty years later, the nucleus of the ideas proposed by Lakatos seem to have retained an appreciable degree of their original dynamic.[3]

The past two decades have seen an unprecedented growth both in gravitational as well as elementary particle physics, and this growth was followed by a peculiar change in the "total status" of these branches of physics.[4]

Gravitational physics has been transformed from a largely theoretical enterprise to one in which meaningful experiments can now be performed. It is no longer merely an admirable mathematical edifice, and its testability is no longer confined to the famous three tests. A whole range of new

K. Gavroglu, Y. Goudaroulis, P. Nicolacopoulos (eds.), Imre Lakatos and Theories of Scientific Change. 123–133.

possibilities for testing (rival) theories of gravitation have become realizable not only because of technological advances and refined experimental techniques, but because the demand for further testability (a demand of a methodological–heuristic character) redefined the whole framework within which one formulated theories of gravitation.

High energy physics, on the other hand, has been transformed from an essentially phenomenological enterprise preoccupied with the mathematization of the classificatory procedures – stemming from an incredible wealth of experimental information concerning the properties of particles – to an activity dominated by the impressive success of the renormalizable gauge theories.

High energy physics is no longer, to use an expression dear to the high energy physicists, the zoo it used to be. It was a peculiar zoo in which you were never sure for how long the same group would remain in the same cage, and the theoretician who was strolling in its grounds was always under the threat of meeting a hitherto unknown beast roaming around free and defying all attempts to capture it. The success, however, of the theoreticians was to eventually capture all the new beasts and put them in the right cages. Our admiration, though, for high energy theoretical physics has been primarily at the ingenuity of the classificatory schemata, at the success of the computational techniques carried over from quantum electrodynamics, at the possibilities that mathematical properties offered for physical interpretations and not only for calculations, and at the swiftness with which research programs could come to terms with the ever present anomalies in the form of unexplainable experimental results. What was lacking, however, was a theory of high energy physics and the various attempts made to formulate one were more of a programmatic character.[5]

The remarkable heuristic role of the gauge transformations and their symmetries gave a boost to certain programs and physicists succeeded in proposing a series of schemata which provide, at least, some convincing breakthroughs for a unified account of – almost – all the interactions.

In order to probe some aspects of Lakatos' proposals, I shall make use of "super-string" theories.[6]

II

Interestingly, the super-string theories were developed during a period when never before "has so much brilliant mathematics [been] done by physicists with so little encouragement from experiment"[7]

Despite the remarkable success of the so-called standard theory which unified the weak and electromagnetic interactions, there is a series of problems which could not be resolved within the context of the standard theory: The present theories contain too many arbitrary parameters and unexplained patterns to be complete. They do not satisfactorily explain the dynamics of symmetry breaking. There is no way that strong interactions can be unified in a relatively straightforward manner. One also has to include gravity in any satisfactory theory.

It appears likely (in view of recent developments) that all of the[se] problems ... can be resolved in the context of superstring theory. This involves introducing three new ingredients into the framework of quantum field theory. The first is to abandon the idea that elementary particles are mathematical points, and to allow them instead to be one-dimensional curves called "strings". The second ingredient is supersymmetry, a remarkable gauge symmetry principle with a 3/2 gauge field, that relates bosons to fermions. The third is the addition of six extra spatial dimensions giving altogether, ten dimensional space-time.[8]

It turns out that such a program contains the gravitational interactions since "string theories are inherently theories of gravity", it includes the so-called grand unification group for the electroweak group for the electroweak as well as the strong interactions, and it may solve the Hierarchy and the Family problems.

In order, however, to make contact between the string theories and the real world one is faced with a formidable task.[9]

These theories are formulated in ten space–time dimensions, they are super-symmetric, they have no candidates for fermionic matter multiplets etc.

These are not characteristic features of the physics we observe at energies below a TeV. If the theory is to describe the real world one must understand how six of the spatial dimensions compactify to four, how the gauge group is broken ... how supersymmetry is broken, how families of light quarks and leptons emerge etc.[10]

What is remarkable is that such mechanisms have been recently proposed and seem to be quite promising.

The reformulation of the whole strategy for the experimental testing of these theories resulted in the redefinition of what we usually accept as being the role of the limiting cases. Instead of devising approximation procedures for low (energy, mass or whatever) limits, and, thus, being able to provide an explanation for all the data and evidence for both the validity of the new theory together with that of the approximation procedure used, there is an attempt to single out some of the already known phenomena to be predicted in the limiting cases, as being "crucial" despite the fact that the crucial new phenomena predicted by the theory are for unimaginably high energy regions. One has to be careful here, lest he is confused with the following situation: When we routinely use limiting approaches, among the data we want to "reproduce" there is some, that our past experience has convinced us as being "more crucial" than the rest, and it is usually *that* specific class of phenomena which have a precedence in our testing. But those phenomena are professed because of our past experience, and if they do not result as predictions of the theory after the application of the approximation procedures, then they cannot be considered as reflecting on the effectiveness of our reduction strategy.

A second strategy concerns the decisions involved in the more subtle questions concerning the interpretative context of the various programs. Both in gravitational physics and high energy physics one is confronted with the task of defining the interpretative *context* and not so much the (no doubt very difficult) task of interpreting the new theoretical entities and the predicted phenomena.

Let us again consider the string theories. We can fairly confidently consider the history of high energy physics as the history of two rival approaches, those of Quantum Field Theory and S-matrix.[11] Quantum Field Theory assumes that it is possible to have a detailed space–time description of high energy physics in terms of quantities which are more fundamental than a particle. S-matrix theory, on the other hand, denies this by asserting that there can be no detailed space–time description of high energy physics with something more fundamental than a particle.

> String theory grew out of S-matrix theory. But in a sense it has some of the features of both S-matrix and quantum field theory – the experts have not yet settled down in their view of what string theory really is. Indeed, this is one of the things that makes the theory hard to learn; not everyone will tell you the same thing about what it is you are supposed to be learning.[12]

III

In high energy physics we are confronted with the formidable task of constructing a new interpretative framework since we can no longer use the relatively well known frameworks of quantum field theory or S-matrix theory.

To understand why the situation in high energy physics necessitated the search for alternative strategies, one also has to realize the following point.

Every successful theory predicts a series of new phenomena and involves a number of more or less ad hoc parameters. Some of them are determined solely by the predicted new phenomena. Hence an additional strength of a theory is expressed by a situation whereby existing data can be retrodicted after the new parameters are determined from the new phenomena. There are, of course, cases where one has the opposite situation: The new parameters are not determined uniquely by the new phenomena, and therefore, their determination from the existing data also leads to (more or less exact) quantitative predictions concerning the new phenomena. There is also a third case which is methodologically acceptable even if a little dubious on ethical grounds., It is to find the values of the new parameters by *demanding* that the theory should be able to explain existing data and *thus* predict the quantitative aspects of the new phenomena as well. And, if the possibility of detecting the new phenomena is removed into the distant future, and the total number of parameters involved is not small, and the errors involved in the existing data are of the usual kind, then this procedure is where you cannot go wrong! But physicists, on the whole, are an honest lot, and quite demanding of a convincing methodology in order to make contact with reality.

Criticism consists in being continually on the *qui vive* and being suspicious of whether a theory is successful because too much has been put in at the beginning and of *the danger that this "excess" has been successfully protected*.

Lakatos after acknowledging that the hard core of a program "develops slowly by a long, preliminary process of trial and error"[13] chooses not to discuss this process, and the hard core is permanently presented as that aspect of the program that is not touched and, what is more, is irrefutable by a methodological decision of its proponents.

All this is fine once the hard core has been satisfactorily constructed and a consensus reached among the proponents of the program that any changes and substitutions will not refer to the hard core. What has, however, been the situation during the development of the hard core through "a long, preliminary process of trial and error?" And the question becomes even more crucial if we realize that there is no process and methodology whereby the hard core is developed first and foremost, *then* the consensus of various people is guaranteed, and *then* there is the beginning proper of a research program. A hard core is developed during the initial stages of a research program, and, therefore, *its* development is inextricably related to and influenced by the processes which form and continuously re-adjust the protective belt of the program. One of the merits of rationally reconstructing research programs which are not too much in the distant past, is that one is facilitated to further explicate the relationship between the hard core and the protective belt.

IV

What emerges as an interesting feature of the recent developments in physics is that they allow us to pose a "series" of new questions thereby redetermining the strategy of problem solving, and we can claim that the more these questions are "allowed" to refer to the foundational aspects of the program, the more progressive the specific program is.

Being able to pose more questions in a (progressive) research program is an additional expression of the increased empirical content of the specific program, and, in that respect, the actual formulation of the questions can indicate directions of research and/or growth of the program complementing the overall strategy drawn out by the positive heuristic. What, however, is quite interesting is that such a criterion can be used to evaluate a series of programs which do not predict anything which is different from the predictions of the program which has become dominant. Yet one cannot dismiss these programs as programs which duplicate known facts, if within their contexts one is allowed to pose a series of questions which were not allowed to be posed within the context of the dominant program. Furthermore, these programs should not be snubbed even in situations where the solution of the problem(s) implied by the(se) question(s) may not be different from the one provided by the dominant program, exactly because being allowed to pose new questions is an indicator of a (peculiar, no doubt) increase in the empirical content of the propositions (or models) of a program.

Lakatos in a comment about the "creative shift" that may give a boost to a degenerating program remarks that

> One should not forget that two specific theories, while being mathematically (and observationally) equivalent, may still be embedded into different rival research programs and the power of the positive heuristic of these programs may well be different. *This point has been overlooked by* proposers of such equivalence proofs (emphasis added).[14]

The assertion that in considering rival programs one has to take into account the kinds of questions which are now posable introduces an indespensable new criterion for seking and establishing "equivalence proofs". Such a criterion, however, exactly because it eases the observability criterion makes the establishment of equivalence more difficult and, hence, "more" real since there are, I think, quite serious epistemological, methodological (and sociopsychological) problems associated with the notion of proving the equivalence of alternative programs. It can only be considered as a technique which establishes trivial correspondences (with very sophisticated techniques many times) and a process which "takes away" from the programs the richness of their individuality, that Lakatos is at pains to defend.

The existence of many different theories which provide equally satisfactory explanations for the phenomena, and with "comparable" empirical contents, should not prompt us to think of them as equivalent modes of explanation and seek a formal way to prove it. Instead we should attempt at constructing metatheoretical frameworks in order to be able to "accommodate" the different programs, and by "accommodating" them to comprehend not their equivalence, but the deeper reasons for the complementarity of the explanatory possibilities they provide. And for this purpose the investigation of the spectrum of the questions which can now be formulated becomes a crucial undertaking.

V

In his intervention during the 1965 London Conference, John Watkins remarked that

> It seems that a dominant theory may come to be replaced, not because of growing empirical pressure (of which there may be little), but because a new and incompatible theory (inspired perhaps by a different metaphysical outlook) has been freely elaborated: A scientific crisis may have theoretical rather than empirical causes.[15]

Both in gravitational physics and high energy physics, various theoretical schemata have become rivals to each other (rather than dominant, as Watkins suggests) and to "more established" theories, because the metaphysical outlook which inspired them proved to have a heuristic value which, in turn, made the proposed metaphysical outlook "less repulsive" – since almost always this new metaphysical outlook is "more unphysical" and counterintuitive. The choices, for example, for what constitutes elementarity entailed a radical departure from a metaphysics which considered the notion of "building blocks" as the underlying as well as guiding metaphysics of the research programs aimed at understanding the structure of matter. The proposal that gauge fields and their symmetries are the fundamental entities is, in fact, the expression of a rival metaphysics and not a change of paradigm, especially since there preceded no crisis. The changes of paradigm in high energy physics within the context of Normal Science, never disputed the metaphysics associated with the "building blocks" attitude. Similar comments can be made for gravitational physics, but, there, the arguments involving the nature of gravity and the choices for the space–time structure which, in fact, lead to different metaphysical commitments, are slightly more technical.

Let us however, be a little analytical on this issue concerning the change of metaphysics in high energy physics. In dealing with any aspects of high energy physics, the researchers in the field are bound to a sissyphian methodology: On the one hand, their efforts are aimed at understanding the behaviour of particles, which are considered to be elementary, and, on the other, every major development, in effect, reformulates and changes the meaning of the notion of elementarity and provides further evidence to support the claim that the study of particles as elementary may have its intrinsic limitations.

One way out of this vicious circle is indicated by the fact that the new "building blocks", discovered in the past 20 years, are there, but they cannot exist freely. The idea of confinement is not only imposed by the null results of the experimental searches, but because such a proposition turns out to be a very fruitful one theoretically. The confinement of the new "building blocks" becomes an (inherent) feature of their elementary status and an indispensable aspect of the new and redefined notion of elementarity. The idea of confinement together with the attempts to unify all interactions is, then, considered by some to elevate the gauge fields together with their symmetries, to the status of fundamental entities.

This situation, which implies a radically new metaphysics, has been anticipated (in a slightly different context) in 1972 by Heisenberg

Nowadays many experimental physicists – even some theoreticians – still look for *really* elementary particles . . . I think that this is an error. . . . What, then, has to replace the concept of a fundamental particle? I think we have to replace this concept by the concept of a fundamental symmetry. . . . But this is again a matter which should be decided by the experiments. I only wanted to say that what we have to look for are not fundamental particles, but fundamental symmetries. And when we have actually made this decisive change in concepts which come about by Dirac's discovery of antimatter, then I do not think that we need any further brakthrough to understand the elementary – or non-elementary – particles. We must only learn to work with this new and unfortunately rather abstract concept of the fundamental symmetries.[16]

The positive heuristic which incorporated the implied metaphysics of such an approach involves aspects which are not readily acceptable within the existing conceptual framework, but these aspects appear in a mathematical form and they comprise a methodololgy whereby a series of constraints become important criteria in our attempts to "simulate reality".

> Nature can(not) be described by independent, individual constituents and forces acting between them. We might have to replace this concept by more general ones. A concept that seems to be more fundamental while at the same time more abstract is that of gauge fields . . . The practically infinite number of mathematically possible gauge fields, is *restricted* by the symmetry properties of these fields. It seems, therefore, that the first principles of describing and understanding Nature may not be little indestructible bricks of matter, but rather abstract concepts, symmetries.[17] (Emphasis added.)

However, the "practical" implications of all these considerations are in the formulation of the positive heuristic, which thus acquires an additional regulative aspect: Such positive heuristics provide a remarkable constraining mechanism, whereby the multitude of possible theories that the mathematical framework allows, is dramatically reduced, not by the limitations introduced by empirical limitation which are inherent in the mathematical instantiations of the (new) metaphysical choices in the way they are specified in the positive heuristic.

And this is why I tend to believe in Lakatos' suggestion that "it is better to separate the "hard one" from the more flexible metaphysical principles expressing the positive heuristic".[18]

Hence, a regulative function during the development of a program is implemented not only by the all too obvious regulative principles mentioned by Lakatos[19] (which are, of course, present independent of the details of the different research programs), but also by the positive heuristic itself. This outstanding feature of the positive heuristic, even though is implied in Lakatos' position, is not singled out and its dynamic is not sufficiently expounded.

Lakatos states that the "positive heuristic consists of a partially articulated set of suggestions or hints on how to change, develop the "refutable" variants of the research program . . ."[20] He then goes on and emphasizes that the positive heuristic in a research-program determines the "strategy both for predicting (producing) and digesting"[21] the expected "refutations". The positive heuristic, however, cannot have this function, if it does not also have the regulative aspect we talked about in the form of continuously imposing a series of constraints and limitations on what, on purely mathematical grounds, can be proposed.

> From the beginning it seemed to me to be a wonderful thing that very few quantum field theories are renormalizable. *Limitations of the sort are, after all, what we most want, not mathematical methods which can make sense out of an infinite variety of physically irrelevant theories but methods which carry constraints*, because these constraints may point the way toward the one true theory. . . . I thought that renormalizability might be the key criterion, which also in a more general context would impose a precise kind of simplicity on our theories and help us pick one, the one true physical theory out of the infinite variety of conceivable quantum field theories.[22] (Emphasis added.)

The change in metaphysics is almost always followed by a new set of constraints. It is this complex of constraining procedures which delimits mathematical physics from theoretical physics. In fact, we can claim that a program in mathematical physics turns into a program in theoretical physics whenever these constraints are explicitly spelt out. And it is this kind of change (which does not necessarily hinder mathematical developments) that should be further studied in order to gain an appreciation of the "relative autonomy of theoretical science", and to relate it with the whole framework of the methodology of research programs and not consider it as a historical fact corroborated by all those historical instances where theoretical work is far ahead from that of the experimentalists.

I think that there is much more to the relative autonomy of theoretical science than that, and Lakatos seems to be well aware of the shortcomings of his initial exposition. "Let us remember that in the positive heuristic of a powerful program there is, right at the start, a general outline of how to build the protective belts: This heuristic power generates the autonomy of theoretical science".[23] I do not think, however, that this general outline is so well spelt out right at the start as Lakatos suggests and studying the relationship between mathematical and theoretical physics during the development of various programs becomes a necessary prerequisite for understanding the non-trivial features of the relative autonomy of theoretical science.

NOTES

[1] See, for example, K. Gavroglou, 'Theoretical Frameworks for Theories of Gravitation', *Methodology and Science,* **19** (1986) 91–124.

[2] A comprehensive account of these developments can be found in the Nobel Prize acceptance speechs of S. Weinberg 'Conceptual Foundations of the Unified Theory of Weak and Electromagnetic Interactions', *Rev. Mod. Phys.,* **52** (1980) 515–524; A. Salam, 'Gauge Unification of Fundamental Forces', *Rev. Mod. Phys.,* **52** (1980) 525–538 S. L. Glashow, 'Towards a Unified Theory: Threads in a Tapestry', *Rev. Mod. Phys.,* **52** (1980) 539–543.

[3] 'Falsification and the Methodology of Scientific Programmes' by Imre Lakatos in *Philosophical Papers* volume I (Cambridge University Press, Cambridge, 1978) ed. by J. Worrall and G. Currie.

[4] The Brinkman Report 'Physics Through the 1990s'. The report, made public in March 1986, has been initiated by the National Science Foundation of the U.S.A.

[5] K. Gavroglou 'Research Guiding Principles in Modern Physics: Case Studies in Elementary Particle Physics', *Zeit. fur allgemeine Wissenschaftstheorie,* **VII** (1976) 223–248 and 'Popper, Tetradic Schema, Progressive Research Programmes and the case of Parity Violation in Elementary Particle Physics 1953–1958', *ibid.,* **XVI** (1985) 261–286.

[6] *Superstrings: The First 15 Years of Superstring Theory*, volumes I and II, edited by J. Schwartz and M. Green (World Scientific Publishers, Singapore, 1986); P. van Nieuwenhuisen 'Supergravity' in *Physics Reports,* **68** (1981) 189–398.

[7] S. Weinberg, 'Particle Physics: Past and Future', *International Journal of Modern Physics,* **1** (1986) 143.

[8] See J. Schwartz, 'Introduction to Supersymmetry', Scottish Universities Summer School.

[9] J. Schwartz and M. Green, ref 6, p. xi.

[10] D. J. Gross, 'Heterotic String Theory', in *Particles and the Universe*, edited by G. Lazarides and Q. I. Shafi (Elsevier Science Publishers, New York, 1986) p. 15.

[11] K. Gavroglou ref. 5.

[12] S. Weinberg, ref. 7, p. 144.

[13] I. Lakatos, ref. 3, p. 48, footnote 4. For a very interesting discussion of these points see A. Musgrave, 'Method or Madness?', in *Essays in the Memory of Imre Lakatos,* edited by R. Cohen, P. Feyerabend and M. Wartofsky (Reidel, Dordrecht, 1983).

[14] I. Lakatos, ref. 3, p. 77, footnote 2.

[15] J. Watkins, 'Against Normal Science', in *Criticism and the Growth of Knowledge,* edited by I. Lakatos and A. Musgrave (Cambridge University Press, Cambridge, 1970) p. 31.

[16] W. Heisenberg, 'Development of Concepts in Quantum Theory', in *The Physicist's Conception of Nature*, edited by J. Mehra (Reidel, Dordrecht, 1973) p. 273.

[17] H. Schopper, 'Elementary Particle Physics: Where is it Going?', CERN Yellow Reports 82–08, 1982, p. 11.

[18] I. Lakatos, ref. 3, p. 51.

[19] *Ibid*., p. 57–58.

[20] *Ibid*., p. 50.

[21] *Ibid*., p. 51.

[22] S. Weinberg, ref. 2, p. 517.

[23] I. Lakatos, ref. 3, p. 88.

National Technical University of Athens

YORGOS GOUDAROULIS

MANY-PARTICLE PHYSICS: CALCULATIONAL COMPLICATIONS THAT BECOME A BLESSING FOR METHODOLOGY

Almost any physical system consists of many interacting atomic particles. Outside the realm of nuclear and high-energy physics, the forces holding these particles together are basically electrical and have been quantitatively understood since the end of the eighteenth century. The laws describing the motion of the atoms and their electrons are also very well known; they are just the laws of quantum mechanics familiar to us for many years. In other words, the atoms in all many-particle systems can be described completely in terms of the well-known electromagnetic forces and the familiar laws of quantum mechanics. But these systems may differ very appreciably in their properties and they often exhibit quite remarkable behavior.

In principle, the behavior of such systems should be completely understandable in terms of known physical laws as applied to the atoms; but, we often realize that the claims of such a statement are practically untenable and understanding "in principle" is meaningless when the systems are so complex that their behavior cannot be predicted from these laws and forces.

The basic problem associated with any system consisting of very many particles is then the following: How can one proceed from the known laws governing the simple atoms in the system to derive significant predictive statements about the observed behavior of this system? So, *we are facing the general challenge of predicting observed behavior on the basis of our knowledge of atomic laws*.

The difficulties are quite sharply expressed by Reif (1970, p. 4)

> The key difficulty is clearly the complexity arising from the fact that the system of interest consists of very many interacting atomic particles. It is worth pointing out, however, that this complexity involves much more than questions of quantitative detail. Complexity may readily lead to unexpected *qualitative* features in the observed behavior of the system . . . and may involve intricate forms of organization of simple atomic constituents. It is also naive to believe that the complexity can be handled merely by resorting to bigger and better computers. The number of interacting atoms in any large-scale system is so enormous that

K. Gavroglu, Y. Goudaroulis, P. Nicolacopoulos (eds.), Imre Lakatos and Theories of Scientific Change. 135–145.

even the most fanciful of future electronic computer could not tackle calculations designed to solve the equations of motion of all these atoms. Furthermore, reams of computer output tape are likely to be meaningless unless one has formulated the right questions.

Feynman also stressed (1955, p. 17) that

We cannot expect a rigorous exposition of how these properties arise. They could only come from complete solutions of the Schrödinger equation for the 10^{23} atoms . . . ,

and Fröhlich concluded (1966, p. 540) that

What is desirable, therefore, is not to solve the equations of motion exactly but to find answers to certain questions in terms of concepts which usually will refer simultaneously to many (or all) of the particles.

Thus, *the task of handling the complexity of the many particle problem is mainly the task of bridging the chasm between microscopic simplicity and macroscopic complexity, and this task is primarily conceptual rather than computational*.

The fact that most of the concepts used to understand the behavior of physical systems on the atomic scale cannot be used when it comes to explaining the behavior of macroscopic systems is a well known difficulty in constructing models. This is especially so when one attempts to explain the macroscopic behavior as resulting from properties at the atomic scale. One has, then to devise "intermediate" concepts that *may* make possible the sought for extrapolation from the simple atomic structure and behavior in order to understand the macroscopic characteristics of the system. These intermediate concepts will necessarily have to play the role of "intermediaries" between the two extremes, especially since it is the two extremes that bring about the paradoxical situation.

(Let us stress that we do not talk of the paradoxical situation that is tautologically inherent in the observation of a new phenomenon which is not predicted by an established model or theory. We do not talk of the lack of a readily available explanation or the fact that in most instances the existing theories, when applied to the new physical situation, give results other than the observed phenomenon. In our case, the paradox expresses the frustration resulting from the deadlock in attempts to make use of tested methodological rules for a problem situation comprised of elements that gave no indication *a priori* that an extrapolation of the known laws would not be effective).

The gap is bridged only through a continuous process of reinterpretation whereby the resulting *intermediate intermediaries* facilitate the creation of a theoretical framework with a relatively autonomous status with respect to the framework within which the two extremes are accommodated.

The process of formulating appropriate intermediate intermediaries should not be confused with providing an algorithm for deriving desired results. It operates under strict rules and one does not have the freedom to introduce concepts in an *ad hoc* manner. In fact, the intermediate intermediaries should:

(a) be firmly based on the known atomic laws,
(b) make it possible to think in simple terms about the system under consideration,
(c) provide quick insights into the behavior expected of the system,
(d) allow quantitative predictions to be made about it.

All the above may, in fact, sound trivial since the possibility of explaining the behavior of many-particle systems from the known behavior of the underlying atomic structure is a well known difficulty for classical systems. Furthermore, the methodological problems involved in the approaches which do provide an explanation of the many-particle systems with the use of statistical methods or with the proposition of a series of concepts signifying collective behavior are quite thoroughly studied, despite the fact that there is no consensus concerning the significance of specific methodological novelties. I will not be concerned with the classical cases.

Quantum systems present a greater interest, and the methodological problems involved are much less understood. Obviously, the quantum mechanical behavior of the atomic structure, induces all kinds of computational as well as conceptual complications in understanding the behavior of many-particle systems and in definition of the motions which express collective behavior. What, however, is the most intriguing problem to my mind is that quantum mechanics does not provide us with any direct methodological and conceptual tools for the quantum mechanical description of a macrosystem as a whole. We may be able to describe successfully a macrosystem if it exhibits certain phenomena which are the result of the quantum mechanical behavior of its underlying structure; but, *we are extremely ill equipped both conceptually and methodologically* as I mentioned, *to formulate the quantum mechanical behavior of a macrosystem taken as a whole, without the assumption of an underlying structure. Unless one understands the various facets of such a system one cannot claim, that the study of the many-particle systems is complete*.

Fortunately, there exists such a case. It is the case of the remarkable properties of liquid helium 4 below 2.9 K, and more generally, the case of quantum liquids.

Here we have a system which has the following characteristics:

(1) It consists of many very simple atoms
(2) It has many extraordinary properties, and
(3) displays a macroscopic quantum behavior.

We will reconstruct the story of low temperature liquid helium[1] as a research programme[2] characterized by:

Its *initial problem*, reformulated as the riddle of how we can bridge the chasm between atomic simplicity and macroscopic behavior;
Its "*hard core*" consisting of the known quantum mechanical atomic laws, and
Its "*positive heuristic*": the task of bridging the chasm is primarily conceptual rather than calculational; which means, devise the appropriate intermediate concepts that can play the role of intermediaries between the two extremes.

The helium atom is, next to hydrogen, the simplest atom (since it contains only two electrons). The helium molecule is just a single helium atom. The forces between atoms are very weak. Liquid helium consists thus merely of a collection of such simple atoms, all exactly alike.

Let us now consider the observed behavior of liquid helium: Helium gas (at atmospheric pressure) becomes a liquid below 4.2 K, and it remains a liquid down to the absolute zero of temperature. From 4.2 K down to 2.2 K, the liquid behaves very much like an ordinary liquid, but, below the critical temperature of 2.2 K, it changes into a new form endowed with very extraordinary properties. It has an uncanny ability to flow through extremely small holes without any apparent friction but, its viscosity is not zero when measured in an experiment involving the drag force acting on an object moving through the liquid; hence liquid helium II (in contrast with the ordinary helium I above 2.2 K) exhibits a "viscosity paradox". Liquid helium II exhibits some other remarkable properties as well. For example, the liquid is able to form a very thin film on the walls of a container and is able to creep up the walls of the container and go over its top, thus gradually emptying the container. I do not intend to list further properties. I believe that I have convinced you by now that liquid helium displays, indeed, most unusual behavior. It exhibits in particularly trenchant form the paradox of atomic simplicity and remarkable behavior.

Thus, we encounter the general challenge of the many-particle problem in the following extreme form: *How can such extreme atomic simplicity lead to the occurrence of extraordinary macroscopic behavior?*

When in 1938 there began a systematic investigation of liquid helium II we find that there coexisted two attitudes to tackling the problem, differing in their epistemological premises and their implied methodology:[3]

(1) One would expect the behavior of helium II to be easily explicable because of the very simple and well understood atomic structure; furthermore, the low temperatures minimize all the disturbing perturbative factors. However, the emerging experimental situation turned out to be quite different: it could not be explained, accounting for and justified, and it was found to be in conflict with all the known laws and their natural extrapolation. The fact that the phenomenon defied such an accepted approach led to considering the problem situation as a paradox between the extreme atomic simplicity of helium and its extremely unusual behavior when cooled below 2.19 K.

(2) A second attitude arose after the failure of the attempts to tackle the phenomenon either within the dominant paradigm or by use of mechanical concepts. It did not regard the above paradox as the basic feature of the situation but rather the chasm between the two extremes; it believes that the bridging of this chasm is primarily a conceptual task.

In a letter to *Nature*, F. London (1938) argued for the first time that the transition from He I to He II was an example of Bose–Einstein condensation and suggested that

> an understanding of a great number of the most striking peculiarities of liquid helium can be achieved *without entering into any discussion of details of molecular mechanics*.[4]

L. Tisza (1938), using London's idea as a basis, developed the phenomenological two-fluid model, in which He II is regarded as a mixture of two components the normal and the superfluid.

Of course, the proposal of a scheme whereby one finds the normal and abnormal situations coexisting and using each one separately to explain the various properties, may seem very naive and lacking strict epistemological justification. It has, however been used extensively in physics, and has turned out to be a particularly fruitful method. There have been cases where just before a change of paradigm, the natural thing to do was to account for the peculiarities by incorporating them by hand in a sea of an otherwise normal state of affairs.[5]

In Tisza's two fluid model the entire liquid is supposed to be, at absolute zero, a *superfluid* consisting of "condensed" atoms, while at the transition temperature this component vanishes. The *normal* component is identical with helium I. It was in this model that Tisza explained the experiments of Allen, Kapitza and their collaborators on flow phenomena of He II and predicted the mechanocaloric effect and the temperature waves which subsequently became known as "second sound". But *this model was forbidden by the hard core of the programme*. Indeed Tisza's model was based on the admission of the physical impossibility of two fluids made up of the same species of atoms, which *by definition* must be distinguishable. Therefore, the *model had to be replaced by one which was able to recover the qualitative explanations of Tisza's two-fluid model without dividing the atoms into superfluid and normal components*.

L. Landau (1941) defined the two fluids as motions, not identifiable portions of matter. This feature was consistent with the spirit of quantum mechanics. He stated his own position quite firmly.

> It follows unambiguously from quantum mechanics that for every slightly excited macroscopic system *a conception can be introduced of "elementary excitations", which describe the "collective" motion of particles* . . . It is this assumption, indisputable in my opinion, which is the basis of the microscopic part of my theory. On the contrary, every consideration of the motion of individual atoms in the system of strongly interacting particles is in contradiction with the first principles of quantum mechanics.[6]

Thus, according to Landau there exists just one fluid: liquid helium, and one might picture it as a background fluid in which excitations move.[7] He moreover showed that a liquid system with excitations as described will behave in many ways like a mixture of two fluids, providing the much needed physical reason for the unreasonable success of Tisza's two-fluid model.

Whereas Landau had postulated the existence of elementary excitations, Bogoliubov (1947) proposed to derive them for a nonideal Bose–Einstein gas. He found that the energy spectrum could be expressed in terms of "quasi-particles" representing the elementary excitations. Thus the problem of dealing with the interdependent motions of almost countless numbers of helium atoms can now be treated by the well understood theories of gases applied to quasi-particles.[8]

Bogoliubov's work was a major step towards unifying the London–Tisza and Landau theories but, although the interaction between helium atoms is comparatively weak, liquid helium cannot be considered as a "nearly perfect" Bose–Einstein gas and the problem of the justification of the Landau excitation or quasi-particle model on the basis of a microscopic theory remained open. It was Feynman who showed that one can justify the Landau model from quantum mechanics applied at the atomic level and remedied, to a large extent, its major failures.

It is clear by now that *Tisza and Landau developed the intermediate concepts of "superfluid" or "background liquid" and of "elementary excitations", or "quasi-particles" and that Feynman justified their role as intermediaries between the two extremes*.

It is easy to see now how these concepts were able to scale the barrier that separated the known atomic laws, on the one hand, from the observable behavior of He II on the other.

Starting from an atomic point of view it is clear that the individual helium atoms in the liquid do not move independently of each other for two reasons: Every atom affects the atoms in its neighborhood by virtue of the forces which it exerts on them; in addition, quantum mechanics imposes certain symmetry requirements to guarantee the complete indistinguishability of all the helium atoms. Furthermore, the mutual effect of the motion of one atom on another becomes particularly significant at low temperatures, because the degree of random thermal agitation is then so small that it does not destroy the interdependence of the atomic motions. Since it is important at low temperatures to consider the fact that helium atoms do not move independently of each other, *the entire liquid should really be treated as if it were a single gigantic molecule*. This idea functioned as an important heuristic principle which governed the construction or modification of the auxiliary hypotheses which, in turn, led to the appropriate concepts and to novel facts.

Indeed, near absolute zero the liquid must be in configurations with energy close to its state of lowest possible energy (its so-called "ground state"), and quantum mechanical ideas ought to be as significant in discussing the liquid as they are in discussing molecules or atoms, which are in states of low energy. Thus, at absolute zero, the liquid should be found in its ground state of lowest possible energy. This is a single well-defined state in which the liquid is as well-ordered and as simple as possible, a "perfect fluid" characterized by uniform density throughout. When the temperature is slightly above absolute zero, the liquid can be found in those of its possible states having energy somewhat greater than that of its ground state. Each such state is less ordered than the perfect fluid and results from some kind of "disturbance"' in the perfect fluid. Each such disturbance involves a characteristic motion of groups of atoms moving together and has associated with it a definite energy and momentum. Thus, as the temperature is increased, the superfluid contains an increasing number of disturbances of various kinds and the liquid, criss-crossed by them, becomes increasingly disorderly as the temperature is raised.

Starting from our knowledge of atomic properties, we have introduced the concepts of perfect fluid or *superfluid* and disturbances or *elementary excitations*, which they describe the liquid in terms of the motion of *groups* of atoms moving jointly and thus circumvent a difficult description in terms of individual atoms which do not move independently.

The question is now if these intermediate concepts (which fill the first two requirements) can play the role of intermediaries. In other words, do these concepts account for the properties of liquid helium II (third requirement)?

Let us see, for example, how they account for the viscosity paradox. It is the superfluid which flows without friction through small holes (the excitations are irrelevant since they do not pass through the holes). But when an object is dragged through liquid helium, the elementary excitation in the liquid collide with the object and are thus responsible for the frictional effects.

One thing is left: If our intermediate concepts about liquid helium are correct and really to be taken seriously, then, do they lead to further predictions (fourth requirement)? The answer is Yes.

The two-fluid model had spectacularly early success in predicting a number of phenomena. One very startling prediction is the quantization of the superfluid circulation (Onsager, 1949) in helium. The idea was essentially developed by Feynman (1955). The reasoning is quite simple. We based the development of our intermediate concepts on the premise that the entire liquid must be treated as if all of its atoms constituted a single gigantic molecule. If quantum mechanical laws do indeed apply to this fluid as they would to any ordinary molecule, then discrete quantum effects will become apparent on the large scale of the entire liquid. Indeed, let us think of a very simple experiment in which the fluid is initially stirred and then left to itself. As is well known, there results then a "vortex". The strength of a vortex in liquid helium ought, therefore, to be quantized. This prediction implies that, although the vortex involves a large-scale motion of the liquid, it should, nevertheless, exhibit discrete quantum effects. Subsequent work by Vinen (1961) and Rayfield & Reif (1963) established that this was the case.

Thus, our intermediate concepts fill the four requirements and they are the appropriate *intermediate intermediaries*. They provide a methodological tool for understanding how the extreme atomic simplicity of helium can lead to the occurrence of its unusual behavior by circumventing the difficulties associated with a description in terms of individual atoms.

The case of liquid helium is not the only example of a successful use of such a methodology. The whole field of quantum liquids is such an example.

Quantum liquids may be Fermi or Bose, normal or superfluid charged or neutral. Key steps in the emerging theory of quantum liquids included a systematic theory of the quantum plasma, an exact theory of Fermi liquids in the limit of long wavelengths and low temperatures, and a markedly improved microscopic understanding of the physical behavior of superfluid helium. Quantum liquids are not only of considerable intrinsic interest, but have made possible key insights into the behavior of almost all solids, as well as finite nuclei, nuclear matter, neutron stars, symmetry-breaking in particle physics, and, possibly quark confinement and quark matter. For a number of model problems it proved possible to calculate not only the ground state energy, but also the low-lying excited states, which for each case could be shown to be a gas of weakly interacting elementary excitations.

The concept of elementary excitations, implicit in the work of many theorists in the 1930s, and made explicit by Landau in his classic 1941 paper on superfluid helium, has turned out to provide a unifying theme in the treatment of the low-lying excited states of almost all condensed matter systems. It is, from this point of view, an analogous concept to that of "particles" in particle physics.

I hope that the preceding remarks have given you a slight appreciation of the power of intermediate intermediaries in furthering our understanding of liquid helium and condensed matter systems.

Similar methodologies are also useful in other fields. Indeed, only in the somewhat esoteric field of subnuclear particles are the fundamental laws of interaction between particles not understood. In most of physics, and in all of chemistry the atomic nucleus is never disrupted. Hence no knowledge of nuclear forces is required. The atoms in all of these systems can be described completely in terms of the well-known electromagnetic forces and the familiar laws of quantum mechanics. Thus, practically all the systems with which we deal in the world around us, systems varying in complexity, consist of very many atoms whose interactions are well understood. *If this statement of understanding "in principle" is to be meaningful so that it can lead to the prediction of observed behavior, then the fundamental intellectual challenge to be faced is that of developing appropriate intermediate intermediaries*. As Fröhlich (1966, p. 540) said:

> such concepts might be of great value in connecting physics with chemistry and biology both of which deal with materials which in terms of physics seem very complicated. Finding new concepts within the realm of very simple materials might, however, also be of importance. For they might open a way in pointing to methods by which such concepts can be found.

Let me finish by saying once again that if the above considerations can really help us to understand the various facets of many-particle systems with a macroscopic quantum mechanical behavior, we can claim that such studies are absolutely necessary in order to provide further techniques for the reconstruction of the story of *all* many-particle systems.

NOTES

[1] For details see Gavroglou & Goudaroulis (1986, 1987) and Brush (1983) 4.9.

[2] See Lakatos (1980).

[3] However, the full understanding has been achieved through their interplay.

[4] F. London, as quoted in Pines (1981) p. 119. Emphasis added.

[5] The rule is quite simple but whether it will work or not, is not always given. It runs: solve the apparent difficulties by postulating the existence of a mechanism with the normal state that "carries away" these difficulties; this coexistence is to be achieved in such a way that the unusual behavior is not just a simple perturbation of an otherwise normal behavior. An instance of a successful use of such a method is Pauli's proposal for the existence of the neutrino where a particle with properties that defied all the accepted criteria for elementary particles was used to account for the missing energy. An attempt years later to use the same method in order to account for parity violation was not successful.

[6] L. Landau, as quoted in Brush (1983), p. 187.

[7] These "elementary excitations" can make their way from one place to another, collide with the walls and with each other, and give to helium some properties associated with the so-called normal fluid component, such as viscosity.

[8] What makes this view so fruitful is that, at low temperature, there are not too many modes of motion to deal with, and hence not too many quasi-particles.

REFERENCES

Bogoliubov, N. N., 'On the theory of superfluidity', *Journal of Physics* (USSR), **11** (1947) 23–32.

Brush, S. G., *Statistical Physics and the Atomic Theory of Matter, from Boyle and Newton to Landau and Onsager*. Princeton: Princeton University Press, 1983.

Feynman, R. P., 'Application of quantum mechanics to liquid helium', in *Progress in Low Temperature Physics*, vol. 1, C. J. Gorter, ed. pp. 17–53, New York: Interscience, 1955.

Fröhlich, H., 'Superconductivity and the many body problem', in *Perspectives in Modern Physics, Essays in Honour of Hans A. Bethe*, ed. by R. E. Marshak, pp. 539–552. New York: Interscience, 1966.

Gavroglou, K. and Goudaroulis, Y., 'Some methodological and historical considerations in low temperature physics II: the case of superfluidity', *Annals of Science*, **43** (1986) 137–146.

Gavroglou, K. and Goudaroulis, Y., 'From *Physica* to *Nature*: the tale of a most peculiar phenomenon', submitted for publication in *Janus* (1987).

Lakatos, I., 'Falsification and the methodology of scientific research programmes', in *Imre Lakatos, Philosophical Papers*, Vol 1, ed. by J. Worrall and G. Currie, pp. 8–101. Cambridge: Cambridge University Press, 1980.

Landau, L. D., 'The theory of superfluidity of helium II', *Journal of Physics* (USSR), **5** (1941) 71–90.

London, F., 'The λ-phenomenon of liquid helium and the Bose–Einstein degeneracy', *Nature*, **141** (1938) 643–644.

Onsager, L., 'Discussion Remark', *Nuovo Cimento*, supplement, **6** (1949) 249.

Pines, D., 'Elementary excitations in quantum liquids', *Physics Today*, **11** (1981) 106–131.

Rayfield, G. W. and Reif, F., 'Evidence for the creation and motion of quantized vortex rings in superfluid helium', *Physical Review Letters*, **11** (1963) 305–308.

Reif, F., 'Superfluidity: the paradox of atomic simplicity and remarkable behavior', in *Quantum Fluids* (proceeding of the Batsheva seminar, Haifa, 1968), ed. by N. Wiser and D. J. Amit, pp. 1–14. New York: Gordon and Breach Science Publishers, 1970.

Tisza, L., 'Transport phenomena in Helium II', *Nature*, **141** (1938), 913.

Vinen, W. F., 'The detection of single quanta of circulation in liquid Helium II', *Proceedings of the Royal Society of London*, **A260** (1961) 218–236.

Aristotle University of Thessaloniki

T. M. CHRISTIDES AND M. MIKOU

THE RELATIVE AUTONOMY OF THEORETICAL SCIENCE AND THE ROLE OF CRUCIAL EXPERIMENTS IN THE DEVELOPMENT OF SUPERCONDUCTIVITY THEORY

1. INTRODUCTION

There are experiments in the History of Sciences which have a particular weight for the development of scientific theories. And there is of course an ongoing dispute in the philosophy of science about the importance of these experiments. The dispute concerning the character of the crucial experiments constitute the axis of the problematic of this controversy and determine in many respects the role that such experiments play in the development of physical theories.

As important as the theoretical issues are, the study of many case studies is always very useful to clarify a host of questions concerning the character of these "crucial" experiments.

There are many well known crucial experiments in the history of Physics. We think, however, that the study of lesser known experiments will illuminate the whole problematic. In this respect we have chosen to study two interesting cases from the history of low temperature physics: the Meissner–Ochsenfeld experiment (1933 – known as the Meissner effect) and the Maxwell–Serrin experiment (1950 – known as the isotopic effect).

These experiments were performed in a period in which there was a dispute also concerning the theoretical considerations and the strategy to be chosen for the further development of the programme. During our rational reconstruction of this historical period we realise that these experiments, particularly the first, are indeed crucial not so much for falsifying or justifying versions of the programme, but for convincing the scientific community that some of their beliefs were the result of deeply ingrained prejudices.

Before the Meissner–Ochsenfeld experiment there was a strong belief that superconductivity was an electrical phenomenon (infinite conductivity) so that scientists had an excessive confidence in the extrapolation of

K. Gavroglu, Y. Goudaroulis, P. Nicolacopoulos (eds.), Imre Lakatos and Theories of Scientific Change. 147–153.

Maxwell's equations to the range of low temperatures. But we think there was no sound reason to do the low temperature extrapolation of the Maxwell's equations and stick to this theoretical result without experimental corroboration. The only reason was prejudice in favour of Maxwell's equations leading to a strong belief that they are valid under basically all conditions (Casimir, 1980).

Similarly in the case of the isotopic effect scientists have stuck to the dictum that the ions of the lattice, in view of their large mass (compared with electrons) could play no important role in the establishment of the superconductive state. This belief, we think, is responsible for the delay of the interpretation of the phenomenon of superconductivity.

In the case of the isotopic effect we point out the double crucial character of the relative experiment both in the development of the London–Fröhlich programme as well as in the predominance of this programme over its competitors.

Finally, our study of low temperature phenomena, and especially of superconductivity, shows that these experiments provided the necessary boost for the programme, during a period when there appeared additional theoretical reasons for a creative shift.

2. THE MEISSNER EFFECT

The phenomenon of superconductivity was discovered by Kammerlingh Onnes in 1911, when he was studying the change of electrical resistance of mercury at low temperatures. The phenomenon consisted in an abrupt drop of the value of the resistance at a critical temperature of 4.2 K. Below this value of temperature the resistance was almost zero. At that time there was no satisfactory theory for the phenomenon of conductivity and quantum theory was scarcely formulated and applied only to very specific problems. Under these conditions superconductivity had been considered as a phenomenon of infinite conductivity to which Maxwell's classical electrodynamics should be applied.[1] This consideration led necessarily to the conclusions that there was no phase transition and that the magnetic field inside the superconductor could not change when the external field was altered. These conclusions did not favour the thermodynamic treatment of the phenomenon.

During this period (1911–1933) the younger generation of researchers working on superconductvity, among other fields, were able to forsee the deadlock that an insistence on this approach implied and to orient their research towards a new direction. Nevertheless the dominant idea of this period was still guiding the work of many researchers.

In 1928 Bloch gave a quantum mechanical interpretation of the conductivity of metals. In the light of this interpretation the consideration of superconductivity as a phenomenon of infinite conductivity led to some new ideas. Especially Bloch and Landau had given two contradictory proposals concerning the most stable state of electrons mechanism in the superconducting state, which resulted in a tendency to refute any new attempt to interpret the phenomenon of superconductivity.[2] This feeling of deadlock pushed probably the younger scientists to search for new paths outside the frame of the established ideas. In attempting this new orientation two factors played a decisive role: (a) The planning and performance of new experiments. (b) The enrichment of their ideas with propositions and theories which had been formulated by that time in other fields of research.

We may point out two of the main achievements of this period: the proof that in the transition to the superconductive state we have a change of phase; and the application of thermodynamics although it was not yet known that this transition was reversible.

Thus this group of young scientists were working under the guidance of a positive heuristic that comprised both the idea that the phenomenon was reversible as well as the belief that in the superconductive state magnetic and thermodynamic properties were equally relevant as electrical properties did (Gavroglou and Goudaroulis, 1984). As Casimir comments (1980):

> By that time we of the younger generation began strongly to doubt the frozen-in field picture and experiments were set up to put it to a direct test. They were anticipated by Meissner and Ochsenfeld, who showed towards the end of 1933 that when a superconductor is cooled in a magnetic field the field is expelled. Our reaction at Leiden was somewhat ambivalent: we were glad to see our ideas confirmed and sorry that we ourselves had been a few weeks too late.

Thus Meissner effect did not emerge as a sudden discovery, at least for those scientists who were working in such a perspective. On the contrary it seems to be the result of their theoretical work.

Now, the Meissner–Ochsenfeld experiment had revealed something very important, namely that the phenomenon of superconductivity was reversible and that superconductors behave like perfect diamagnetic materials. Schematically, the problem of superconductivity was transformed from a predominantly electrical problem into a magnetic one.

The crucial character of the Meissner effect consists mainly in the throwing down of the following prejudice: that superconductivity was rather an electrical irreversible phenomenon to which Maxwell's equations should be applied.

The impact of this experiment on the formation of the ideas in the scientific community was very strong. It supplied scientists with such a heuristic boost that oriented their groping efforts to some concrete choices which led to new theoretical considerations and to the formation of a sound strategy.

F. London working in this framework of ideas had at first elaborated a phenomenological programme and then formed the fundamental steps towards a microscopic programme by formulating explicitly the positive heuristic of this programme.

3. THE ISOTOPIC EFFECT

At the very start of the fifties the interpretation of superconductivity enters a new period, which results in the BCS theory in 1957. An open problem, which was of great urgency, was the determination of the kind of interactions which were responsible for superconductivity. This conviction seems also to be responsible for the delay of the interpretation of superconductivity.

In 1950 Fröhlich enters on the scene. Working within the frame of London's programme, he elaborates theoretically the interaction problem. Guided by the same positive heuristic ("A superconductor is essentially a quantum structure on a macroscopic scale") he comes to a conclusion about the kind of interaction which predominates superconductivity. In doing so he extends the positive heuristic creating a heuristic boost based on his knowledge of both the solid state physics and field theory. The use of this theoretical background, which led to the introduction of new methods, was not very common at that time. Fröhlich says (1966): "Solid state physicists were so selfcentred that the concept of field had been connected with elementary particles and electromagnetic theory but nothing else".

A conclusion of Fröhlich's theoretical work was that the critical temperature depends on the isotopic mass of the superconductor. This conclusion resulted from Fröhlich's conviction that the interaction between the electrons and the lattice was mainly responsible for superconductivity. In the same year (a few months before the publication of

Fröhlich's paper) the results of an experiment are presented according to which the critical temperature varies inversely to the square root of the isotopic mass. This experiment, known since then as the isotopic effect, was the empirical corroboration of the theory of Fröhlich, who ignored its occurrence. According to Bardeen (1963):

> At that time Fröhlich visited the Bell telephone laboratories, where I was working. He told me about his own work on a theory of superconductivity based on electron–phonon interactions, which he had done at Perdue in the spring of 1950. Fröhlich's work was done without knowledge of the isotopic effect. He was greatly encouraged when he learned, just about the time he was ready to send his manuscript to the Physical Review, about this strong experimental confirmation of his approach.

We point out that in this case theoretical discovery and experimental confirmation were quite simultaneous. We could, however, allege rather plausibly that the London–Fröhlich programme would have been continued by Bardeen, Cooper and Schrieffer even if this experimental confirmation either had never occurred or was to be performed some years later.

We shall now examine the role played by Maxwell–Serrin's experiment from two points of view: the way it influenced the development of the London–Fröhlich programme and the conflict between this programme and its competitors. As we have mentioned before, London considers Coulomb interactions between electrons to be responsible for superconductivity while Fröhlich introduces the interactions between electrons and the lattice. If London's view is the nth scientific version within the programme and Fröhlich's view the $(n + 1)$th version, then the Maxwell–Serrin experiment played the role of a minor crucial experiment between these subsequent versions.

About the same time other scientific programmes also develop. In these programmes the interaction between electrons and the lattice is not taken into account. Thus, isotopic effect constitutes an anomaly, which was ignored by the scientists working within these programmes, guided by their positive heuristic. As Fröhlich comments (1961):

> Discovery of the isotopic effect was, of course, an important feature in this change of opinion. Yet it appears that all this did not affect some of those physicists who had previously formed an opinion in a different direction. A striking example can be found in the article by W. L. Ginsburg published in 1953. This author, it appears, had formed the opinion that the free electron model can never succeed in accounting for the properties of superconductors but that the electrostatic interaction between electrons – modified by the high dielectric constant of metals – should be of primary importance.

It is worth noticing the sharpness Ginsburg uses in reacting to Fröhlich's view as well as his dogmatic adherence to his own programme.

In this case, we see that an experimental result could not decide instantly for the victory of one programme over its competitors. The final issue of the war is achieved after a long period during which all the available data (experimental evidence, explanatory power, etc.) lead to the predominance of the one over the others. Thus, only after the emergence of the BCS theory the isotopic effect can be characterised as a major crucial experiment that was decisive for the superiority of London–Fröhlich's programme over its competitors.

4. CONCLUSIONS

It is worth mentioning an additional aspect that these crucial experiments accomplish: they clarify the conceptual framework by eventually preventing the scientists from an adherence to prejudices favoring specific explanatory schemata.

Thus, an experiment would not have such a crucial character if it was designed to answer a question which disputes the specific belief system. In our case the Meissner and the isotopic effects played such a role, as we have seen.

The Meissner effect, in addition, strongly affected the situation in another way: It provided London's programme with a heuristic boost during a phase that such a boost could not have been provided only by theoretical means.

Regarding the crucial character of some experiments we may point out the double role that some of them have played in the development of scientific research programmes. On the one hand they have provided a confirmation of a version within a research programme; on the other hand have proved crucial with hindsight, in favouring the predominance of a programme over its competitors.

NOTES

[1] The scientific community had reason for such a strong belief in Maxwell's equations in that period (1911 and later) especially since Einstein's special relativity theory "confirmed" the superiority of Maxwell's framework.

[2] This tendency found its expression in Pauli's remark, "Theories of superconductivity are wrong", and in Felix Bloch's theorem, "Theories of superconductivity can be disproved".

REFERENCES

Bardeen, J., 'Developments of concepts in superconductivity', *Physics Today,* **1** (1963) 24.

Casimir, H. B. G., 'Superconductivity in the Thirties', in Cohen, E. G. D. (ed.), *Fundamental Problems in Statistical Mechanics V*, North Holland Publ. Co., 1980.

Gavroglou, K. and Goudaroulis, Y., 'Some methodological and historical considerations in low temperature physics: The case of superconductivity, 1911–57', *Annals of Science,* **41** (1984) 144.

Fröhlich, H., 'The theory of the superconductive state', *Rep. on Progr. in Phys*., **24** (1961) 20.

Fröhlich, H., 'Superconductivity and the many body problem', in Marshak, R. E. (ed.) *Perspectives in Modern Physics*, *Essays in Honour of Hans A. Bethe*, Interscience, New York, 1966.

Aristotle University of Thessaloniki

PART III

NIKOLAOS AVGELIS

LAKATOS ON THE EVALUATION OF SCIENTIFIC THEORIES

Central to, and of paramount importance in the course of any critique of scientific reason, is the question concerning the justification and validity of scientific statements and theories. Any attempt to clarify the general issue and offer a respected solution to the problem inevitably involves some objective standards or criteria whose evaluative function is indispensable. In fact, theory-appraisal and use of such standards in a systematic and austere way can be identified since the days of early logical empiricism up to contemporary philosophy of science. Indeed, it was a leading and provocative theme around which rigorous argumentation and counter-argumentation has developed. The vast literature pertinent to the problem is rich and fruitful and deserves a thorough exploration. However, thematic and time limitations oblige me to trace and critically outline the several developmental stages of the problem, focusing more on these aspects of the issue in question that can prove very useful for the understanding of Lakatos's methodology. My discussion will begin with the consideration of the logical-empiricist point of view.

In the logical-empiricist argumentation the concept of confirmation plays the decisive role in theory-appraisal: those theories (hypotheses) should be accepted if and only if they are most highly confirmed: to be confirmed means in this context to be in agreement with observable facts that can be logically derived from the theory. To be precise, in Carnap–Hempel's sophisticated account, confirmation takes the form of a confrontation between the theories in question (hypotheses) and the observation reports, i.e. the propositions we have taken as evidence. This confrontation is considered to be of a logical character meaning that the rules of evaluation can be completely formulated in the language of pure logic. To put it another way, theory-appraisal is a matter subject to a logic of confirmation that Carnap, as it is well known, formulated in his theory of inductive probability. According to Carnap, the degree of probability in the light of evidence is our guide to theory acceptance.

I shall now try to point out that further developments in the problem of

K. Gavroglu, Y. Goudaroulis, P. Nicolacopoulos (eds.), Imre Lakatos and Theories of Scientific Change. 157–167.

theory evaluation up to Lakatos are closely connected with important changes in the aforementioned criteria of theory-acceptance. It should be mentioned at this point that analysis and evaluation of all the new perspectives opened to us in the postpositivistic interpretation of science cannot obviously be undertaken in the context of this paper. Here I shall only restrict myself to some main lines of the argumentation. To start with, unavoidably, I shall refer to Popper's philosophy of science – the uncontested landmark in these developments.

Popper, as is known, has challenged the logical empiricist criterion of acceptability and choice between competing theories in his *Logic of Scientific Discovery*. He argues that no number of confirming instances will show that a universal generalization is true, whereas a simple disconfirming instance will show that it is false. Scientists, in fact, test theories by trying to find ways in which they might be falsified. Consequently, falsification, not confirmation, plays the important role in theory-acceptance. This point of view is immediately associated with his major assumption that there is no logic except deductive logic. Therefore, any inference in scientific theories is legitimized only as deduction from them of observable implications and the consequent possibility of their falsification by *modus tollens*. The Popperian method of deductive testing can possibly lead to the rejection of a theory if an inconsistency appears either in the very same set of premises or in the relation of the consequences of the theory with the system of accepted basic propositions. If, however, no inconsistency appears, this does not mean that the theory is verified, but it becomes corroborated.

Despite Popper's different approach to this problem, falsification shares with confirmation the status of a logical relation between theory and evidence. From this point of view theory-appraisal through falsification, like appraisal through confirmation, is not time-bound, i.e. no time element enters into the theory-evaluation. But unlike the confirmation theory of the logical-empiricists the emphasis here is on the degree of corroboration of a theory, not on truth or on a precise probability measure. For the degree of corroboration is according to Popper our rational guide to theory choice. Popper's account of science, however, involves historical considerations to the extent that the progressiveness of scientific theories are taken to be relevant to their cognitive assessment.

Now the degree of corroboration of a theory is to be understood as "a concise report evaluating the state (at a certain time *t*) of the critical discussion of a theory with respect to the way it solves its problems", hence it is "an evaluating *report of past performance*" and "says nothing about future performance".[1] The main question that arises at this point is: if according to Popper the degree of corroboration pertains to the past performance of the theory and has no future implications, it is difficult to see how it should be a reason for preferring theory A to theory B. For in choosing between A and B we want to choose the more reliable theory, the theory with the better explanatory and predictive power in the future. Several critics of Popper, notably Lakatos, have stressed this point and underlined the need for a link between corroboration and verisimilitude that would enable us to assert that a highly corroborated theory B is likely to be more reliable than a poorly corroborated rival A. As long as corroboration cannot be treated as an indication of verisimilitude and reliability, it is difficult to see how it could be serve as a rational guide to theory choice. Popper needs obviously an inductive argumentation to establish the required link between corroboration and verisimilitude, otherwise he cannot justify our choice of a theory on the basis of its evidential support. But this would imply abandoning what is unique and important in his methodology, i.e., the rejection of induction. Popper was in fact faced with this dilemma. In what follows I intend to interpret Lakatos's methodological proposal as an attempt to find a way out of this dilemma.

Let me start mentioning some common lines of argumentation. Lakatos shares Popper's central position that every synthetic proposition is both fallible and theory-laden; consequently, there is no observational language which preserves its empirical meaning despite change in scientific theories, as logical-empiricists maintain. This point has been stressed further by Feyerabend. "Theories are meaningful independent of observations; observational statements are not meaningful unless they have been connected with theories . . . It is therefore the *observation sentence* that is in need of interpretation and *not* the theory".[2] According to Popper, even basic statements are interpreted in the light of theories, hence they cannot be verified or falsified by sense-experience, they are hypotheses as well as the theoretical ones. To the question which arises here, due to the rejection of the theoretical-observational distinction, how would, indeed, the truth-value of scientific statements be tested, if all the propositions are hypothetical, Popper gives the following answer: the deductive inter-subjective testing of theories becomes possible, if we assume by convention, i.e. without further scientific foundation, that certain basic statements are true. Lakatos shares this special status of basic statements rested on a conventional decision. What counts as scientific evidence is, therefore, dependent on the consensus of the scientific community. However, in Lakatos argumentation three objections against Popper can be discerned:

The *first* objection is connected with the Duhem–Quine problem. As is well known, Duhem correctly pointed out that any conflict between the derived statement and the observation statements indicates only that something is wrong with the entire theoretical system but not in what hypothesis the error lies, hence no isolated hypothesis can be falsified. Lakatos is right in claiming that the falsifying instance reflects negatively upon the whole body of our knowledge, hence some negative outcomes of tests do not constitute refutations, for they can be ascribed to incorrect auxiliary assumptions. Consequently, crucial experiments, which are important for Popper, play no role in Lakatos's sophisticated falsificationism, i.e. they cannot decide in a moment between competing theories. According to Lakatos, "we cannot learn from experience the falsehood of any theory".[3]

The *second* objection concerns the role of rival theories in tests, for in case of a conflict between theory and observation a modification of the protective auxiliary belt is more reasonable than dropping the theory, on the condition of course that the scientific research programme, of which it is a part, is not a degenerating one. According to Lakatos, tests are not two cornered fights between theories and experiments but three cornered fights between rival theories and experiment, where "some of the most interesting experiments result, *prima facie* in confirmation rather than falsification'.[4]

The *third* objection concerns the fact that falsification is not a necessary condition for the elimination of theories. One finds evidence for this in a common phenomenon in the history of science; theories are retained by the scientific community well after their decisive falsification and contra Popper's falsificationism principle. According to Lakatos, "*there is no falsification before the emergence of a better theory*".[5]

Now my point is that Lakatos's views can be adequately understood as a further systematic development of Popper's insights. Popper himself, as time passes, becomes more and more conscious of the fact that experience by itself cannot falsify a scientific theory once and for all, for a further element should be taken into consideration and this is the comparison between rival theories.[6] This idea, latent in Popper, was developed systematically by Feyerabend and Lakatos. If experimental data are necessarily interpreted by existing theories, then falsification of a theory is meaningful only in the light of a rival theory; for it is precisely from the point of view of its rival theory that data can be seen as counterevidence of a certain theory. For example, it was only in the light of Einstein's theory that Michelson's experiment was interpreted as counterevidence for the Newtonian theory. Lakatos suggests that we should regard a theory T as falsified if there is a rival theory T^1 that makes new predictions, some of which are eventually confirmed. Unlike Popper, Lakatos considers as a necessary condition for the acceptance of a theory not only falsifiability but also confirmability. Consequently no theory should be abandoned except in favour of a better theory. This epistemological fact has important consequences for scientific research: if the formulation of rival theories is a fundamental presupposition for the falsification of a theory, then pluralism concerning the use of theories in scientific research should be the basic policy in the pursuit of knowledge.

This idea of theoretical pluralism is most sharply formulated by Feyerabend: "Invent and elaborate theories which are inconsistent with the accepted point of view, even if the latter should happen to be highly confirmed and generally accepted".[7] Feyerabend, however, gives no guiding principles nor any indication of rules to be followed throughout the process of proliferation. But with Lakatos this process obtains some definite criteria in the context of "research programmes" and in particular by rules of a "positive heuristic" which guide us in the formulation of variants of a theory (and eventually of alternative theories). Thus Lakatos, by shifting the whole problem of the evaluation of scientific theories from an evaluation problem of a single theory to an evaluation of a series of connected theories (so called by Lakatos "research programme"), suggests a better criterion of demarcation and falsification than the Popperian one. The empirical character of theories is determined by their being set against empirical propositions accepted as basic, as Popper demands, but in the context of the competitive relation of theories where the empirical base simply mediates the rivalry of theories. Given this new historical dimension in the concept of theory and its priority against facts the problem of the criteria employed for the appraisal of scientific theories has turned out to be the central problem in contemporary philosophy of science. Lakatos proposes a criterion of progress to be considered as a criterion of the validity of scientific theories, where progress is always to be decided with respect to some accepted basic propositions. A progressive series of theories or a scientific research programme has to satisfy two conditions: it must be theoretically and empirically progressive, i.e. it must have some excess empirical content over its predecessor, some of which must be corroborated.

The critical questions encountered at this point concern the kinds of evidence which may be considered to advocate real progress of our theories, that is their getting closer to truth. The problem is how to bridge the gap between the conventional game of science and the demand that science, if it is to be something more than a simple game, should search for truth. Popper hesitates to seek the connecting link between scientific method (the game of science) and its claim for truth, for he accords no value to the theory of knowledge in the context of his logic of discovery. In his *Objective Knowledge* he stresses this point: "*No theory of knowledge should attempt to explain why we are successful in our attempts to explain things*".[8] Consequently he restricts himself to an analytical appraisal of scientific theories, whereas Lakatos is fully conscious of the necessity of some speculative genuinely epistemological theory connecting scientific standards with verisimilitude. Lakatos's proposal is that we should recognise the signs of the growth of knowledge on the basis of "an inductive principle which connects realist metaphysics with methodological appraisals, verisimilitude with corroboration, which reinterprets the rules of the scientific games as a – conjectural – theory about *signs* of the *growth of knowledge*, that is, about the signs of *growing verisimilitude of our scientific theories*".[9] According to Lakatos, "only such an inductive principle can turn science from a mere game into an epistemologically rational exercise".[10]

At this point it is obvious that the Popperian dilemma reemerges. Lakatos's way out of the dilemma leads to the acceptance of a metaphysical principle at the cost of calling in question the strict Popperian deductive pattern, for he leaves space in his methodology for the principle of induction. Whereas on the one hand we can indeed agree with Lakatos, when he points out the lack of an appropriate theory of knowledge in Popper's philosophy which could bridge the gap between the game of science and its claim for truth, we can on the other hand seriously doubt that his metaphysical-synthetic principle is able to correct the Popperian deficiencies and surmount the relevant difficulties. Lakatos's weak principle of induction, which he accepts "without believing it'[11] in order to explain the technological reliability of scientific theories, cannot take the place of an appropriate theory of knowledge that is lacking, in the context of which this problem could be tackled satisfactorily. Lakatos himself is fully conscious of the severe difficulties that go along with his approach to theory appraisal, when he writes that "only God could give us a correct, detailed estimate of the absolute reliability of *all* theories by checking them against his blueprint of the universe".[12] But God does not give us His plan so we can perform this appraisal by ourselves; that is to see how close our theories get to the truth. Here the question arises: is there no way out of the Popperian dilemma?

My concern is not to challenge Lakatos's conception of what counts as a unit of appraisal in science, for I think he is right in claiming that the growth of knowledge has a dynamic dimension which is best determined within the framework of scientific research programmes. The problem is, given that scientific method is our only guide to nature, how can the hypotheses of reason have any bearing on the empirical world; on what grounds can they be accepted? There is a tension regarding this problem in Lakatos's thought. Lakatos oscillates between realistic metaphysics and methodology. On the one hand he claims that scientific method should be considered as a way to a pre-existent truth and reality and stresses the necessity of the concept of verisimilitude in theory appraisal, whereas on the other hand he tries to develop a methodological account of the theory evaluation in the context of which the assumption of a realistic metaphysics seems to have no place. Lakatos points out that "as long as the Leibniz–Whewell–Popper requirement is met" (i.e. "*the – well planned – building of pigeon holes must proceed much faster than the recording of facts which are to be housed in them*"), "it does not matter whether we stress the instrumental aspect of imaginative research programmes for finding novel facts and for making trustworthy predictions, or whether we stress the putative growing Popperian verisimilitude (that is the estimated difference between the truth and falsity content) of their successive versions".[13]

I interpret this passage as trying to reduce the meaning of a realistic-metaphysical assumption that Lakatos, for want of an appropriate theory of knowledge has introduced *ad hoc* without justifying it in order to explain the reliability of scientific theories, and as stressing the role of methodological procedures in the growth of knowledge. Lakatos assumes that there has been a growth in scientific knowledge and that this growth has been achieved through methodological procedures. But failing to provide any good reason for thinking that there has been progress towards greater verisimilitude, Lakatos seems sometimes to toy with the idea of substituting truth for method. Certainly this would relieve the tension which has been mentioned. But we ought not to take seriously the possibility of such an interpretation of Lakatos's thought, for one might cite Lakatos's claim that the rationality of scientific activity need "some extramethodological inductive principle to relate . . . the scientific gambit of pragmatic acceptances and rejections to verisimilitude".[14] According to Lakatos, "only such an 'inductive principle' can turn science from a mere game into an epistemological rational exercise, from a set of light-hearted sceptical gambits pursued for intellectual fun into a – more serious – fallibilist venture of approximating to Truth about the Universe".[15]

Looking at Lakatos's theory through Kantian spectacles, I will subsequently utilize basic concepts and arguments derived from Kant's Transcendental Methodology (B 737 f.) intending to find a way out of the Popperian dilemma. In the first place, Lakatos realises that unless we appraise a scientific theory with respect to its truth-status, scientific enterprise could be taken to be a mere game. Consequently the question arises for him why scientific theories arrived at through procedures involving a specific methodology (in Lakatos's terms the methodology of scientific research programmes) is the sort of thing that can be true of nature and hence can be accepted. For it is obvious that we employ methodological procedures which are guided by our own interests, that is the systematic unification of empirical knowledge. Needless to say, hypothesis formation is to be taken as one element in a methodology that seeks to unify empirical knowledge.

The important thing to notice for our purpose is that the hypothetical use of reason is guided by a transcendental principle, the principle of the systematic unity of all physical phenomena. Lakatos has not fully realized the regulative function in theory building of the "hard core" and the "positive heuristics" in scientific research programmes. As a result he has stressed inductive elements (successful prediction of novel facts) rather than systematic properties as grounds for theory choice. But theory choice will have a rational foundation, if we view the sequence of theories as guided by the transcendental idea of the systematic unity of nature; that is, scientific theories are to be constructed so as to conform to the ideal of the systematic organization of experience. For, unless we assume a systematic unity of nature in a Kantian sense, it would not make sense to treat theories as capable of truth or to speak about real progress in science. It seems to me that the rationality of Lakatos's methodology cannot be established at the pre-critical level of realistic metaphysics, i.e., on the basis of the Popperian notion of verisimilitude. Popper did not realise that his "new meta-logical idea" was in fact a transcendental principle.

What I hope to have in fact established is that one needs to use further presuppositions in order to justify the claim that scientific theories, the products of a specific methodology, should be capable of experimental confirmation or corroboration. Moreover, in order to show the rationality of the scientific activity one needs to know the possibility of achieving one's goal, i.e., the establishment of true empirical theories, through the application of the rules of the game. Thus, unless we presuppose a system of nature that science will enable us to uncover, such a rationality would not be established. Rational scientific inquiry is a goal-directed activity. To achieve the goal one needs to adopt a particular strategy that will regulate the necessary movements in the scientific game. But first we must know *a priori* that, following the rules of the game, it is possible to achieve the gaol: otherwise, scientific activity seems to be irrational. On the other hand, if we come to achieve our goal, that is to know the structure of natural phenomena, we need to proceed in accordance with a specific methodology. As is well known, on the grounds of case studies Lakatos argued that scientists proceeded in accordance with the methodology of scientific research programmes.

What I am trying to suggest, in short, is that the evaluation of scientific theories involves two aspects that should not be regarded as being unrelated. On the one hand, to the extent that evaluation is directed to the achievements of the theory, the emphasis will be on the criterion of corroboration, since verification is beyond the reach of scientific method. On the other hand, inasmuch as theory evaluation takes place when a choice must be made between rival theories, the contribution of a rival theory to the systematic organization of knowledge is more important than instance-confirmation. Implicit in this systematization of empirical knowledge is the possibility of unifying hitherto diverse areas or of giving rise to interesting extensions to further regions of experience. To evaluate a theory or a research programme would mean in this case to explore the dynamic character or the fertility of the relevant theory or research programme by reference to their prior historical development. Lakatos is right, not only in thinking that the historical dimension of science affects theory evaluation, but also in stressing that the requirement of continuous growth is of considerable significance for a rational reconstruction of the unity of science. But in formulating criteria of acceptability he failed to give due consideration to the prominent feature of systematization of experience implicit in scientific research programmes, although the unifying character of scientific theorizing is manifest in his account of the historical development of the Newtonian research programme. This may help to explain why Lakatos does not reflect on the principles that guide the ordering of experience as a goal to be sought by the knowing subject and stresses instead inductive elements as grounds for theory acceptance. His commitment to a realistic view of science, however, does not help him to surmount the difficulties regarding the required link between the game of science and the goal to be sought, i.e. truth or verisimilitude. In conclusion, I should emphasize that it would be more in keeping with Lakatos's methodological intentions, if we considered the growth of knowledge, not as a process of approximating to a pre-existent truth, but as a process that is guided by regulative principles towards a systematic description of nature.

NOTES

[1] Popper (1972), p. 18.
[2] Feyerabend (1965a), p. 213.
[3] Lakatos (1978b), p. 211.
[4] Lakatos (1970), p. 115.
[5] Lakatos, *op. cit.*, p. 119.
[6] See Popper (1963), p. 232.
[7] Feyerabend (1965b), pp. 223–4.
[8] Popper (1972), p. 23.
[9] Lakatos (1978a), p. 165.
[10] Ibid., p. 156.
[11] Lakatos (1978b), p. 187.
[12] Lakatos (1978a), p. 185.
[13] Lakatos (1978a), p. 100.
[14] Lakatos (1978a), p. 113.
[15] Ibid., p. 113.

REFERENCES

Feyerabend, P. K., 'Problems of Empiricism', in R. Colodny (ed.): *Beyond the Edge of Certainty*, 1965a.

Feyerabend, P. K., 'Reply to Criticism', in *Boston Studies in the Philosophy of Science*, **2**, 1965b.

Lakatos, I., 'Falsification and the Methodology of Scientific Research Programmes', in I. Lakatos and A. Musgrave (eds): *Criticism and the Growth of Knowledge*, 1970.

Lakatos, I., *The Methodology of Scientific Research Programmes* (ed. by J. Worrall and G. Currie, Philosophical Papers, Vol. 1), 1978a.

Lakatos, I., *Mathematics, Science and Epistemology* (ed. by J. Worrall and G. Currie, Philosophical Papers, Vol. 2), 1978b.
Popper, K. R., 'Truth, Rationality and the Growth of Scientific Knowledge', in K. R. Popper, *Conjectures and Refutations*, 1963.
Popper, K. R., *Objective Knowledge*, 1972.

Aristotle University of Thessaloniki

MARCELLO PERA

METHODOLOGICAL SOPHISTICATIONISM: A DEGENERATING PROJECT

1. LAKATOS'S PROJECT

Lakatos's project is here taken as consisting in the attempt at looking for a universal methodology which fits scientific practice, especially as witnessed by the history of science.

In the context of the philosophy of this century this project is new. Logical positivists had equated the philosophy of science with the logic of science and this with the anatomy of scientific achievements considered as finished linguistic products. They had neglected practice to the benefit of logical substitutes. Carnap, for example, wrote that "philosophy is the theory of science" and that "philosophy deals with science *only* from the logical viewpoint. Philosophy is the logic of science, i.e., the logical analysis of the concepts, propositions, proofs, theories of science, as well as of those which we select in available science as common to the possible methods of constructing concepts, proofs, hypotheses, theories" (Carnap 1934, p. 6). In the same vein, Reichenbach wrote that "epistemology is interested in internal relations *only*" (1934, p. 4). Popper widened the scope of philosophy. He agreed that "the logic of knowledge is concerned *only* with logical relations" (1959, p. 30), but he observed that logical analysis is not enough, for "if we characterize empirical science merely by the formal or logical structure of its statements, we shall not be able to exclude from it that prevalent form of metaphysics which results from elevating an obsolete scientific theory into an incontrovertible truth" (1959, p. 50). Accordingly, he added to the logic of science a "methodological supplement" (1959, p. 54) containing instructions on how to treat scientific theories in order to ensure their perennial revisability. In Popper's view this supplement was so important that he went so far as to maintain that "epistemology, or the logic of scientific discovery, should be identified with the theory of scientific method" (1959, p. 49).

Popper, however, was not very clear on the question of the warrant or justification of methodology. Although he resolutely rejected the logical positivist view that methodology is a descriptive, empirical science, for

K. Gavroglu, Y. Goudaroulis, P. Nicolacopoulos (eds.), Imre Lakatos and Theories of Scientific Change. 169–187. © 1989 by Kluwer Academic Publishers.

"science" and "scientist" are (also) value-predicates (1959, p. 52), he proved to be ambiguous on its foundation. On the one hand, he maintained that a set of methodological rules is introduced as a "proposal for an agreement or convention" depending on a choice of purposes or values (1959, p. 37); from this point of view, epistemology has an axiological warrant whose suitability requires a discussion which "is only possible between parties having some purpose in common". On the other hand, he affirmed that his own proposal for a falsificationist method depends on the consequences of this method, and especially on "how far it conforms to his [the scientist's] intuitive idea of the goal of his endeavours" (1959, p. 55); taken in this sense, the warrant of methodology is intuitionistic or, perhaps, descriptive, for methodologists can resort to the scientists' intuitive ideas of scientific goals and their greatest achievements only through empirical scrutiny of their practice and, therefore, through historical examination.

Lakatos took the second view as the only legitimate one and embarked upon his project. But at the very beginning he made an unexpected, disappointing discovery; he noticed (or was sure he had noticed) that "history 'falsifies' falsificationism (and any other methodology)" (1971a, p. 213). Faced with this state of affairs, he envisaged only two possible alternatives. As he wrote:

> But if – as seems to be the case – the history of science does not bear out our theory of rationality, we have two alternatives. One alternative is to abandon efforts to give a rational explanation of the success of science. Scientific method (or 'logic of discovery'), conceived as the discipline of rational appraisal of scientific theories – and of criteria of *progress* – vanishes. We may, of course, still try to explain *changes* in 'paradigms' in terms of social psychology. This is Polanyi's and Kuhn's way. The other alternative is to try at least to *reduce* the conventional element in falsificationism (we cannot possibily eliminate it) and replace the *naive* version of methodological falsificationsim ... by a *sophisticated* version which would give a new *rationale* of falsificationism and thereby rescue methodology and the idea of scientific *progress*. This is Popper's way, and the one I intend to follow (1970, p. 31).

For the sake of simplicity, let us call these alternatives the *methodologistic thesis* and the *sophisticationist thesis*, respectively. The methodologistic thesis maintains that scientific rationality depends on method; the sophisticationist thesis maintains that scientific method has to fit scientific practice, at least its best examples. Both these alternatives are based on various presuppositions which need to be clarified and examined if we wish to evaluate Lakatos's project.

The following presuppositions are those most relevant to the sophisticationist thesis:

(a) Sophisticationism holds the view that methodologies have to be tried at the bar of the history of science to prove their worth, just as theories have to be brought before the court of facts to show their truth-value. The best methodology is that which withstands the severest historical tests represented by paradigmatic cases of scientific performance, or which takes more "basic appraisals of the scientific élite" into account, or which makes more actual science internal or rational (1971a, p. 123).

(b) In a sophisticated methodology historical adequacy and analytical precision can be combined to any desired or, at least sufficient, degree. Sophisticationism assumes that in the whole history of science (or at least the history of science of the "last two centuries") there is a methodological L.C.D. which has a non-generic content. A historical scrutiny would show, for example, that scientists work with a rejection rule which says not simply "Reject those theories which conflict with contrary evidence" but "Reject a theory T which is in conflict with the evidence if and only if another theory T' has been proposed with such and such well-defined characteristics" (see 1970, p. 32). If one methodology turns out to be inadequate, sophisticationism also assumes it is always possible to find another one which fits the history of science better, both extensively, that is, with reference to the amount of history saved, and intensively, that is, with reference to the precision acquired.

Two relevant presuppositions also lie behind the methodologistic thesis. They are:

(c) A methodology is, or offers, a theory of rationality or a demarcation criterion or a criterion of progress or a code of scientific honesty (1971a, p. 103 and 140; see also 1974a, p. 315n. and 1976, p. 168). Thus, the scope of rationality is conceived as being coextensive with the domain of rule-governed activity. Outside this domain there may be only personal, irrational decisions and behaviour. In this sense, methodology is like Kant's universal moral law: obeying its commands is esteemed as moral and honest, whereas violating it and acting according to personal maxims is condemned as dishonest.

(d) Accordingly, the aim of a theory of rationality is that of eliminating or reducing to a harmless minimum the personal factors which may enter into the scientific enterprise, especially into the appraisal of scientific theories (1970, p. 31). To the extent that such factors still persist, science is not rational. In this sense, methodology acts like Orwell's Big Brother. His supreme eye vigilates in order to keep law and order, to discover deviant tendencies which might be the outcome of subjective freedom, to condemn them as dishonest attacks on the universal norms which ensure rationality and progress for the commmunity, to eradicate them, and to deprive their authors of funds and journals, until they admit their faults and show they are willing to make amends (1971a, p. 117).

Elsewhere (Pera, 1986), I have criticized (a) and (b). I do not think that methodology can be tested against history; in my view, what history offers is, at the very best, no more than consolations for methodologists; for conservative methodologists, I should add. Nor do I believe that the sophistication of methodological rules can be extended *ad infinitum*; on the contrary, precisely the history of science can be used to prove something like a "methodological indeterminacy theorem": what we gain in scope we lose in accuracy. But here I wish to focus on (c) and (d). In my view, these presuppositions amount to a more general assumption which I have called the *Cartesian Syndrome*. I shall, then, proceed in two stages. I shall first describe this syndrome and I shall try to show that Lakatos was really affected by it. Then I shall discuss Lakatos's project and its failure. What seems to be a promising alternative to it is a question which cannot be examined here (see Pera, 1987 and 1988).

2. THE CARTESIAN ROOTS OF LAKATOS'S PROJECT

A quick look at history may help to clarify what I mean by "Cartesian Syndrome".

Descartes emphasized the idea that science and, in general, knowledge, may be pursued, acquired and elaborated upon, only if our mind is bound by rigid norms. According to him, science or knowledge possesses a specific method which, as Rule 4 says, consists of "certain and easy rules such as whoever has employed them exactly never supposes anything false as true, and without uselessly consuming his mental effort but rather always gradually increasing his knowledge, will arrive at a true cognition of all those things of which he will be capable" (Descartes, 1966, pp. 371–72). To further explicate this point, the title of the same Rule says that "*necessaria est methodus ad rerum veritatem investigandam*".

Descartes was certainly not the first to maintain that science is a public game, a rule-directed activity. Bacon had already held a similar opinion, comparing his method to an instrument which "levels man's wits", like a rule or compass (Bacon, 1860, p. 63). And the idea of method dates back at least to Plato's *Phaedrus*, which also contained two other ideas that proved to be influential in Western thought, namely that method is a demarcation criterion for what is scientific (268a–c), and that method consists of a fixed sequence of intellectual steps (270d–271c). Descartes, however, seems to have been the first to maintain that, outside method, there can only be irrational passions. His Rule 4 explicitly says that methodology "should contain the primary rudiments of human reason" (1966, p. 374; see Schouls, 1980, Chapter III.1). As a consequence, any controversy on scientific matters must, in principle, be settled in a clear-cut way. For Descartes, controversy is the negation of science; as he says in Rule 2, "whenever the judgements of two people on the same subject are opposite, it is certain that either is wrong and that neither possesses science" (Descartes, 1966, p. 363).

While Spinoza introduced the Cartesian method in ethical matters, Leibniz found that, as it stood, it was terribly lacking. "I almost feel like saying that the Cartesian rules are rather like those of some chemist or other", he wrote, "take what is necessary, do as you ought to do, and you will get what you wanted" (Leibniz, 1961, IV, p. 329). It should be noted, however, that this irony was not directed towards Descartes' project as such, but towards its *technical* poverty. Leibniz, too, was of the opinion that controversy is the hall-mark of ignorance; his "calculus" and "scales of proofs" were intended to offer precise tools for proving truth or for weighing the degree of probability of any rival opinion. Thanks to these tools, as he wrote, "all truths can be discovered by anybody and with a secure method (*methodo certa*), to the extent that, making use of reason, they may be obtained by data available even to the greatest and trained talent, with the only difference of readiness, whose importance is greater in action than in meditation and discovery" (Leibniz, 1961, VII, p. 202).

The underlying ideas common to Descartes and Leibniz and to the tradition to which both of them belong are that science is an intellectual activity submitted to universal and impersonal rules and that violating such rules leads to irrationality. It is the combination of these ideas that I call the *Cartesian Syndrome*. For those affected by this syndrome science becomes an impersonal body of knowledge in the sense that its intellectual content and growth may be, if not constructed, at least reconstructed in terms of rules which are independent of the personality of scientists and of the cultural and social conditions in which they perform their research.

The logical positivists' notion of a "context of justification" to be analyzed in merely logical terms is a contribution to this idea, and Carnap's probabilistic logic of confirmation may be regarded as an up-to-date version of Leibniz's scales. But perhaps the most eloquent example is offered by Popper's view of science as "knowledge without a knowing subject". In the following passage this view is clearly stated:

> Scientific knowledge may be regarded as *subjectless*: it may be treated as a system of *theories-in-themselves* (in Bolzano's sense) on which we work like workers on a cathedral, though we should try to do so in a critical spirit: the belief or faith of a scientist, though it may inspire him, poses no problem for the theory of 'scientific knowledge' (that is, for the theory of theories-in-themselves); though of course this faith may cause as much trouble as most inspirations do (Popper 1968, p. 295).

Popper's early comparison of science to a game governed by rules belongs to the same line of thought. "Just as chess might be defined by the rules proper to it, so empirical science may be defined by means of its methodological rules" (1959, p. 54). Granted, to *win* a game requires experience, talent, ingenuity, luck; but to *play* chess only requires strict observance of the fixed rules. In the same way, experience, talent, etc. are necessary for a scientist to invent "happy guesses"; but once the invention has been made, all that is needed is deductive logic and a "methodological supplement". In World$_3$, where science properly lies, there is no room for personal factors.

Lakatos took his dread of these factors from Popper. Just as Popper took deductive logic as the "organon of rational criticism" (Popper, 1963, p. 64), so Lakatos maintained that "without deductive logic there can be no genuine criticism" (1978d, p. 243). And just as Popper had been looking for a (methodological) "*clear* line of demarcation between science and metaphysical ideas" (1959, p. 39),[1] so Lakatos entrusted methodology with the task of providing "*sharp* criteria" or "*universal* definitions" of science (1976, p. 192; 1974a, p. 315n.). His view amounts to saying that either we create an exact logic of discovery, or science irremediably degenerates into "mob psychology" (1970, p. 91).

It is precisely this alternative that proves that Lakatos was affected by the Cartesian syndrome. He could not conceive of any middle land, at least any *rational* middle land, between these two extremes. The passage quoted above in Section 1 proves how firmly he stuck to this view. In another passage he wrote:

> Often these rules, or systems of appraisal, also serve as 'theories of scientific rationality', 'demarcation criteria' or 'definitions of science'. Outside the legislative domain of these normative rules there is, of course, an empirical psychology and sociology of discovery (1970, p. 103; see also 1978d, pp. 224–225).

Accordingly, Lakatos saw the purpose of methodology as being primarily aimed at reducing or mitigating the conventional factors any scientific decision involves, the underlying assumption being that such factors would destroy scientific rationality and that scientific rationality coincides with rule-domination (that is assumptions (c) and (d) above).

Feyerabend criticized Popper and Lakatos on the ground that their methodologies did not work for they are violated by scientists in many cases. But Feyerabend's arguments prove how the Cartesian syndrome affects even those who seem to be farthest away from Cartesian ideas. Consider, for example, the following objection to Popper's and Lakatos's emphasis on tests and criticism:

> Now this reference to tests and criticism which is supposed to guarantee the rationality of science and, perhaps, of our entire life may be either to *well-defined procedures* without which a criticism or test cannot be said to have taken place, or it may be purely abstract so that it is left to us to fill it now with this, and now with that concrete content. The first case has just been discussed. In the second case we have but a verbal ornament, just as Lakatos' defence of his own 'objective standards' turned out to be a verbal ornament (Feyerabend, 1970, p. 151).

The Cartesian alternative is at work here. Descartes had maintained the view that *if* science had no method, knowledge would be impossible; Feyerabend reaches the conclusion that *since* science has no method or, to be more precise, no universal and impersonal rules of method, then "science is much closer to myth than a scientific philosophy is prepared to admit" (1975, p. 295), and scientific ideas are defended and accepted by "means other than arguments" or by "irrational means, such as propaganda, emotion", etc. (1975, pp. 153–54; see also 1970, p. 150). From this point of view, Feyerabend is the latest coherent Cartesian; his criticism of Lakatos is tantamount to reminding Lakatos an alternative he himself had stated: if the rules of the methodology of scientific research programmes cannot be rendered precise and operative without a massive introduction of personal factors, and if the weight of such factors proportionally reduces the rationality of science, why then not frankly admit that the aim of science is "inaccessible to those who decide to rely on arguments only"? (Feyerabend, 1970, p. 151).

Kuhn had remarked to no effect that, when no algorithms or precise rules exist, the acceptance of a new theory is a question of "argument and counterargument" (1962, p. 152); the Cartesian syndrome was too deeply-rooted in the philosophical community for these remarks to be noted. Thus, quite typically, Kuhn was labelled as an "irrationalist" (and

Toulmin as an "élitist"). Scientific argumentation different from logic and its "methodological supplement", on the one hand, and from mere propaganda, on the other, is something that Cartesian patients cannot even conceive of.

This, then, is the project; let us now take a look at the way Lakatos carried it out.

3. THE FAILURE OF LAKATOS'S PROJECT

In order to attain his goal of submitting mob psychology to the closest possible inspection by the police of impersonal method, Lakatos had to transfer science from (to use Popper's terminology) $World_2$ to $World_3$, which is the proper domain of method, where it fully shows its strength, and look at science through "Popperian spectacles" (Lakatos 1970, p. 91). But for this operation to be successful, one needs to be a coherent Popperian; in particular, one needs to maintain Popper's three main theses of "epistemology without a knowing subject"; that is:

(i) $World_2$ is "irrelevant to the study of scientific knowledge" (Popper, 1972, p. 111);
(ii) $World_3$ contains objective thoughts (in Frege's sense) which are "independent of anybody's claim to know" and also "independent of anybody's belief, or disposition to assent; or to assert, or to act" (Popper, 1972, p. 109);
(iii) between $World_3$ and $World_2$ there is a bridge, a "principle of transference" by which "what is true in logic is true in scientific method and in the history of science" (Popper, 1972, p. 6).

All these theses are disputable. Let us consider them briefly.

Thesis (i) is true in the sense that negating it gives rise to what textbooks of logic call "genetic fallacy", and in the sense that truth is different from what is actually accepted by people, including communities of experts. Granted, considerations about the proponent of a cognitive claim, for example considerations about how he arrived at it or why he holds it, are irrelevant to the validity or truth of the claim; nor do psychological convictions justify it. But this does not mean that personal (or pragmatic) factors do not enter into the appraisal of a claim. Take just a few cases. The confirmation of an hypothesis h on evidence e depends, among other things, on the prior probability of h, and the evaluation of this probability is largely subjective. The degree of confirmation of h given e depends, in

Carnap's inductive logic, on a choice of the value of the parameter λ and this choice, too, involves personal decisions. Popper's degree of corroboration of h given e is a function of the "severity" of e, that is of the "sincere attempts" at falsifying h, which, in spite of Popper's efforts, are to a great extent subjective. In all these cases an examination of the conditions ($World_2$) in which a claim ($World_3$) is put forward is constitutive of the evaluation of that claim.

Thesis (ii) is also true in the sense that expressed thoughts are different from the process of thinking and may be analyzed in different terms. However, from this it does not follow that objective thoughts are analyzable in strict logical terms. Popper seems to presuppose that $World_3$ contains only propositions and arguments, but it certainly also contains value-concepts and evaluations or discussions. Such discussions can be said to be "logical" in the broad sense that arguments supporting them follow some pattern of reasoning, not in the more narrow sense that these patterns are deductive.

Lastly, thesis (iii) also holds good in a certain sense. For example, if an argument is contradictory, one can say (following Hume) that it is not even psychologically conceivable. In this sense, what is impossible in logic is impossible in psychology. But if T contradicts T', it is not impossible for one and the same person to work on both of them; and if T' turns out to have a certain relation (for example, inclusion or smaller content) to T', this does not mean that this relation is the reason why upholders of T *preferred it to* T', nor that it is the reason why they should have done it. Logical possibility has no normative bearing on factual (historical) reality.

Lakatos apparently accepted (i) and (ii), but he did not feel like going so far as to accept (iii). On the one hand, shaken by Kuhn and Feyerabend, he never refused to look at real science and to note how far its actual course is from its reconstructions in $World_3$. On the other hand, probably under the influence of Hegel, he never stopped attempting to show that what is rational (in $World_3$) is also real (in $World_2$). Lakatos had probably handled too many Hegelian "explosives" (1978b, p. 61) to leave praxis without the guidance of the "cunning of reason" and thus to let scientific practice grow without revealing an ideal design behind itself.

But this is a conjecture. What happens in actual fact is that, in order to let scientific method descend from the heaven of $World_3$ to the earth of $World_2$, that is in order to fill the gap between received methodological rules and real scientific practice, Lakatos had to sophisticate Popper's falsificationism. Two questions then crop up: does Lakatos's sophisticated falsificationism contain essential novelties with respect to Popper's methodology? Does Lakatos's sophisticated falsificationism really reduce or mitigate the conventional elements present in Popper's methodology, as his project demands?

To start with the first question, it is true that Lakatos's methodology corrects Popper's falsificationism to some extent. For example, the distinction between falsification and rejection helps in clarifying certain ambiguities of Popper's; and Lakatos's attention to the heuristics or logic of discovery may be considered as a step further with respect to Popper's homologous logic. It is to be doubted, however, whether these amendments and additions are genuine novelties and constitute a real improvement or a "creative shift" (Worrall, 1978, p. 65) or a "subtle refinement" (Zahar, 1978, p. 72). This holds, in particular, for the core of Lakatos's methodological sophisticationism, that is his demarcation criterion. This criterion is expressed by the following rule:

L: That research programme is to be considered scientific whose series of theories is both theoretically and empirically progressive, that is it is such that: (1) T' has excess empirical content over T; (2) the unrefuted content of T is included in the content of T'; (3) part of the content of T' is corroborated (Lakatos, 1970, p. 32).

Lakatos remarks that this criterion "shifts the problem of how to appraise *theories* to the problem of how to appraise *series of theories*. Not an isolated theory, but only a series of theories can be said to be scientific or unscientific: to apply the term 'scientific' to one single theory is a category mistake" (1970, p. 34); he also criticizes Popper for having conflated these two categories. But if this is doubtful for the young Popper,[2] it is certainly not true for the later Popper.

If we look at Popper's 'Truth, Rationality, and the Growth of Knowledge', a paper which "contains some essential further developments of the ideas of the *Logic of Scientific Discovery*" (Popper, 1963, p. 215), we find the following rule:

P: T' is better than T if it fulfils three requirements, namely (1) T' is testable; (2) T' has new and testable consequences; (3) part of these consequences are corroborated by new and severe tests. (Popper 1963, pp. 240–41).

Although Popper's three requirements are the same as Lakatos's, this is not enough to prove that P = L. To prove this point, we must check whether P is a demarcation criterion as L is and if it too applies to series of theories. Now, this is actually the case. Popper explicitly says that P applies to a "sequence of theories" (1963, p. 244), that, "if the progress of science is to continue", this sequence has to satisfy all three requirements, especially the third (p. 243), and that the empirical character of science depends on the progressiveness of the sequence of theories. This last point is made clear in the following passage:

> Earlier I suggested that science would stagnate, and lose its empirical character, if we should fail to obtain refutations. We can now see that for very similar reasons science would stagnate, and lose its empirical character, if we should fail to obtain verifications of new predictions (1963, p. 244).

Since losing *empirical* character amounts to losing *scientific* character, Popper's remark proves that his P is the same as Lakatos's L, and thus that the core of methodological sophisticationism is included in Popper's methodology. But not only does Popper's methodology anticipate Lakatos's; it anticipates other variants as well. Take Larry Laudan's methodology of scientific research traditions. Laudan maintains that his approach offers a better understanding of scientific rationality for it makes it parasitic upon progressiveness and not the other way round: "to make rational choices is, on this view, to make choices which are progressive" (Laudan, 1977, p. 125). Although Popper has a different idea of scientific progress, he has a similar view about the relationship between progress and rationality: "it is my thesis" – he wrote in the paper quoted above – "that it is the *growth* of our knowledge, our way of choosing between theories, in a certain problem situation, which makes science rational" (Popper, 1963, p. 248).

If Popper's "sophisticated" methodology anticipates Lakatos's, then the latter cannot be expected to reduce or mitigate the conventional elements of the former. This is our second question.

Lakatos's Cartesian project requires personal factors to be eliminated for they would destroy the rationality of science. And Lakatos was of the opinion that he had gone a step further in this direction for his methodology would contain "fewer methodological decisions' (1970, p. 40). But this claim, too, does not correspond to facts. Let us suppose that the five kinds of decisions mentioned by Lakatos do really exhaust the decisions which are needed for Popper's methodology to apply to scientific practice, however disputable this may be.

The fourth-type decision concerns which element of a theoretical whole consisting of a theory, laws, initial conditions and auxiliary assumptions, is to be replaced if the whole clashes with some contrary evidence. Lakatos's solution to this problem, that one should try to replace *any* part of the whole provided it does not lose its progressive character (1970, p. 40), does *not* render the decision "completely redundant" as he maintains, but simply liberalizes it. A decision is redundant when there is no need for it or, if there is, when it is disciplined by a universal, impersonal rule, as it was in the Baconian (and Cartesian and Leibnizian) dream of levelling man's wits. But Lakatos's methodology contains no rule to this purpose; it leaves scientists to themselves, precisely as Popper's methodology does.

The fifth-type decision is about syntactically metaphysical theories, that is theories expressed by statements containing both universal and existential quantifiers. Lakatos's advice is that a theory of this kind has to be eliminated if "it produces a degenerating shift in the long run and there is a better, rival, metaphysics to replace it" (1970, p. 42). Here, too, the decision is not rendered redundant but only liberalized; and here, too, the advice is too liberal for no criteria are fixed which specify the limits of the "long run". Lakatos himself admits that "the very choice of the logical form in which to articulate a theory depends to a large extent on our methodological decision" and this, in its turn, on "the state of our knowledge" (1970, p. 42). But the progressive (or regressive) character of the state of our knowledge is a diachronical property and thus Lakatos's advice amounts to no more than a poor "wait and see".

As for the first, second and third-type decisions, regarding, respectively, the acceptance of basic statements, the separation of those accepted from the others, and the rejection of probabilistic hypotheses, Lakatos maintained that, although unavoidable, they may be "reduced" (1970, p. 42). The second-type decision, for example, is "mitigated" by "allowing an appeal procedure". But to mitigate a decision is to make it less dramatic and Lakatos's methodology renders it less dramatic simply because it puts it off. The advice "Put off till tomorrow what you have to do today" may sometimes be wiser than the standard one, but it is not wise to forget that, sooner or later, tomorrow will come.

Summing up. Following his Cartesian project, Lakatos aimed at a methodology which were more articulated than Popper's and which gave room to fewer personal decisions. Actually, the core of his methodological sophisticationism did not improve on Popper's methodology, nor was it less troubled by the problem of decisions.

4. LAKATOS'S SURRENDER

Was Lakatos aware of his failure? Two views he maintained in his last period seem to show that he was.

The first view has already been examined by Toulmin (1976). Lakatos ended up admitting that Popper's view that "there must be the constitutional authority of an *immutable statute law* (laid down in his demarcation criterion) to distinguish between good and bad science" (1971a, p. 136) is untenable. He also recognized that the idea of finding a methodological L.C.D. at least for the best examples of scientific practice is not viable. "Is it not then *hubris* to try to impose some *a priori* philosophy of science on the most advanced sciences? Is it not *hubris* to demand that if, say, Newtonian or Einsteinian science turns out to have violated Bacon's, Carnap's or Popper's *a priori* rules of the game, the business of science should be started anew?" (1971a, p. 137).

That Popper's methodology is completely *a priori* may be disputed on the grounds that, according to one of Popper's own requirements, it explicitly aims at capturing the intuitive ideas scientists have of their own work. Rather than *a priori*, Popper's methodology is *universal* with reference both to history and disciplines; but this holds good for Lakatos's methodology as well, at least for the original plan of it. Thus, the admission that no immutable statute law is possible is contrary to the idea that methodology has to offer "sharp criteria" or "universal definitions" of science and, therefore, is an admission of failure.

The second view consists in Lakatos's separation (not only distinction) of methodology from heuristics. As he wrote, "methodology is separated from heuristics, rather as value judgements are from 'ought' statements" (1971a, p. 103 n.; see also 1971b, pp. 174 and 178). But, as A. Musgrave (1976) and others have remarked, this is contrary to Lakatos's original project, and, in any case, this is contrary to the very idea of methodology, for if methodology offers criteria for distinguishing good science from bad, or progressive programmes from regressive ones, then these criteria also have practical consequences and can be used to give advice.

Trying to defend Lakatos's separation, J. Watkins holds the view that "while methodologists or philosophers of science may have something worthwhile to say about the comparative appraisal of the products of scientific research, it is not their business to advise scientists how to go about their research. In particular, it is not for them to tell scientists what they should or should not work on" (Watkins, 1984, p. 156). But this view relies too much on the principle of the division of labor; it forgets that methodology (provided it exists) is an *internal* component of science. Scientists may not have a fully articulated and explicitly professed methodology, but when they work, they follow, consistently or not, some methodology, for example, they apply certain standards and argue for them. Admittedly, the articulation and justification of a methodology is a *philosophical* job, but the fact that this job has mainly been done by professional philosophers does not matter; scientists sometimes do it better than philosophers. What does matter is that a methodology (or something akin to it) is always at work in scientific practice and that scientists refer to it in taking their own decisions, that is in advising themselves as to what they should do.

Nor does Watkins's distinction between "accept" and "work on" (1984, Chapter 4.51; 1987) bring grist to Lakatos's mill. Watkins maintains that these questions

(1) Which of these theories should I *accept*?
(2) Which of these theories should I *work on*?

are very different. Indeed they are. But this does not mean that if we have a solution to (1) this solution is no guide to (2). Suppose a theory T is to be accepted because it fulfils what Watkins calls OAS (the "Optimum Aim for Science"); also suppose (for sake of discussion) that scientists agree on what this aim is and on how to reach or approach it (for example, through some measure of corroboration in Watkins's sense). Then the value judgement

(3) T fulfils OAS better than T'

together with the descriptive statement

(4) X is a scientist

and the ought-statement

(5) As a scientist X should pursue OAS

implies the ought-statement

(6) X should work on T.

Thus, if methodology offers value judgements like (3), it also offers advice like (6). The fact that working scientists sometimes do not follow the latter may be used, if one sticks to Lakatos's history-oriented methodology, as an argument against the former, not as an argument for the separation of methodology from heuristics.

And here the last point comes in. Why did Lakatos take these views? Three reasons may perhaps explain his change of mind.

The first is that Lakatos found out that the history of science resisted his rational reconstructions. As the reconstruction of episodes and the examination of case studies went on, Lakatos and his followers tried to further sophisticate their methodology in order to make it suitable for the new fields. Do not scientists reject a theory when it is contradicted by contrary evidence? Here is a distinction between "disproof" and "rejection" (or "stopping working on") (Lakatos, 1970, p. 25; Watkins, 1984, Chapter 4.51 and 1987). Were Planck's modification of the Lummer–Pringsheim formula, Ptolemy's introduction of equants and Lorentz's contraction hypothesis unsatisfactory? A new concept of an *ad hoc* hypothesis ("ad hoc_3") is coined to help (Lakatos, 1970, pp. 80, 88; 1971a, p. 112; 1974b, p. 149; 1976, p. 182; Zahar, 1973). Is not the acceptance of Copernican theory explainable in terms of the sophisticated rules, for all its consequences were already known? Here is a new conception of "novel fact" (Lakatos, 1976, p. 185; Zahar, 1973, p. 103).[3] And so on. But in spite of these *ad hoc* methodological manoeuvres the phenomenology of the history of science still proved to be richer than its reconstructions. At a certain point Lakatos was tempted to put real history "in the footnotes" (1971a, p. 120), but eventually he probably realized that his attempts at reconstructing history with well-fitting methodological rules were like trying to dress Proteus with ready-made clothes or like running after one's own shadow, and frankly admitted that they were due to *hubris*, gratuitous violence.

The second reason stems from the first. Even if it were possible to reconstruct history, this would not provide any warrant that the resulting methodology is the best one, for "great scientific achievements may change scientific standards" (Lakatos, 1978c, p. 201). But if the standards are changing, what is the use of going on to look for "universal definitions" of science?

This introduces the third reason. Not only did Lakatos realize that scientific standards of different traditions may be different; he also acknowledged that good science may be pursued even through the violation of the best or more reasonable standards. Contrary to his own original criticism, he then came to admit that "Oakeshott's and Polanyi's position has a great deal of truth in it" (1971a, p. 136); that, at least in Newton's case, "Kuhn is very near to the historical truth when he says that 'science starts when criticism stops'" (1978c, p. 207); and that, still in this case, "Feyerabend rightly blames Newton's Rule IV for 'theoretical monism'" (1978c, p. 207). But if this is so, if scientists violate even their own standards and if these standards may block scientific progress, then not only can there be no universal definitions of science and constitutional methodological authorities, but methodology cannot even provide any advice for working scientists.

Faced with these difficulties, Lakatos retreated until he surrendered. He had started as a soldier of the Popperian army with the aim of routing conventionalists, elitists, historicists, sociologists, etc.; actually he ended up as a (disguised) "fellow in anarchism". Many of his shots missed their target and many of his troops passed into the hands of his "irrationalist" enemies. In this he was a victim of his Cartesian syndrome, for he was not equipped to conceive of rationality in terms other than strictly methodological. In spite of his many subtleties and refinements, his project turned out to be a degeneration of falsificationism rather than an improvement on it.

Popper's project at least had the fascination of plans for wide-scale reform; Popper aimed at improving not only our image of science but science itself. Accordingly, he introduced new aims (fallibilism and progress) and new means to such ends (bold conjectures and refutations), and, like a true reformer, he did not care much about the history of science. Lakatos and the sophisticationists wanted more: they wanted to maintain the core of falsificationism while, at the same time, making past and present science rational. But, in these matters, one cannot have it both ways. Lakatos's generous and courageous attempt at letting World_3 descend on World_2 encountered all the difficulties of combining the ideal with the real: however great an effort we may make and whatever results we may obtain, the gates of Paradise remain irremediably locked for the inhabitants (even though they are scientists) of this earth.

NOTES

[1] Later on, Popper lost part of his original faith. See (1963), p. 257 ("the criterion of demarcation cannot be absolutely sharp"); (1974), p. 981 ("any demarcation in my sense *must* be rough"; "our criterion must not be too sharp"); (1974), p. 984 ("such a rule of method is, necessarily, somewhat vague – as is the problem of demarcation altogether").

[2] I say "doubtful" for in his *Logik der Forschung* Popper had *two* demarcation criteria, one logical and the other methodological, depending on the logical or methodological sense in which the predicate "falsifiable" is taken. Unlike the former, the latter is *not* an instant criterion, for it allows certain immunizing manoeuvres (for example, introduction of auxiliary hypotheses) provided these do "not diminish the degree of falsifiability or testability of the system in question" (Popper, 1959, pp. 82–83). This means that the methodological criterion applies not to *theories* as such, but to *modifications of theories*, that is to theoretical *sequences*. On this point see my 1981, Chapter 5.3.

[3] That this idea of "novel fact" is new may be disputed. Indeed it is not different from one of Whewell's views of the "consilience of inductions". Zahar (1973, p. 103) says that "a fact will be considered novel with respect to a given hypothesis if it did not belong to the problem-situation which governed the construction of the hypothesis". Whewell (1847, vol. 2, pp. 67–68) had said that "when the hypothesis of itself and without adjustment for that purpose, gives us the rule and reason of a class of facts not contemplated in its construction, we have a criterion of its reality, which has never yet been produced in favour of a falsehood". On Whewell's view, see Laudan 1981; on the adhocness of Lakatosian adhocness, see Nickles 1987.

REFERENCES

Bacon, F., *Novum Organon*, in Spedding, J., Ellis, R. L. and Heath, D. D. (eds.), *The Works of Francis Bacon*, London, Longman, Vol. IV, 1860.

Iarnap, R., 'On the Character of Philosophic Problems', *Philosophy of Science,* **I** (1934) 5–17.

Cohen, R. S., Feyerabend, P. K. and Wartofsky, M. (eds.), *Essays in Memory of Imre Lakatos*, Dordrecht, Reidel, 1976.

Descartes, R., *Regulae ad directionem ingenii*, in *Œuvres de Descartes*, publiées par C. Adam et P. Tannery, Paris, Vrin, Vol. X, 1966.

Feyerabend, P., 'Consolations for the Specialist', as reprinted in Feyerabend (1981), Vol. 2, pp. 131–161, 1970.

Feyerabend, P., *Against Method*, London, New Left Books, 1975.

Feyerabend, P., *Philosophical Papers*, 2 vols, Cambridge, Cambridge University Press, 1981.

Kuhn, T., *The Structure of Scientific Revolutions*, Chicago, Chicago University Press, 1962.

Lakatos, I., 'Falsification and the Methodology of Scientific Research Programmes', in Lakatos (1978a), Vol. 1, pp. 8–101, 1970.

Lakatos, I., 'History of Science and its Rational Reconstructions', in Lakatos (1978a), Vol. 1, pp. 102–138, 1971a.

Lakatos, I., 'Replies to Critics', in Buck, R. C., and Cohen R. S. (eds.), *PSA 1970*, Dordrecht, Reidel, pp. 174–182, 1971b.

Lakatos, I., 'The Role of Crucial Experiments in Science', *Studies in History and Philosophy of Science,* **54** (1974a) 309–325.

Lakatos, I., 'Popper on Demarcation and Induction', in Lakatos (1978a), Vol. 1, pp. 139–167, 1974b.

Lakatos, I., 'Why Did Copernicus's Programme Supersede Ptolemy's?' in Lakatos (1978a), Vol. 1, pp. 168–192, 1976.

Lakatos, I., *Philosophical Papers*, 2 vols, edited by J. Worrall and G. Currie, Cambridge, Cambridge University Press, 1978a.

Lakatos, I., 'What Does a Mathematical Proof Prove?', in Lakatos (1978a), Vol. 2, pp. 61–69, 1978b.

Lakatos, I., 'Newton's Effect on Scientific Standards', in Lakatos (1978a), Vol. 1, pp. 193–222, 1978c.

Lakatos, I., 'Understanding Toulmin', in Lakatos (1978a), Vol. 2, pp. 224–243, 1978d.

Lakatos, I. (ed.), *The Problem of Inductive Logic*, Amsterdam, North-Holland Publishing Company, 1968.

Laudan, L., *Progress and its Problems*, Berkeley, University of California Press, 1977.

Laudan, L., 'William Whewell on the Consilience of Inductions', in *Science and Hypothesis*, Dordrecht, Reidel, 1981, pp. 163–180.

Leibniz, G. W., *Philosophische Schriften*, herausgegeben von C. I. Gerhardt, 7 vols, Hildesheim, G. Olms, 1961.

Musgrave, A., 'Method or Madness?' in Cohen *et al.* (eds.), 1976, pp. 457–491.

Nickles, T., 'Lakatosian Heuristics and Epistemic Support', *British Journal for the Philosophy of Science*, **38** (1987) 181–206.

Pera, M., *Popper e la scienza su palafitte*, Roma-Bari, Laterza, 1981.

Pera, M., 'Narcissus at the Pool: Scientific Method and the History of Science', *Organon*, **22–23** (1986/1987) 79–98.

Pera, M., 'From Methodology to Dialectics. A post-Cartesian Approach to Scientific Rationality', in *PSA 1986*, Vol. 2, edited by A. Fine and M. Forbes, East Lansing, Philosophy of Science Association, 1987, pp. 359–374.

Pera, M., 'Breaking the Link between Methodology and Rationality: A Plea for Rhetoric in Scientific Inquiry' in *Theory and Experiment*, edited by D. Batens and J. P. van Bendegem, Dordrecht, D. Reidel, 1988, pp. 259–276.

Popper, K., *The Logic of Scientific Discovery*, London, Hutchinson, 1959.

Popper, K., *Conjectures and Refutations*, London, Routledge & Kegan Paul, 1963.,

Popper, K., 'Theories, Experience and Probabilistic Intuitions', in Lakatos (ed., 1968), pp. 285–303.

Popper, K., *Objective Knowledge*, Oxford, Clarendon Press, 1972.

Popper, K., 'Replies to My Critics', in *The Philosophy of Karl Popper*, edited by P. A. Schilpp, La Salle, Ill., 1974, pp. 961–1197.

Radnitzky, G. and Andersson, G. (eds.), *Progress and Rationality in Science*, Dordrecht, Reidel, 1978.

Reichenbach, H., *Experience and Prediction*, Chicago, University of Chicago Press, 1938.

Schouls, P. A., *The Imposition of Method. A Study of Descartes and Locke*, Oxford, Clarendon Press, 1980.

Toulmin, S., 'History, Praxis and the "Third World"', in Cohen *et al.* (eds., 1976), pp. 655–675.

Worrall, J., 'The Way in Which the Methodology of Scientific Research Programmes Improves on Popper's Methodology', in Radnitzky and Andersson (eds., 1978), pp. 45–70.

Watkins, J., *Science and Scepticism*, Princeton, Princeton University Press, 1984.

Watkins, J., 'The Methodology of Scientific Research Programmes: A Retrospect', *this volume*, 1989.

Whewell, W., *Philosophy of the Inductive Sciences*, London, 1847; reprinted 1967.
Zahar, E., 'Why Did Einstein's Research Programme Supersede Lorentz's?', *British Journal for the Philosophy of Science,* **24** (1973) 95–123.
Zahar, E., '"Crucial" Experiments: A Case Study', in Radnitzky and Andersson (eds., 1978), pp. 45–70.

University of Pisa

PANTELIS D. NICOLACOPOULOS

THROUGH THE LOOKING GLASS: PHILOSOPHY, RESEARCH PROGRAMMES AND THE SCIENTIFIC COMMUNITY

In this paper I wish to discuss some relationships between philosophy and science, or rather, between the philosophy of science and the natural sciences, as well as the role of research programmes in these relationships. My interest in the topic rises out of the problems involved in teaching philosophy, and especially epistemology, to science, engineering and technology students. My "field experience" in the last five years has led me to the conclusion that the purpose is better served through a programme of courses of different levels carried out by the joined forces and complementary efforts of philosophers and scientists. However, I do not intend to discuss either this conclusion or my own teaching experiences concerning a particular educational system at a particular period – in fact, one of reformation. If I mention this, it is in order to draw attention to two implicit distinctions, made by teachers and students, philosophers and scientists, but not always in the same manner, which may become the source of dispute when one attempts to make them explicit. The two distinctions are (a) the one between philosophy (of science) and (the natural) science(s), and (b) the one between the view of science presented in the philosophy of science and the view of science created by the scientists and science teachers.

The first distinction is almost a trivial truth for scientists and science students; the latter take for granted that philosophy is completely different from the natural and technological sciences, and do not even ask the question whether philosophy is a science, what the differences really are. Of course, it is not surprising that they do not stop to pose such questions, when they are introduced to science by means of a currently prevailing theory, are taught to believe that science is what scientists do, and are led to identify the concept of a given science, such as physics or chemical engineering, with a certain procedure or a set of problems. The question of what differentiates technology from science is not put to them, so they often assume that technology is simply applied science; if pressed to

K. Gavroglu, Y. Goudaroulis, P. Nicolacopoulos (eds.), Imre Lakatos and Theories of Scientific Change. 189–202.

distinguish chemistry from chemical engineering they will usually say that the problems dealt with by the latter are problems of application.

The second distinction presupposes some knowledge of the history of science. When history of science is absent, as it is usually the case in hard-core pressed-for-time science courses, there cannot be much discussion on the problems that bring about new theories, nor much interest in competing theories, new phenomena, crises, paradigms and the like. Instead, there is a prevailing view that there currently exists one theory which best serves the interests of a given science, and that whatever other theories might have existed are now wrong, or outdated, or holding in borderline cases, or approximations, and do not figure any more in scientific progress. This distinction is essentially a question of conceptual framework; when made by scientists, however, it is often accompanied by the belief that whatever the view of science presented by the philosophers of science, it is of no consequence to the advancement of science, and its merits are to be looked for elsewhere. In the conceptual scheme of things of a practising scientist, the history and philosophy of science is of no greater interest to science and its students than, say, the history of medieval Europe or the political philosophy of Plato.

That the adoption, use, and change of a conceptual framework, though of great importance to the practice and advancement of science, is not a presuppositionless, rational choice of the working scientist, is one of the important findings of the philosophy of science. Despite the disagreements among the various philosophers of science, it is safe to say that, on the whole, they give a different, and more fruitful meaning to the two distinctions above. Let me, then, approach these distinctions, and the problem of the relationship between philosophy and science, by posing and discussing a question that rises out of the work of Imre Lakatos. In doing so, I will also have to chance to present a critical evaluation of some aspects of Lakatos's contribution to philosophy itself.

The question I wish to pose may be phrased in a number of ways: Are there research programmes in philosophy? Can philosophy be organized along the lines of research programmes of the type described by Lakatos? Or are there different types of research programmes in philosophy that cannot be described as scientific? Let me try to provide some contexts for this series of questions.

On the one hand, this series of questions arises out of the twists and turns of philosophy in our century, and especially in the last couple of decades. In a schematic manner, one may point to the following characteristic turns in recent philosophy.

1. There has been an emphasis on "practical" or "applied" philosophy. With respect to the subject-matter of philosophy, one may point to such developments in ethics as theories concerning animal rights or rights of future generations, or as the search for realism in ethics. With respect to the occupation of philosophers, there have been efforts to open up non-academic positions for philosophers, coupled with attempts to train philosophers for participation in interdisciplinary activities that may also involve knowledge of and experience in law, medical ethics, politics and public administration, etc.[1]

2. There has been a re-emergence of some classical philosophical problems. It is, of course, an important characteristic of the history of philosophy that certain key problems are posited again and again, despite numerous efforts at their solution, or even answers accepted by the philosophical community for a substantial length of time, and despite long, intermediate periods during which they are not recognised as genuine or serious problems, and they meet with only marginal treatment by some ill-respected individuals with peculiar and misplaced interests. This characteristic by no means proves that the problems of philosophy are perennial or inherently unsolvable, for it may be plausibly argued that what the historical fact indicates is that the problems of philosophy are, in a sense, dated and expressed within a context which incorporates internal and external conditions and which gives meaning(s) to the problem at hand. Evidence in support of the latter may be found, first, in the striking differences in the ways in which the recurring problems are each time expressed and, secondly, in the recurrence not only of problems but also of philosophical battles, involving antagonisms and disputes about the very meaning and solution of a given problem, or a series of problems, among proponents of competitive theories. In this last case, it becomes more evident that what we are dealing with is the reappearance of a problem expressed in new terms, set in a new context, and resulting in a new theory. The interesting new development this time around, however, is that certain classical philosophical problems, belonging to traditional branches of philosophy such as the theory of knowledge or metaphysics, are put forth again by developments in the philosophy of science. A good example is the re-opening of the realism–empiricism debate by van Fraassen's 'constructive empiricism'.[2]

3. There has been some kind of a dialogue between an Anglo-Saxon and a French philosophy of science. Now, this statement is an

over-simplification; it implies, for example, the existence of two well-defined schools in philosophy of science, which is not the case, and it attempts to make geographical or even ethnic distinctions, which are at the least inaccurate. To qualify it somewhat, let us speak of two trends, each originating from a different, but extremely rich and old philosophical tradition, developing separately and facing a lot of internal disputes and disagreements, which now recognize the existence of each other, develop an interest in each other and attempt to reach, comprehend and influence each other. It seems to me, then, that the objective of this dialogue is the role of philosophy today; an objective which may be interpreted to be as broad as a re-evaluation of the social function and aim of philosophy, but which is specifically focused on the intervention of philosophy in science.[3]

4. At the same time, the Marxist tradition, which had always been interested in the relations between science and society, has turned its attention to the analysis of scientific knowledge and the comprehension of the structure of science. Without giving up the effort of discovering, choosing or interpreting scientific developments as facts or elements of an objective reality that confirm the thesis of marxist philosophy, it has shown a significant interest in current work in the philosophy and history of science it still terms 'positivism'.[4]

5. The last few decades have also witnessed the emergence of some new fields, or at least theories aiming to establish new fields, such as metaphilosophy, presuppositionless philosophy, the archaeology of knowledge and others, which have enriched our approaches not only to the understanding of science but also to the reconstructions we usually term 'the history of ideas'. It should not pass unremarked, however, that certain classical approaches to philosophy, including the construction of speculative systems and the history of philosophy itself have, as a result, been put aside.[5] 6. Nevertheless, the acknowledged and evident predominance in contemporary philosophy of logic, philosophy of language and philosophy of science seems to continue, despite the philosophical turns mentioned in points 1 and 5 above. However, the development of these dominant fields has not always been smooth. Some of their leading theories of the recent past (positivism, for instance) are now considered outdated, if not outright wrong, some trends (such as analytic philosophy) have proved to be too narrow or not sufficiently productive, and there have even been blows against philosophy of science as a whole coming from within (Feyerabend), on the grounds that it has not been at all effective in influencing science.[6]

It seems to me, then, that all these turns and developments, in posing the problem of the need for a new function for philosophy, emphasize methodology; not only the perennial quest for a method, but also, and especially, the role of philosophy as the study of methods. This role, one of the many that philosophy has played throughout its history, appears to be the main, if not the only one, philosophy is capable of playing, or perhaps is allowed to play, if it wishes not only to survive but to critically intervene in today's technological society. This role might provide the identity of the philosophical endeavour in the intermediate period between the end of classical philosophy and the beginning of an era of which there are no concrete descriptions, but quite a few, and quite different, visions – in nuclear colors, I am tempted to say, in a characteristically unscientific manner.

On the other hand, the above series of questions arises out of a problem I encounter in reading Lakatos, especially his 'Methodology of Scientific Research Programmes'.[7] I could state that problem as a question: Does Lakatos (always) distinguish (properly) between science and philosophy? There is a hidden "ought to" statement in this question, and I hasten to acknowledge it, lest it be misinterpreted. I do not mean to suggest that there is a known, agreed-upon and undisputed, "proper" distinction between science and philosophy that Lakatos ignores or fails to accept. I mean, instead, that there ought to be some kind of distinction between science and philosophy, especially in a philosophical text on scientific methodology such as Lakatos's, or at least some hints on distinguishing criteria. Granted, there is no generally accepted criterion to distinguish science from philosophy; yet many such criteria have been proposed, put to use, and have met with some success and acceptance. Lakatos neither uses any of these, nor proposes some of his own; what is more problematic, not only does he not seem to be interested in such a distinction, not even as a side-issue in his argumentation, but he may actually confuse the point by misleading us, here and there, towards some kind of identification between science and philosophy (of science). Let me give some examples. To whom does he intend important (at least for his argumentation) characterisations such as 'justificationists' and 'conventionalists' to apply? Working scientists, philosophers of science, epistemologists, or all of the above?[8] It is very difficult to tell whether his attempt to refute justificationism and 'naive' falsificationism is made on scientific or philosophical grounds, whether, that is, the propositions at stake are scientific or philosophical. Again, does he really mean that Kuhn and Popper have research programmes? He almost interchangeably refers to "Kuhn's philosophy" and "Kuhn's research programme', for example,[9] and there are other similar indications that, for Lakatos, these philosophies are research programmes, or at least that these philosophers have research programmes. If this is so, one may well assume that Lakatos means them to be scientific research programmes, for the only other kind of research programme he distinguishes is the pseudoscientific or degenerating one, and it would be rather difficult to take the position that Lakatos thought of Popper's philosophy as a degenerating one. There is an additional complication at this point concerning Lakatos's very distinction between scientific and pseudoscientific or degenerating research programmes; the distinction is made on the grounds of progressiveness, i.e. the ability to predict novel facts, and it aims to distinguish science (more precisely, scientific methodology) from pseudo-science rather than from non-science. I will take up this point in more detail below.

When my series of questions is placed in the above contexts, it becomes clear, I hope, that what is at stake is philosophical methodology. Lakatos's analysis of scientific methodology poses for philosophy a methodological problem which reflects the uncertain progress of philosophy – indeed, the very notion of progress in philosophy in the age of science – and which refers to the relations between science and philosophy. The courses of science and philosophy have been construed in many ways, some of which are mutually exclusive. One may view the two courses as parallel and mutually beneficial. But there have been understandings of philosophy as the mother, or the heart, or the guiding force of science. Or, one may speak of philosophy as always lagging behind science. Again, one may view philosophy as independent from science, with its autonomy, its own subject-matter and domain. And one may point to attempts at construing philosophy as a special scientific branch.[10] Now, the fundamentally methodological character of this series of questions remains clear, despite the ontological manner in which the questions have been stated. If there is an ontological dimension to them, it is in the direction of examining the present status of the philosophical endeavour, but by no means in the sense of hunting after new entities to provide referents or meaning to philosophical pronouncements. In short, the questions I posed arise in the context of the interaction between two modes of human activity, both of which make claims to discover, produce, examine, or clarify knowledge, and which must be methodologically distinguishable but cognitively interrelated.

Consequently, the questions concern also the relations between two communities, the philosophical and the scientific, if indeed there exist two communities rather than one (or none, for that matter).[11] The problem of the interactions between philosophical and scientific methodology could be translatable, for example, into more specific questions, such as the following: How does each community react to certain criteria – of rationality, progress, validity, acceptance of propositions – set, proposed or used by the other? What are the conventions of the community with respect to making decisions? By whom and for whom are norms set? How does the community use descriptions and prescriptions? Essentially, however, the problem of the methodological connections and differences between the two communities, or even of the very identity of the communities, lies, I think, in the manner in which each community views itself not only as a user of criteria and norms but as a criterion or norm.

Too many questions. Before I proceed to some answers, let me summarize the contexts and backgrounds of the problem at hand by sketching the image I tried to capture in the title of my paper. I see philosophy as being in a transition. We are moving from a period when (i) philosophy used science as a mirror to reflect itself on and "correct" its misgivings, for instance by throwing out metaphysics, (ii) philosophy tried to look good in, and in accordance with, the mirror, by modeling itself after science and attempting to become a branch of science (or perhaps, conversely, philosophy saw itself as the priviledged space of the representation of science), to a period when (i) philosophy wishes to transcend the mirror, go through it, so to speak, and be involved with and participate in the scientific endeavour, (ii) philosophy recognizes that it has been using a mirror and that its reflection in that mirror was not self-reflection. To speak less metaphorically, philosophy of science has now developed the notions of the scientific community and of the scientific research programmes through which, in conjunction with an analysis of the structure of scientific theories, it has not only advanced its reconstruction of scientific methodology and the progress of science, but it has also reached a point from where it can now – indeed, it must – reflect upon itself, and reconsider its own methodology and progress, re-identify its own community, ponder on its future as philosophy.

Let me, then, attempt to give some answers focusing on the question of whether philosophy can be organized along the lines of (scientific) research programmes. My attempt will provide a critical evaluation of some aspects of the Lakatosian methodology, but it is mainly aimed at the sketching of some positions concerning the present function of philosophy. I will begin with some positions that answer the above question in the negative, and I will then proceed to some positive responses to the notion of research programmes in philosophy.

I take it that Lakatos was not seriously interested in the distinction between science and philosophy, and, consequently, he did not belabour the point whether he ought to consider specific philosophies of science such as Kuhn's and Popper's as research programmes – scientific, degenerating or of a different type. I take the position that philosophy can not develop through Lakatosian scientific research programmes because of certain important differences between science and philosophy.

1. There are no 'stunning novel facts' predicted by philosophy, and therefore no 'content increase'. Unlike science, philosophy does not aim to make predictions, especially predictions about facts and events. There can be no rule in philosophy to the effect that a philosophical theory is to be accepted only if it has (corroborated or not) excess empirical content over a predecessor or rival, nor can there be any criterion about philosophical theories concerning the discovery of novel facts they lead to.[12]

2. The object of philosophy is significantly different than that of science, and the predictions of empirical facts is included only in the latter. The foundations and the structure of scientific theories cannot work for philosophy not only because there exist fundamental philosophical concepts with no 'empirical component' (nor reducible to 'empirical terms'), but also because, unlike the case of science, one can point to philosophical theories that have not emerged simultaneously with their object.[13]

3. In particular, philosophy of science may limit itself to reconstructions of science – which are always reconstructions after the fact – or it may also try to be prescriptive for science. But a prescriptive philosophy of science (such as Popper's or Lakatos's) gets its prescriptions from its reconstructions. Meanwhile, science has moved along, and philosophy of science has not been very successful in having its prescriptions followed.

4. Despite Lakatos's attack on conventionalism, he allows some room for conventions in the decision-makings of the scientific community – more in the naive, less in the sophisticated methodological falsificationsim. Generally speaking, philosophy of science, if not conventionalist or at least in favour of certain conventions, has not managed to totally replace conventions by some internal structural elements or relations of scientific theories, and it cannot outlaw conventions from scientific methodology

for that, of course, would be a convention itself. The decisions of the scientific community in setting certain norms concerning not only due procedures but also acceptance of certain propositions or experimental findings is, perhaps, the type of convention most difficult to attack. Be that as it may, it remains a privilege of the scientific community to accept or reject certain theories proposed as scientific and claimed to be valid not on the basis of given criteria – beyond those which test the conformity of the proposed theory to explicit standards of scientific methodology – but merely on the basis of the community's decision to accept or reject them. In that sense, the community uses itself as a criterion. To put it differently, a proposed theory needs the scientific community's acceptance if it is to be used and tested, to live and possibly flourish. A rejection by the community of a proposed theory at an early stage makes it impossible for the theory to compete against alternative theories when it still matters, and a later recognition of the rightness of the rejected theory will interest only the historian of science but will not affect the course of science. Not so in philosophy. Here, theories have a different endurance through time and a different relationship with the community to which they are addressed.[14] The philosophical community's acceptance or rejection is not necessarily a determining factor for a theory's "career", and a theory may survive despite an early rejection, or even a permanent rejection by a part of the philosophical community. In fact, there have been philosophical theories that need the rejection of some philosophical quarters in order to flourish. And though the endorsement of the philosophical community is always desirable for and sought by a theory, such an endorsement need not come at a "crucial" early point. One could go a step further and point out that it is not only theories but also conventions that, unlike the case of science, do not need the endorsement of the community as a whole in order to apply.

5. Philosophy cannot be organized on the basis of scientific research programmes because, as a rule, there are no 'problemshifts' in philosophy and no progress in a cumulative sense, except in occasional periods and with respect to specific problems. We cannot use the concept of a 'series of theories' as a unit of discourse in philosophical methodology. As a rule, the story of philosophy is one of internal, often closed (in the systemic sense) philosophical dialogues that aim at the consistency of arguments and at the replacement or displacement of presuppositions.

6. The unit of philosophical discourse is to be sought elsewhere. It seems to me that despite all the twists and turns, and all the different

approaches and objectives throughout the history of philosophy, the philosophical endeavour has always employed the same unit: a text and its argument. I would go as far as proposing that even science, when becoming a subject-matter of philosophy, is seen as a text, and it is the text of science that philosophy of science talks about.

7. Consequently, instead of rationality, philosophy looks for order; order in experience, order in thought, order in the text. The aim of philosophy of science, hence, is to discover, or perhaps impose, order in the results of science. It is towards that aim that philosophy deals with the methodologies of the scientific discovery and production of knowledge.

8. Science is an object of philosophy, but philosophy is not an object of science.

On the other hand, philosophical methodology has something to gain, if it sets out to build some kind of research programme of its own, and it can, thus, benefit from Lakatos's reconstruction of scientific methodology.

1. The methodology of scientific research programmes is, most of all, a very good proposal for the understanding of science. It was, of course, intended to be something more than that, namely a prescription about how to do science, and as such it is an improvement over previous prescriptive proposals, with a more detailed reconstruction of scientific methodology than the "trial and error" method, and with a more sophisticated analysis than Popper's falsificationism. It has also benefited from non-prescriptive approaches to the philosophy of science, such as Kuhn's and Feyerabend's, and it has managed to retain, despite its fundamental differences from these approaches, a historical dimension in its presentation of the tenacity and the proliferation of science coupled with some room for freedom from rationality. But seen as a proposal for the understanding of science, Lakatos's methodology offers something original and helpful, namely the notion of heuristics (negative and positive).

2. From my point of view – that of a philosopher involved in the educational programme of a science/technology university – the Lakatosian heuristics and the associated notions of a hard-core and of a protective belt form the basis of an excellent method for teaching science, especially for introducing students to science, for two reasons. First, it comes fairly close to the richness of the history of science (a richness no reconstruction can actually capture), and it warns the newcomer that in the on-going processes of science nothing is to be taken as fixed and settled; for what is at stake is not only the possible application of a theory but the theory itself, and what is to be discovered is not only the outcome of an

experiment but a world according to a theory, or rather, a series of worlds according to a series of theories. Secondly, it prepares the student of science for participation in a battle – that of defending the fundamental nucleus of a theory, the hard-core, for as long as it remains productive, and of giving it up when and only when it can be replaced by the next one in a progressive series – rather than for becoming a proselyte, a dedicated follower of a given theory who might wish to go down with it.

3. What is more, the methodology of scientific research programmes is a good proposal for the teaching of philosophy, though it was not actually proposed for that purpose. It constitutes a model after which one may construct "pseudo" research programmes, for example through the use of philosophical theses to make reconstructions of the history of science.[15] I call such research programmes "pseudo" ones to distinguish them from both scientific and pseudo-scientific ones, because I am not interested here in whether they are progressive or degenerating; I see them as teaching devices. I still call them research programmes, however, because I wish them to retain not only a likeness to, but also some of the essence of the Lakatosian model; I see them as heuristic devices.

4. Finally, Lakatos's reconstruction of scientific methodology may actually turn out to be helpful not only in teaching but also in doing philosophy. What we are faced with is the problem of constructing philosophical research programmes that will retain the Lakatosian notion of heuristics but will redefine the notions of progressiveness and degeneration. That involves an examination of the conception of progress in philosophy (and a comparison with the conception of progress in science), as well as a determination of the hard-core of a philosophical text. It may also require a criterion other than that of progressive vs. degenerating in determining whether a research programme is 'genuine' or 'pseudo' philosophical. Essentially, however, the search for philosophical research programmes opens up the problematic of the philosophical methodology appropriate for the participation of philosophy in interdisciplinary research, which I consider to be a demand of the times, and which cannot but alter the identities and the relationships between the philosophical and the scientific community. Perhaps Lakatos's failure (or lack of intention) to distinguish between philosophy and science – which is reflected in the original but problematic complex position I tried to describe, according to which some philosophies (of science) are research programmes, research programmes are distinguished as of their being scientific or pseudo-scientific in terms of their progressiveness or degeneration, and yet no

other type of research programme is acknowledged – is, after all, a sign of the ambiguity that characterizes the relationships between the two communities and the two disciplines. Ambiguity is something science tries to get rid of, but philosophy has always thrived on ambiguities.[16]

NOTES

[1] Instead of mentioning books and articles, let me refer to different sources. With respect to the subject-matter of philosophy, one might take a look at the sections of the 1988 World Congress of Philosophy (Philosophical problems of artificial intelligence, Masculine and feminine in philosophy, The practical role of moral values in human life, Ethical issues in the treatment of animals, Philosophical problems about genetic engineering, Dangers of nuclear war, Ecology and the future of life on earth, etc.). With respect to the occupation of philosophers, one might take a look at the literature of the American Philosophical Association (advertisements of non-philosophical and non-academic jobs for philosophers, Association Committees, Congress Grants for philosophers, etc.).

[2] See van Fraassen (1980), Churchland and Hooker (eds.) (1985).

[3] A central element in this dialogue is the work of the historian of science Alexandre Koyré, which has influenced both Kuhn and Althusser. But one should point to the work of D. Lecourt with respect to an intentional effort for a dialogue. Althusser and his "school" is what I mainly have in mind when I speak of a French philosophy of science, and I think that Kuhn is the representative of the Anglo-Saxon philosophy of science who has raised the interest of the French. If we do not limit ourselves to philosophy of science, we may also point to the interest in the work of Ricoeur, Derrida or even Piaget in the U.S.A., and in the work of Wittgenstein or Dewey in France.

[4] Here, I have the situation currently existing in Greece in mind; it may not be typical but it is suitable. It would take too long to give my reasons here, but I tried to do so in public once, in a seminar on French and Anglo-Saxon Epistemology at L.S.E. in November 1985, by talking on the case of Greece under the title 'Imported Epistemology vs. Footnotes to Plato'.

[5] See, for example, Lazerowitz (1964), Lefebvre (1965), Lübbe (1978), Rorty (1980), Foucault (1971) and (1976).

[6] See Feyerabend (1975) and (1978).

[7] Lakatos (1970).

[8] Lakatos (1978a), pp. 10–47.

[9] *Ibid*., pp. 90–93, 31 fn. 2.

[10] See Veikos (1983).

[11] See Nicolacopoulos (1985), pp. 217–223.

[12] Compare Lakatos (1978a), pp. 31–32.

[13] See the paper by A. Baltas in this volume.

[14] I am leaving aside, at this point, a rather interesting problem. Philosophical theories are not necessarily addressed to the philosophical community alone, though in our age of professionalised philosophy that seems to be the case. For the sake of my present argument, and in order to make a specific distinction between the relationship of scientific theories with the scientific community and that of philosophical theories with the philosophical com-

munity, I take theories to be addressed to their respective professional communities. It seems to me, however, that if I did not take this narrow view, I would find myself before further distinctions, or at least differences, between the scientific and philosophical communities, with respect to their relations with the "general public".

[15] An example of what I mean by the application of philosophical theses to make reconstructions of the history of science is the research of V. Kalfas of the University of Crete, which aims at examining the history of astronomy against the Platonic demand of "saving the phenomena". However, I do not have the slightest intention of attributing to it the characterization "pseudo-research programme". Kalfas's extremely interesting research is still in progress and for the most unpublished, and attributing any label to it would be not only unfair but also against academic principles. So, this is *not* an example of what I mean by a proposal for the teaching of philosophy, although I believe that useful educational ideas could be derived from it.

[16] See also Cohen *et al.* (1976) and Nicolacopoulos (1983).

REFERENCES

Churchland, P. M. and Hooker, C. A. (eds.), *Images of Science. Essays on Realism and Empiricism with a Reply from B. C. van Fraassen*, Chicago, University of Chicago Press, 1985.

Cohen, R. S., Feyerabend, P. K. and Wartofsky, M. W. (eds.), *Essays in Memory of Imre Lakatos* (BSPS Vol. 39), Dordrecht, Reidel, 1976.

Feyerabend, P., *Against Method*, London, New Left Books, 1975.

Feyerabend, P., *Science in a Free Society*, London, New Left Books, 1978.

Foucault, M., *The Order of Things. An Archaeology of the Human Sciences*, New York, Random House, 1971.

Foucault, M., *The Archaeology of Knowledge*, New York, Harper and Row, 1976.

Kuhn, T. S., *The Structure of Scientific Revolutions*, Second Edition, Chicago, The University of Chicago Press, 1970.

Lakatos, I., 'Falsification and the Methodology of Scientific Research Programmes', in Lakatos and Musgrave (eds.) 1970. Also in Lakatos 1978a.

Lakatos, I., *The Methodology of Scientific Research Programmes*, Philosophical Papers Vol. 1, edited by J. Worrall and G. Currie, Cambridge, Cambridge University Press, 1978a.

Lakatos, I., *Mathematics, Science and Epistemology*, Philosophical Papers Vol. 2, edited by J. Worrall and G. Currie, Cambridge, Cambridge University Press, 1978b.

Lakatos, I. and Musgrave, A. (eds.), *Criticism and the Growth of Knowledge*, Cambridge, Cambridge University Press, 1970.

Lazerowitz, M., *Studies in Metaphilosophy*, London, Routledge and Kegan Paul, 1964.

Lazerowitz, M., *The Structure of Metaphysics*, Third Edition, London, Routledge and Kegan Paul, 1968.

Lefebvre, H., *Metaphilosophie. Prolegomenes*, Paris, Edition de Minuit, 1965.

Lübbe, H., *Wozu Philosophie?*, Berlin, 1978.

Nicolacopoulos, P., 'Science as Artifact, Knowledge as Praxis', in *Abstracts of the 7th International Congress of Logic, Methodology and Philosophy of Science*, Vol. 3, Salzburg, 1983 (pp. 188–191).

Nicolacopoulos, P., 'The Methodology of Mr. Feyerabend', in *Deucalion*, **38**, April 1985, Athens, 1986 (pp. 211–228).
Rorty, R., *Philosophy and the Mirror of Nature*, second printing with corrections, Princeton, Princeton University Press, 1980.
van Fraassen, Bas C., *The Scientific Image*, Oxford, Clarendon Press, 1980.
Veikos, Th., *Prolegomena sti Filosofia*, Athens, Themelio Editions, 1983 (in Greek).

National Technical University of Athens

EMILIO METAXOPOULOS

A CRITICAL CONSIDERATION OF THE LAKATOSIAN CONCEPTS: "MATURE" AND "IMMATURE" SCIENCE

The aim in this paper is simply to sketch some basic characteristics of the concepts of mature and immature science. In other words, my aim is to show the importance of these concepts for two purposes that every historian of science has sometime or other to deal with: firstly, the aforementioned concepts are useful for the understanding of great scientific changes or revolutions. (This is particularly true for the historian concerned with the transition from medieval science to the new scientific universe of the 17th century); secondly, it seems possible to me that these concepts might give us the key to the understanding of the interaction between internal and external factors in science's historical development. Furthermore, I think that we could define mature and immature science in a way that permits us to include external factors in our rational reconstructions without having to give up a criterial theory of scientific rationality.

According to my view there are two important elements that our science and that of the 17th century have in common. These elements can thus be taken as a basis for a rational distinction between mature and immature science.

(a) Despite some considerable differences there exists a minimal common corpus of methodological procedures and strategies of rational choice. This can easily be proved if we study the 17th century's philosophies of science. Virtually all the philosophies of science of our own century have been anticipated in some form or other by the great philosophical systems of the 17th century. But even if this were not the case, there exists at least a common understanding about what is to be considered a rational scientific strategy based on rational methodological principles.

(b) Independently of the *Weltanschauungen* within which they were first expressed, there are particular problems and specific solutions that form the minimum basis of content-continuity between Galileo's science and our own.

The problem of science's maturity should, indeed, be treated in a way combining methodological considerations and specific content-instances.

K. Gavroglu, Y. Goudaroulis, P. Nicolacopoulos (eds.), Imre Lakatos and Theories of Scientific Change. 203–214.

Furthermore, we should take into account two additional, independent (in the sense of being external) parameters; (i) the idea of progress together with epistemological realism, that both have initially external, sociological and political roots but they achieve gradually a methodological status, and (ii) a parameter which is purely social for it concerns science's position in the social division of labour.

The concepts of maturity and immaturity cannot be understood in a purely theoretical way. They are, above all, historiographical categories with theoretical and historical–empirical components. Consequently, we cannot understand the difference between mature and immature science if we don't resort to a methodological scheme containing some definition of science, some criteria of scientific rationality and rational evaluation. But, once more, theory is not enough: in order to understand these concepts and their difference we should also take into consideration some external factors. In fact it seems rather obvious that mature science is not only an epistemological phenomenon. It is also a social phenomenon; and this observation brings us to the very heart of the problem, because internal and external factors in the development of science move in structurally different historical times. Internal growth is certainly not the product of social events, though it can be accelerated or even stopped by them. At the same time internal growth cannot have a direct influence on external events; the mediation of technology is also needed. As a result, a comparative analysis of mature and immature science is an extremely interesting field for those who want to clarify, within a rational and prevalently internalist model, the interactions between external and internal factors.

Lakatos' concept of maturity and immaturity are opposed to the respective Kuhnian ones. For Kuhn the existence of paradigms is the hallmark of science's maturity. According to Lakatos, monolithical domination or dogmatic acceptance of a paradigm is a symptom of pseudoscience etc.[1] But, if things are quite simple in Kuhn's rather simplistic model, they are not in the Lakatosian philosophy of science. Lakatos never rigorously defined mature and immature science.[2] Sometimes mature science seems to be a genuine 17th century product, depending on the appearance of conflicting research programmes. Nevertheless, Lakatos also refers to Ptolemy's research programme, inconsistently since in other occasions – as we've already noticed – he himself relates the appearance of research programmes to the scientific revolution of the 17th century.

By way of putting forward a tentative, nominalistic definition, I'd say that for science to be mature, from an internal point of view, the following three conditions must hold:

(1) systematic construction of its propositions and rigorous inferential rules (the highest expression of systematicity is mathematization and axiomatization),
(2) procerdures of empirical control (no matter what type of control: verification, inductive confirmation, falsification etc.), and
(3) pluralistic conflictuality between different, well articulated and antagonistic research programmes.

However, it is a sign of maturity that the battles are conducted according to some rules of the game, that is, according to some commonly accepted criteria of rationality. Ancient and medieval science obviously do not fulfill all three conditions at the same time. In particular they do not satisfy the last one. We can, for example, think of Aristotelian and Galenic medical science in terms of research programmes but it is quite evident that their conflict is not mediated by a common theory of rationality: their respective conceptions of empirical control, their models of explanation are very far apart. Furthermore, some sciences are, from a specific point of view, systematicity for instance, more advanced than others. Astronomy provides us with a good example: from Ptolemy to Copernicus it exhibits a high degree of systematization and its geometrical tools are considerably sophisticated. Nevertheless with minor exceptions, during the whole period that goes from Hellenic years to the Renaissance, astronomy is dominated by a monolithic paradigm[4] (or, to put it differently, a lonely research program with no rivals). This is the reason why astronomy, by far the most advanced science during Antiquity and the Middle Ages, cannot be considered to be a mature science before the 16th century. I might add at this point that the aforementioned nominalistic definition applies to concrete scientific branches. There is no such entity as "science". It would consequently be erroneous to speak about science's maturity in general terms. Not all sciences achieved maturity during the 17th century, though some extremely important scientific branches certainly did. A further clarification is perhaps needed here. I disagree with Lakatos when he refers to a "Ptolemaic research program". Immature science consists of paradigms. Thus we shouldn't speak about research programmes when there is only one, or, even, when there exist only research traditions in Laudan's sense[5] and not research programmes with their specific characteristics as such. The term "research programme" implies the parallel existence of a multitude of conflicting programmes.

Reformulating the preceding observations we may conclude that a scientific branch achieves maturity when (i) it starts growing on the basis of a regulative theory of rationality. In other words when a demarcation criterion and some criteria of acceptability based on empirical testability and sometimes on conceptual problem-solving are incorporated to its methodological principles. In the context of ancient science such criteria are simply non-existent. There are no experiments and the concepts of empirical testability, despite Aristotelian empiricism, play a rather poor role. Medieval nominalists tried, instead, to formulate some principles of empirical testing. Indeed, an extreme nominalist, such as Nicholas of Autrecourt, virtually anticipated a Humean demarcation principle. But this was all philosophy. There are only few concrete scientific researches or theories in specific branches that follow these methodological rules. I am convinced that nominalistic late Middle Ages offered to modern science its basic methodological tools but medieval science gave no signs of maturity in any field.

(ii) Rival research programmes appear which, although based on opposed metaphysical cores, share some methodological criteria, a sort of minimum theory of rationality. It is on the basis of this theory that judgements about the cognitive content, and the conceptual fertility of the competing programmes are formulated. Immature science is, on the contrary, characterized either by the monolithic domination of a sovereign programme or by non-systematic, random discoveries or even, as I have already mentioned, by the parallel existence of rival research traditions in Laudan's sense, which do not share a common theory of rationality. When, for instance, Aristotle criticizes Leucippus or Democritus he does not oppose his own theory's major content, which could be verified on commonly-accepted methodological grounds, to their atomism. That is: from its superior resistance to empirical control or its greater predictive power. Preference for Aristotelian or atomistic science was purely arbitrary because it depended entirely upon previous arbitrary acceptance of their philosophical assumptions. No other arguments were given in favour of one or the other scientific theory. I think that ancient science was lacking in all criteria of rational choice between theories. Ancient rationality coincides with the metaphysical presuppositions of each school of scientific thought. For modern rationality, the requirement of empirical control is absolutely essential. And this is not of course to say that Galileo was an inductivist or that we are all disguised Popperians. It only means that almost all scientists, from 17th century to the present day, share the same conviction: when a theory is repeatedly contradicted by "hard" facts – and nowadays everybody knows that even "hard" facts are in some way theory-laden – the theory must be rejected. When, on the contrary, a theory has no significant counter-evidence, then it is, usually, worth insisting upon.

Closely connected with the methodological requirement of comparative evaluation is the idea of linear progress, which is also virtually unknown during antiquity and the Middle Ages.[6] Although we considered the idea of progress as an independent, non-internal, parameter, mature science in fact appears only when the regulative ideal of progress invades science and sets up the rules of the game. The concept of progress would not be intelligible independently of the concept of empirical control and it undoubtedly represents a typical ideal of modern science, an essential ingredient of the new image of science as a finalized process of natural selection.[7] Furthermore, it is only in the context of modern science that all the profound interconnections between the idea of progress and epistemological realism become evident. Science is progressive because it can be seen as a process of approximation to truth. In conclusion, the generalized acceptance of the requirement of empirical testability and the consideration of empirical content (sometimes also conceptual richness) as the basic criterion of rational choice are *sine qua non* pressuppositions of mature science. That is why the greater part of science up to the 17th century seems closer to myth than to authentic science.

(iii) As a third instance, science achieves maturity when the axiomatic, mathematical elaborations of theories begin. But as we have already noticed, this third requirement might be expressed in a more elastic way: instead of mathematization we should perhaps speak of systematization whose aim is to render empirical controls easier. Indeed, some branches of natural science were not, until recently, mathematized and this is still the case, to a considerable extent, for social science.

Having briefly stated the three basic internal conditions of science's maturity, we may now pass to another consideration which may prove useful in understanding the concept of continuity. If there were not a common skeleton, that is elements of continuity in the history of science, there would simply not be any history of science. The concepts of specific problem-posing and problem-solving could, perhaps, give us the key to the understanding of the continuity of scientific knowledge independently of the history of research traditions and research programmes. These concepts are related to the historically continuous dealing with categories of problems which, though in the long run deeply transformed, they nevertheless conserve some common elements independently of the frameworks within which they are perceived and conceived in any era. I think that the concept of specific problem-posing could prove quite useful for the analysis of questions such as the continuity between medieval and modern, or ancient and modern science. It would be, thus, possible to identify some elements of linear continuity from ancient to medieval science, on the one hand, or from medieval to modern science on the other, though there is no such continuity, in any specific field, between ancient and modern science. Any atomistic *Weltanschaung* deals with problems, such as gravity, space-time relationships, movement etc. compatibility with its own premises. Nevertheless, though both coherently atomistic, the theories of Democritus and Gassendi present, in specific problem-posing and problem-solving, few elements in common. There is accordingly an interruption of continuity at the level of specific problem posing within the same research tradition. There exists, instead, a clear continuity, though the research traditions are sometimes profoundly different between Heytesbury and Galileo or Oresme and Galileo: it is a continuity of specific problem posing and problem solving.

Going back now to the study of the basic conditions of science's maturity we might agree that the first in order among the two aforementioned independent parameters is realism. During the 17th century realism had already become an almost universally accepted conviction. Nevertheless, during the 16th century the question for a realist epistemology did not grow out of some internal needs of science. Studying the history of renaissance philosophy one readily concludes that realism is externally transfused into science. Indeed, it mainly represents an important weapon of the ideological revolution against academic and religious culture.[8] I think that during the 16th century the ingenuous compromise of science and theology, known under the label "theory of the double truth", demonstrates all its limits. Truth gradually becomes simply the truth of science. However, if we tried to evaluate the epistemological arguments that Bruno, and even Galileo, later, put forward in defense of their realism, we would conclude that they are quite inconsistent, if not non-existent. From a strictly methodological point of view Osiander and Bellarmino have stronger arguments. The reason is not a very complicated one: rigorously speaking the Copernican system in its original form does not present any clear epistemological advantages compared to ptolemaic astronomy. In my opinion the concept of "heuristic power" that Lakatos used in order to show the superiority of Copernicus' research program does not help us sufficiently.[9] Furthermore the "objective promises" of a research programme as they have been analysed by Peter Urbach consist of such an informal and fuzzy set of criteria that they do not solve any problem. In order to understand what "objective promises" really mean, we have to reduce them to acceptability 1 and 2.[10] In all other cases the lack of precision that affects their definition combined with Lakatos insistence, in some texts, to the retrospective nature of our methodological evaluations would bring us dangerously close to anarchism. Anyway, it is a fact that when Bruno or Galileo defend the truth of the heliocentric system they substantially give an ideological, political and social battle. But at the same time this idea of truth in connection with the new ideal of progress is one of the most typical characteristics of the transition to maturity. Indeed, since the third decade of the 17th century, realism ceases to be a purely ideological arm and gradually assumes the form of an internal paramater. From now on in order to be convincing it must be given a methodological foundation. Descartes tried to offer one. Hobbes or Mersenne's circle another. But it seems to me that the majority of the 17th century scientists agree on one fundamental point. The theories which are closer to truth are those which, compared to their predecessors, represent a progress in the double sense of being more profound and better-tested. Realism as a symptom of maturity is closely connected to the abandonment of the apodictical patterns of explanation (despite Descartes' view). It is also connected to the idea of progress, to the requirement of empirical testability (whatever this could mean and it means actually quite different things) and to the appearance of rival research programmes, the comparative evaluation of which requires a minimal consensus of the scientific community about the rules of the game.

Needless to say, the same concept of scientific community introduces us to the second independent and genuinely external parameter of science's transition to maturity. I have in mind the new social division of labour, formed during the Renaissance. The constitution of scientific communities functioning according to some formal or even informal rules; a novel form of ideological justification of science's utility as well as a radical defense of science's autonomy in the search of truth; all these are the products of the new capitalist division of labour. I think that mature science's historical appearance is related, (i) to the social autonomization of the scientific activity within the system of division of labour, (ii) to the professionalization of scientists – though this is a process which is far from being entirely realized during the 17th century, (iii) to the scientific specialization – which includes the separation of science from philosophy, the constitution of new scientific branches etc, and finally, to the formation of what Merton called the "ethos of science".[11] Indeed, it is this corporative "ethos" that renders the methodological requirement of empirical testability and the resort to some "objective" criteria of rational choice, effective. We can take for granted that the specific cognitive contents of science are not externally determined. To reduce them to extrascientific social matrices would be, in any case, a dangerous mistake. (Though it is indisputable that an interaction of science and ideology, mediated by philosophy is always present.) But mature science cannot exist as a socially legitimate activity independently of some concrete social context. A history of science ignoring this would be blind. Thus, if history of science without philosophy of science is blind, history and philosophy of science without sociology of science are also blind.

Let us now try to sketch a possible application – in a schematic manner, indeed – of the preceding observations to an interesting case of the history of science: the continuity from medieval to modern science through that of Renaissance. Medieval epistemological ideas anticipate some methodological characteristics of science's maturity. Furthermore, even at the level of specific problem-posing an indubitable continuity between medieval elaborations and modern scientific achievements exists. Nevertheless from an internal point of view this continuity is interrupted during Renaissance which is, to my opinion, the antiscientific age *par excellence* – with the exception of some Aristotelian-minded philosophers, namely the ones belonging to the school of Padua. This is to say that medieval ideas reveal their potentialities, after a curious interruption of a century and a half, during the 17th century. This seems to be the conclusion of any possible internalist analysis. Internal continuity is interrupted during Renaissance because of the revival of magical, hermetic, occultist philosophies which virtually set aside the progressive dynamics of the 14th century's scholastic crisis. We might express this in the following scheme:

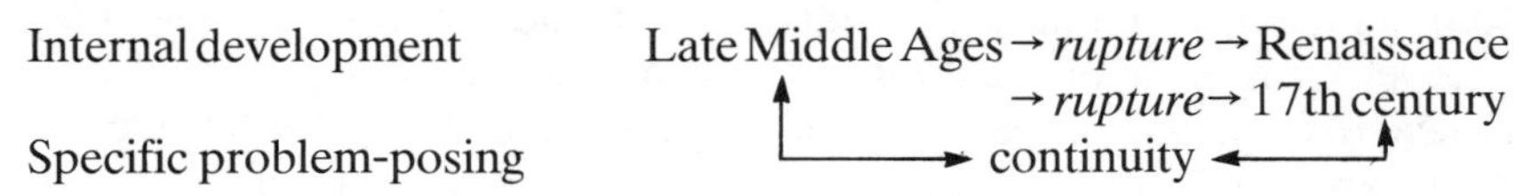

Still, should that be true, the Renaissance, considered from an external point of view, has been a necessary condition of the Scientific Revolution. Though there has been a rupture in science's growth, internally, the external conditions of the transition to maturity were all created in the course of Renaissance's turbulent ideological, social, political events. Renaissance's thought is, according to rational principles, to a considerable extent, magical, animistic, antiscientific. But in non scientific fields, Renaissance, with the refusal of authority, the divorce between science and theology, the gradual professionalization of the scientific activities and the creation of the first scientific societies, opens the way to mature science. Thus, from an external point of view, our previous scheme could be exactly the reverse:

External development	Late Middle Ages → *rupture* → Renaissance →
Social conditions	*continuity* → 17th century

Only if we could combine these two, apparently contradictory schemes, in a unified and coherent explanation, would we properly understand the 17th century scientific revolution. I am convinced that exclusively internalist histories as well as exclusively externalist ones are quite useless. And I also confess that if a third type of approach were to be simply an eclectical combination of the other two, we wouldn't really need it. It would be erroneous to adopt a model with insufficient theoretical foundation. Consequently – if I may do so – I would like to suggest the adoption of a prevalently internalist point of view, that is an epistemological approach to the history of science. Nevertheless, for most internalists, external factors seem to represent irreductible anomalies, an invasion of irrationality in the kingdom of rationality. Externalist explanations, to the mind of a rigid internalist, represent at most necessary but undesirable solutions of some historical anomalies. According to Lakatos, for instance, the ideal historiographical program would be the one enabling the comprehension of the greatest part of the real history of science in purely internalist schemes.[12] Lakatos was proud because the MSRP succeeded in explaining rationally, that is internally, a greater part of real history of science than other rival methodologies. In this consisted its superiority which was proved when the MSRP, raised to a second degree, was applied firstly to other methodologies and then to itself. Lakatos was well aware of the role played by external factors. He was convinced that the "internal skeleton of rational history defines the external problems" but he also affirmed that "the question of the motives and the reception of

Copernicus' achievement is an important one and cannot be answered in strictly internal terms.[13] To his opinion, however, though "appraisal alone does not logically imply acceptance or rejection" it is "our crucial ("internal") 'third-world' premise" . . . that . . . "as a matter of fact defines the problem situation for the 'externalist' ".[14]

Two objections might be raised at this point. The first concerns our "third-world" premise. I don't think there is any such thing as a third world neither do I think that our epistemological premises, that is our theory of scientific appraisal, actually define the premises of our external explanations. The hard core of a historiographical research program, attempting a rational reconstruction of the history of science, is always a philosophy of science. But this philosophy of science need not always be identical to the theory of history that leads our sociological reconstructions of the external conditions of the growth of science. What I mean by this is that Lakatosian methodology, for instance, has nothing to do with Marxist social theory. Nevertheless, I think that the best history of science would be one based on a Lakatosian theory of rational choice, as far as internal problem situations are concerned, and on a Marxist approach to the social conditions that externally determine science's growth. Lakatos speaks of psychological premises of the scientific discovery of acceptance. I think that premises which are not logical are simply social. And I also think, contrary to Lakatos' view, that according to the rational criteria of MSRP the best sociological reconstruction of the history of science would be a sophisticated Marxist one. I thus endorse the basic theses of a Marxist sociology but I am fairly convinced that Marxism is not a philosophy of science. More than this: Marxism does not imply a philosophy of science. The second objection now. When Lakatos holds that our internalist premises define the problem situation for the externalist he seems to have in mind two different kinds of history. And sometimes he goes still further in the wrong direction as when he writes:

> All these examples show how the methodology of scientific research programmes turns many problems which had been external problems for other historiographies into internal ones. But occasionally the borderline is moved in the opposite direction. For instance there may have been an experiment which was accepted instantly . . . as a negative crucial experiment. For the falsificationist such acceptance is part of internal history; for me it is not rational and has to be explained in terms of external history.[15]

The internalist prejudice is particularly clear in these lines. Not only have we two distinct kinds of history: rationally reconstructed history and external history; not only are externalists only concerned with eternal facts; but also, external facts are "not rational". Though this is not always the case – as we have already seen – Lakatos has a strong tendency to classify every event that resists internal explanation under the label "not rational'. Of course there is not such a thing as rational or irrational actual history. There are only rational or irrationalist reconstructions of history and events which appear rational or irrational in the light of these reconstructions. What I would like to point out is that a rational reconstruction of the history of science which would also include rational external explanations is possible. So though I believe in the rightness of Lakatos' basic views, I think that non-reducible external factors could be integrated in a rational, prevalently internalist historiography. I agree with Lakatos that the best methodology is the one which, applied to itself, proves to be, as a historiographical research program, more progressive than rival methodologies. Let us suppose that the MSRP is indeed, from a metamethodological point of view superior to its rivals. Then the MSRP explains, in rational terms, a considerable amount of the basic value judgements of the scientific community; but it can also fallibly predict a programmes future success – though Lakatos never really decided on this point – if we use a metaphysical inductive principle according to which a program's achievements, its major corroboration, its technological applications etc., measure its degree of versisimilitude. So, were there a scientific research program which, independently of the MSRP, but according to its criteria, could explain in the most promising way the external conditions of science's historical growth, then the MSRP should be completed with this program.

NOTES

[1] This, I think, is true even if we take into account Lakatos' claim according to which his "demarcation criterion between mature and immature science can be interpreted as a Popperian absorbtion of Kuhn's idea of 'normality' as a hallmark of mature science" I. Lakatos, *Philosophical Papers*, Vol. I, Cambridge, 1980, p. 90.

[2] Lakatos defines mature science as follows: "My account implies a new criterion of demarcation between 'mature science' consisting of research programmes, and 'immature science' consisting of a mere patched up pattern of trial and error", *ibid*., p. 87. He also says, in other occasions that "the unit of mature science is a research programme", *ibid*., p. 179, explaining that "mature science consists of research programmes in which not only novel facts but, in an important sense, also novel auxiliary theories, are anticipated; mature science – unlike pedestrian trial-and-error – has 'heuristic power'", *ibid*., p. 88.

[3] I think that Lakatos would not accept the idea that science was always 'mature science'. If this was the case then the demarcation between 'mature' and 'immature' science would be

completely pointless. Actually, the crucial question is: when was mature science born? And this is a question that Lakatos does not answer.

[4] As a matter of fact, Buridan's and Oresme's intuitions about Earth's movement around its axe, were not developed into a coherent research programme.

[5] See L. Laudan, *Progress and its Problems*, London, 1977, p. 78 sq.

[6] See also A. Crombie, 'Alcuni atteggiamenti nei confronti del progresso scientifico: Antichità, Medioevo, inizi dell'era moderna', in E. Agazzi (ed.), *Il concetto di progresso nella scienza*. Milano, 1976, pp. 15–36, and P. Rossi 'Sulle origini dell'idea di progresso', *ibid.*, pp. 37–89.

[7] I have in mind Popper's beautiful and interesting, though controversial description in his article 'The rationality of scientific revolutions', in I. Hacking (ed.), *Scientific Revolutions*, Oxford, 1981, pp. 80–106.

[8] About the distinction between ideological and scientific revolutions see again Popper's article "The rationality of scientific revolutions", *ibid*., p. 98 et seq.

[9] See I. Lakatos 'Why did Copernicus' research programme supersede Ptolemy's' in *Philosophical Papers*, vol. I, *op. cit*., pp. 168–192. It does not help us sufficiently not because it constitutes a step in the wrong direction but rather because it adds nothing to the previously elaborated criteria of acceptability 1 and 2.

[10] See P. Urbach, 'The objective promise of a research programme', in G. Radnitzky and G. Andersson, *Progress and Rationality in Science*, Dordrecht, 1978, pp. 99–113. I have tried to show that "objective promises" can be reduced to the two criteria of acceptability in my introduction to the Greek translation of Lakatos' *Methodology of Scientific Research Programmes*. See E. Metaxopoulos, *Εἰσαγωγή*, in I. Lakatos *Η μεθοδολογία των προγραμμάτων επιστημονικής έρευνας*, Athens, 1986, pp. 1–78.

[11] For the definition see R. K. Merton, 'Science and the Social Order', in *Social Theory and Social Structure*, Glencoe, 1959, pp. 537–549 and in the same volume 'Science and Democratic Social Order', pp. 550–562.

[12] See I. Lakatos 'History of Science and its Rational Reconstructions', in *Phil. Papers*, *op. cit*., pp. 132 et seq.

[13] *Ibid*., p. 189.

[14] *Ibid*., p. 191.

[15] *Ibid*., p. 116.

University of Ioannina

ULRICH GÄHDE

BRIDGE STRUCTURES AND THE BORDERLINE BETWEEN THE INTERNAL AND EXTERNAL HISTORY OF SCIENCE

1. INTRODUCTION

In this paper I want to take up two ideas presented by Imre Lakatos in his article on the 'History of Science and its Rational Reconstructions':[1] the ideas that (1) one ought to distinguish between an internal and an external history of science, and that (2) these two are in a relation of both competition and mutual completion.

Let me first recapitulate the distinction. According to Lakatos, the internal history of science accounts for the "innertheoretical" aspects of scientific development. It presupposes a concept of scientific rationality which serves as a normative methodology. Guided by this methodology, one tries to interpret as much as possible of the history of science as a "rational enterprise" – i.e. as an enterprise, which is in accordance with the normative rules specified previously. These historical studies may, in turn, lead to criticism and subsequent modification of the underlying methodological concept.[2] One major task for this part of the history of science is to provide an adequate explication of the term "scientific progress".

The internal history of science, however, cannot explain everything: It has to leave a certain range of historical phenomena unexplained. Or, as Lakatos puts it: "The history of science is always richer than its rational reconstruction".[3] These remaining phenomena constitute the domain of the "external history of science". It tries to explain them as brought about and controlled by external factors which have to be investigated by economic, psychological, sociological, or other empirical methods.

Lakatos now had the following idea: How much of the history of science we can understand "rationally" depends on how much we know about the structure of empirical theories, and about the normative rules that govern their construction and refinement. If we have but a very rough picture at our disposal, then we can understand only few processes within the boundaries of an internal history of science. Most scientific developments will appear to be externally controlled. If, however, our picture gets

K. Gavroglu, Y. Goudaroulis, P. Nicolacopoulos (eds.), Imre Lakatos and Theories of Scientific Change. 215–225.

increasingly detailed and refined, we will be able to understand more and more processes in the framework of some rational methodology. Thus Lakatos proposed the following strategy: By choosing increasingly refined normative methodologies and increasingly detailed pictures of empirical theories we should try to expand the internal and push back the external history of science.

Some aspects of Lakatos' distinction between an internal and an external history of science remain rather opaque. In particular, the meta-criteria which govern the choice of a certain normative methodology deserve further elaboration. I think, however, that Lakatos is basically right in that (1) the idea of a "rational reconstruction" of scientific changes is indispensable, and that (2) the extent, to which such a reconstruction is possible, depends on our knowledge of the structure of empirical theories. In the following paper, I want to take up these two aspects of Lakatos' position.

The scenario of the philosophy of science has changed since Lakatos' death: The enforced use of formal and semiformal methods led to a significant gain in precision in the metatheoretical description of empirical theories. New, rather filigreed innertheoretical structures were discovered which provide new starting points for theory-dynamical processes. Thus, the following question suggests itself: Does this progress in the description of theories allow for some further expansion of the internal history of science? I think that the answer is "yes". I shall argue this optimistic view by focusing on a special refinement introduced into the metatheoretical picture of empirical theories in recent years: bridge structures which connect different applications of these theories. I shall proceed in three steps. As a first step I will explain what bridge structures are and which role they play in empirical theories. Then, as a second step, I am going to explain how these structures might change and how certain theory-dynamical processes can be traced back to these changes. Finally, I am going to show how new questions for the internal history of science can be derived from these considerations.

II. NEW INSIGHTS INTO THE STATICS OF EMPIRICAL THEORIES: BRIDGE STRUCTURES

Most metatheoretical concepts start from a simplifying assumption: Empirical theories have but one great universal application. According to this view, electrodynamics and mechanics, for example, have the entire physical universe as their one all-embracing application. This assumption, however, is not only simplifying, but fictitious. Some empirical sciences, in fact, have rather extensive applications: Cosmology deals with the structure and development of the universe as a whole, and a lot of recent research work in astrophysics has been directed toward an investigation of galaxies, clusters of galaxies and other large scale objects. Nevertheless, the overwhelming majority of applications of empirical theories focuses on rather limited segments of reality: on some stellar object with time-dependent luminosity, on a certain type of mechanical motion damped by friction, or on some unexpected geological formation on Jupiter's moon Ganymede. Numerous major research problems in the natural sciences consist in trying to fit these individual descriptions of small segments of reality together. The following example illustrates this point.

Stellar atmospheres are, in general, far from being in a state of thermal equilibrium. This would, among other things, presuppose that the gradient of temperature is equal to zero everywhere in the atmosphere – which is, of course, not so. In the case of some stars, however, it is possible to dissect the atmosphere into small volume elements, and to assume that each of them is – at least approximately – in a state of thermal equilibrium. In other words: Each individual element is treated as an application of equilibrium thermodynamics. What causes problems with this so-called 'Local Thermal Equilibrium-approach' ('LTE-approach') is not so much the description of the individual elements, but the task of fitting them together to obtain a picture of the whole stellar atmosphere. A metatheoretical approach which, from the start, accepts the assumption of but one universal application per theory cannot adequately mirror research enterprises of this kind.

It goes to the credit of the structuralist concept that it does not make use of this fictitious assumption: It allows for any number of different applications of one and the same theory. These applications are not necessarily independent of one another. Instead, they may overlap, or be in spacial contact, as in the case of the stellar atmosphere. A variety of relations between different applications of one or more theories may exist. In order to represent these relations adequately on a metatheoretical level one needs certain technical devices: Bridge structures serve this purpose. The structuralist concept distinguishes two types of such structures: (1) innertheoretical bridge structures, which connect applications of one theory only, (2) intertheoretical bridge structures, which connect applications of different theories.

I shall start by discussing innertheoretical bridge structures, the so-called "constraints". A standard example is the "extensivity constraint", which concerns the mass function in classical mechanics. Let us think of three applications of this theory. In the first application there occurs an object p, in the second an object q and in the third an object obtained from p and q by means of some physical concatenation operation. The sum of the masses of p and q has then to be assigned to this concatenated object.

Other important examples of constraints are supplied by the invariance principles associated with physical theories. These principles connect descriptions of identical physical systems as seen from different frames of reference. Since, in structuralist terminology, these descriptions correspond to different applications of the same theory, invariance principles can be interpreted as innertheoretical bridge structures, i.e. as constraints.[4] So far, formal explications of the Galilei and Lorentz invariance principles have been provided.[5]

Let us now turn to a more complex example. It refers to so-called "variable stars". These stars do not emit electromagnetic radiation with constant intensity, but instead show a (in many cases periodic) change in their brightness. An important subclass is formed by the Cephei variables. To be more precise: "to be a Cephei variable" does not denote a lifelong membership in some special class of stars, but merely a transitional stage which is passed by a variety of stars in the process of stellar evolution. Cephei variables are distinguished by the fact that the periods of their light change and their mean luminosities during that period are interrelated by the so-called "period-luminosity equation". In astronomical literature this equation is commonly presented in an abbreviated form, in which all arguments of the occurring functions have been dropped. Let me try to provide a more complete formulation. Let p be the star in question and T the time interval during which the star passes the Cephei stage. Then

$$(1) \quad \exists\alpha\,\exists\beta\,\forall t\,[\alpha \in \mathbb{R}^- \wedge \beta \in \mathbb{R}^- \wedge t \in T \rightarrow M(p,t) = \alpha + \beta \log P(p,t)]$$

Here the function P gives the period of the light change (P in days). The function M is the so-called "absolute magnitude", i.e. the luminosity the star would have if seen from some standard distance (which is 10 parsec, or about 33 light years). The inclusion of t among the arguments of M and P accounts for the fact that both the absolute magnitude and the period P may slightly change while the star passes through the Cephei variable phase of its development. The absolute magnitude M has to be distinguished from the apparent magnitude, i.e. the luminosity the star has when observed from the earth. Let $r(p, t)$ denote the distance of star p from the earth at time $t(t \in T, r(p, t)$ in parsec). Then absolute and apparent magnitude are related as follows:

$$(2) \quad m(p,t) - M(p,t) = -5 + 5 \log r(p,t).$$

In a more precise presentation, both M and m would have to be supplied with some index which denotes the interval of the electromagnetic spectrum, to which, in turn, the measurement of these luminosities relates. Thus, one distinguishes between M_v (="visual magnitude"), M_{pg} (="photographical magnitude"), etc. In general, for different sectors of the electromagnetic spectrum the corresponding values for α and β, respectively, will not coincide. This is why here "the period-luminosity equation" is identified with the rather weak existential proposition (1). By inserting special values for α and β into (1), specialized statements concerning certain sectors of the electromagnetic spectrum can be obtained.

So far, we have dealt with individual Cephei variables only. A connection between different objects of this type is brought about by a constraint which is attached to the period-luminosity equation. According to this constraint, all Cephei variables fulfill this equation – for some sector of the electromagnetic spectrum – with the same values for α and β, respectively.

One further remark: Equation (1) can be interpreted in the framework of a theory of stellar evolution. Observation by astronomers, however, can cover only a time interval which is negligible compared to T. Thus, one might reformulate (1) by referring to a certain time $t^* \in T$ (which is rather common in astronomy textbooks). This modification turns equation (1) into a trivial, mathematically true statement. In that case, the constraint becomes the only carrier of empirical information.

Let us take a second look at equations (1) and (2). They make clear why Cephei variables have played such an outstanding role in modern astronomy and cosmology: They provide the key to a cosmic distance scale. The procedure for determining distances using Cephei variables can be described as follows: As a first step those stars are investigated for which both the period P of their light change and their absolute magnitude can be obtained independently of equation (1). (For some objects, for example, the absolute magnitudes can be derived from an analysis of their electromagnetic spectrum.) Once M and P are known for some stars, the values of α and β can be calculated. Then, as a second step, one proceeds to the analysis of those Cephei variables for which the distance to the earth is to be determined. Even if these stars are rather remote objects, the period P of their light change will, in general, be readily measurable. By making use of the constraint, their absolute magnitudes are calculated via equation (1). Their apparent magnitudes can be obtained by photometric measurement from the earth's surface. Once the absolute and apparent magnitudes are known, the distance from the earth can be calculated by means of equation (2). In other words: Cephei variables can be used as "cosmic milestones".

Note the holistic character of this procedure: Information gained through the examination of some less remote Cephei variables is used for the investigation of other stars of that type, which are more remote or for some other reasons less accessible to observation. Intuitively speaking, these objects form a kind of "holistic unit". The elements of this unit are connected by the above mentioned constraint.

Obviously, this constraint differs essentially from the extensivity or invariance constraints mentioned before: The latter are not attached to any special law. They can connect all models whatsoever. Conversely, the constraint occurring in the Cephei variables example makes sense only in connection with some special law: the period-luminosity equation.[6] For this reason constraints of the first type are called "general constraints', those of the second type "special constraints".

Constraints are innertheoretical bridge structures: They connect different models of one and the same theory only (although, possibly, models of different specializations of this theory). Besides these inner-theoretical there are intertheoretical bridge structures, the so-called "intertheoretical links". Their role can be explained by means of an analogy taken from computer sciences. In general, a computer program uses both "input parameters" and "output parameters" which occur in its headline. Input parameters have to be fed into the program either by the user himself or by some other program. Without them, it cannot be run. By contrast, output parameters are calculated by the program and are put at the user's – or some other program's – disposal. The situation is completely analogous with empirical theories. In general, they make use of functions supplied by other theories. Classical electrodynamics, for example, uses the term "mass density" and "force density" which are supplied by the mechanics of the continuum. Functions of this kind, which are fed into the theory "from outside", correspond to the input parameters in computer programs. By contrast, classical electrodynamics provides values for the intensity of electric and magnetic fields: They correspond to output parameters. Intertheoretical links are bridge structures, which correlate "output parameters" of one theory with "input parameters" of some other theory. (They may also serve further purposes. I shall, however, not go into any detail here.) Links play a decisive role in the reconstruction of intertheoretical relations, and thus for an analysis of the "superstructure" of empirical science.

III. NEW INSIGHTS INTO THE DYNAMICS OF EMPIRICAL THEORIES: CHANGE OF BRIDGE STRUCTURES

So far, I have given no more than a static description of constraints and links: A kind of snapshot which freezes the scenario of different applications of one or more empirical theories connected by certain bridge structures. What do these considerations have to do with theory-dynamics and the history of science?

Theory-dynamical aspects come in as soon as we move on from this static picture to an analysis of how bridge structures might vary with time. By means of some examples from physics I shall try to illustrate that a variety of – major and minor – developments in the history of science can be traced back to changes of bridge structures.

Let us start with an example of a major change, brought about by the change of an intertheoretical bridge structure. The link described above provides an example for this type of theory-dynamical processes. In the 19th century electrodynamics had to confront the following problem: On the one hand, Maxwell's theory was neatly connected with the classical mechanics of the continuum: It made use of both the mass density and force density functions supplied by this theory. On the other hand, both theories did not fit together: The principle of Lorentz invariance which belongs to Maxwell's theory and the principle of Galilei invariance which belongs to the mechanics of the continuum are inconsistent.

The situation was similar to the one described by Lakatos in his paper on the 'Methodology of Research Programmes'.[7] In it he described the early phases of the development of Bohr's theory of the atom as an example for a "research programme progressing on inconsistent foundations": On top of classical electrodynamics Bohr imposed his postulates, which contradicted Maxwell's theory. In complete analogy Maxwell's theory itself made use of a "pretheory", although both theories did not fit together.

The theory-dynamical process by which this anomaly – as Lakatos would have called it – was eliminated can be described by using the concept of "intertheoretical links": The links which connected mass and force density in non-relativistic mechanics and electrodynamics were broken up. They were replaced by analogous links connecting electrodynamics with the relativistic mechanics of the continuum. The antinomy due to confligating invariance principles vanished. This constitutes an example for a rather dramatic process in the history of science which was brought about by a change of intertheoretical bridge structures.

The concept of bridge structures, however, allows for the reconstruction of more moderate developments in the history of science as well: changes which affect rather filigreed structures of empirical theories. The example of the constraint attached to the period-luminosity equation for Cephei variables provides an intriguing example for a theory-dynamical process of this type.

The procedure of determining distances with the help of Cephei stars rests on the assumption that all stars of this type fulfil the period-luminosity equation with the same values for α and β. This constraint, however, broke down.

The fault was noticed, when an increasing number of anomalies occurred. Among others, the absolute luminosities assigned to stars in neighboring galaxies were smaller than statistical considerations would suggest. Photometric measurements, however, reaffirmed the data obtained for the apparent luminosities. In the light of these difficulties, the argument ran as follows: If these galaxies in general and the Cephei variables in them in particular were more remote than previously assumed, then the mean absolute luminosity of these stars could well be the same as that of the stars in the Milky Way. They would seem less bright only due to their greater distance. These considerations suggested that the method for determining distances with the help of Cephei variables contained some fundamental fault. A new analysis by Baade[8] led to the following result: The set of Cephei variables contains two subsets: The subset of the so-called "δ-Cephei stars" and the subset of the "W-Virginis stars". The δ-Cephei variables, however, fulfil the period-luminosity equation with different values for α and β than the W-Virginis variables. Thus the original constraint had to be dropped. It was replaced by two analogous, though weaker constraints: The first requires that all δ-Cephei stars fulfil this equation with the same values for α and β. The second formulates this requirement for W-Virginis stars. By contrast, an analogous requirement for the total of Cephei variables was no longer upheld. This correction forced a revision of the cosmic distance scale. Once this revision was carried through, the anomalies vanished.

Note that at this point a general connection between bridge structures and holistic phenomena becomes evident: Bridge structures connect different applications of empirical theories, thus leading to the formation of "holistic complexes" (in our example: the complex of all Cephei variables). When these bridge structures break down or are replaced by weaker structures, these complexes may disintegrate into sub-complexes (in our example: (1) the subcomplex of all δ-Cephei variables, (2) the sub-complex of all W-Virginis variables). This illustrates that the concept of bridge structures may well serve a further purpose in the philosophy of science: They may supply the key to an investigation of the dynamics of holistic phenomena, thus opening a new field for philosophical research.

IV. SUMMARY: NEW QUESTIONS FOR THE INTERNAL HISTORY OF SCIENCE

Let me summarize the preceding considerations. They refer to a major modification in the picture of empirical theories: The concept of one universal application for each theory was dropped, more-elementary sets of applications have been admitted. As these applications do not have to be independent from each other, formal tools for the metatheoretical description of the relations between them have to be provided. Inner-theoretical bridge structures (=constraints) and intertheoretical bridge structures (=intertheoretical links) serve that purpose. The introduction of these structures reveals new starting points for theory-dynamical processes. The first example I discussed before relates to the change of an intertheoretical bridge structure: The change of links which connect electrodynamics with a mechanical "pre-theory". This process belongs – in Kuhn's terminology – to a revolutionary phase of scientific development. The second example provided illustrates the change of an innertheoretical bridge structure. A constraint, which is attached to some special law (the period-luminosity equation for Cephei variables) is dropped. It is replaced by two weaker constraints of the same type. Although only rather filigreed innertheoretical structures are directly affected, major developments in the history of the natural sciences can be traced back to this change.

These considerations suggest a variety of questions for the internal history of science. Let me list some of them, without claiming completeness:

(1) Bridge structures play an important role in the construction of the "protective belt"[9] for empirical theories. They help to immunize theories against reluctant measuring data. The Cephei variable example illustrates this point. In particular, they account for a procedure one might call the "non-local elimination of anomalies": An anomaly, which occurs in some application of an empirical theory, is eliminated by corrections carried out in a different application.[10] Historical case studies should supply further insights into this mechanism.
(2) As has been mentioned before, invariance principles can be treated as innertheoretical bridge structures (see sec. II). These constraints have played a key role in revolutionary phases of the development of physics. In particular, conflicts between some theory and its pretheories due to inconsistent invariance requirements would deserve much further historical study. Furthermore, recent investigations[11] suggest that in the construction of at least one theory net (the net of electrodynamics) special theory elements with differing invariance principles have been used. Further studies in the history of physics could help to clarify this point.
(3) By connecting different applications of empirical theories, bridge structures induce the formation of "holistic complexes". If some bridge structures are abandoned, these complexes may disintegrate. If, in contrast, new structures are added, several complexes may combine. Furthermore, one may even argue that an important aspect of scientific progress consists in the formation of increasingly extensive holistic units. The dynamical aspects of holistic phenomena, however, have hardly been an object of historical studies so far.

These questions should suffice to illustrate my main thesis: The refinement of the metatheoretical picture of empirical theories – brought about by the introduction of bridge structures – allows for a better and more detailed understanding of numerous developments in the history of science. Thus, they may help to regain terrain for what Lakatos called 'the internal history of science'. Since it has become rather en vogue to stress the role of external factors in science excessively, this should be regarded as a major challenge for the philosophy of science.

NOTES

[1] See Lakatos (1978).
[2] For details see *ibid*., pp. 121–138.
[3] *Ibid*., p. 118.
[4] Another possible way of dealing with invariance principles would be the following: An application of some theory T is no longer identified with a single partial potential model –as is usual in the structuralist framework. Instead, it is defined as an equivalence class of partial potential models, which can be transferred into each other by means of invariance transformations.
[5] The formalization of the Galilei invariance principle is rather trivial. A formalization of the Lorentz invariance principle is given in T. Bartelborth (1988).
[6] This equation serves as a special law in the (rather extensive) theory of stellar interiors and evolution. For an outline of this theory compare Novotny (1973), and Cox (1980).
[7] Compare Lakatos (1970).
[8] See Baade (1952).
[9] See Lakatos (1970).
[10] For a more detailed description of this mechanism see U. Gähde and W. Stegmüller (1987).
[11] Bartelborth (1988).

REFERENCES

Baade, W., 'A Revision of the Extra-Galactic Distance Scale', *Transactions of the International Astronomical Union*, **8** (1952), pp. 397–398.

Bartelborth, T., *Eine logische Rekonstruktion der klassischen Elektrodynamik*, Frankfurt, 1988.

Cox, J. P., *Theory of Stellar Pulsation*, Princeton Series in Astrophysics, Princeton, 1980.

Gähde, U. and Stegmüller, W., 'An Argument in Favor of the Duhem–Quine Thesis: From the Structuralist Point of View', in Hahn, I. and Schilpp, P. H. (eds.), *The Philosophy of W. v. Quine*, The Library of Living Philosophers, vol. 15. La Salle, Il., 1987.

Lakatos, I., 'Falsification and the Methodology of Research Programmes', in Lakatos, I. and Musgrave, A., *Criticism and the Growth of Knowledge*, Cambridge University Press, 1970.

Lakatos, I., 'History of Science and its Rational Reconstructions', in Lakatos, I. *The Methodology of Scientific Research Programmes*. Cambridge University Press, 1978.

Novotny, E., *Introduction to Stellar Atmospheres and Interiors*. Oxford University Press, 1973.

Freie Universität Berlin

PART IV

ILKKA NIINILUOTO

CORROBORATION, VERISIMILITUDE, AND THE SUCCESS OF SCIENCE

1. INTRODUCTION: TWO FRONTS AT THE LONDON COLLOQUIUM

To what extent did the positions articulated in the 1965 London Colloquium become guidelines for the ensuing discussions among philosophers of science? A re-reading of the programme of the International Colloquium in the Philosophy of Science suggests an answer to this question: the menu contained two important topics which have remained as central epistemological and methodological themes for the next two decades.

First, "the problem of inductive logic" was discussed in no fewer than six sessions, with R. Jeffrey, R. Carnap, H. Freudenthal, J. Hintikka, H. Kyburg, Y. Bar-Hillel, M. Hesse, W. Salmon, and I. Hacking as the main speakers. The main papers, which developed different modern variants of the theory of inductive inference, were challenged in the discussions by Popperian anti-inductivist arguments. The Proceedings of this part of Colloquium, *The Problem of Inductive Logic* (1968; ed. by I. Lakatos), contained a long survey paper, 'Changes in the Problem of Inductive Logic', where Lakatos attempted to reveal the "degenerating problem shifts" in Carnap's research programme of inductive logic – but also criticised the measures of corroboration proposed by K. Popper as alternatives to Carnap's probabilistic measures of confirmation.

Secondly, "the growth of knowledge" was discussed in a session with T. S. Kuhn and J. Watkins as the main speakers, and Popper as the chairman, and in R. Suszko's paper on 'Formal Logic and the Development of Knowledge'. While Watkins questioned the rationality of "dogmatic" normal science, Kuhn (especially in his later reflections in the book) challenged with his incommensurability thesis the Popperian emphasis on truth as the aim of science. The Kuhn–Watkins debate, with contributions by other participants,[1] eventually appeared in a book, *Criticism and the Growth of Knowledge* (1970, ed. by I. Lakatos and A.

K. Gavroglu, Y. Goudaroulis, P. Nicolacopoulos (eds.), Imre Lakatos and Theories of Scientific Change. 229–243.

Musgrave), which again contained a long paper by Lakatos, outlining his methodology of scientific research programmes.

The London Colloquium was organized at the main headquarters of the Popperian critical rationalism. The situation may be viewed as a battle that Sir Karl with his general staff were fighting in two fronts (see Fig. 1). In popular terms, the two fronts may be called *the Popper–Carnap controversy* and *the Popper–Kuhn controversy*.[2] In the first case, Popper's "falsificationism" stood up against Carnap's "inductivism": while "weak justificationism" wanted to retain a probabilistic version of the classical concept of knowledge, Popper wished to formulate a weaker concept of conjectural knowledge through his concept of corroboration. In the second case, Popper's "realism" was opposed to the "pragmatist" tendency in Kuhn: while Popper wanted to retain an approximate version of the classical concept of truth with his theory of verisimilitude, Kuhn wished to formulate a weaker account of scientific progress in terms of problem-solving. In both cases, Lakatos attempted to play a mediating role in the controversies by his efforts to find a creative synthesis of the rival positions.

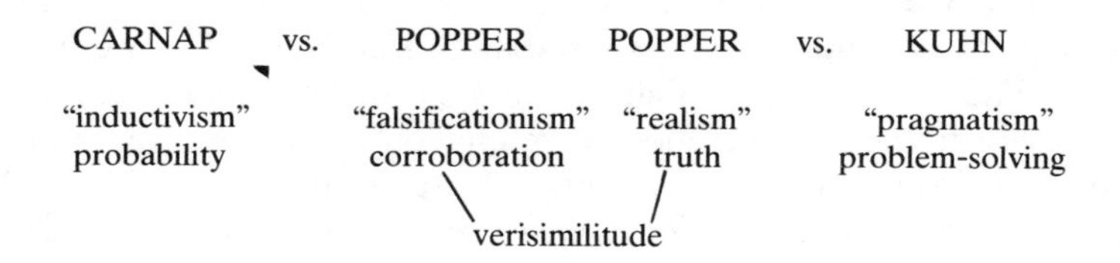

Fig. 1.

In the 1960s it was still possible to think that the two controversies were to a large extent independent of each other: as Bar-Hillel put it, the former concerned the "synchronic" problem of justification, the latter the "diachronic" change in science. Lakatos sharply attacked this independence assumption with his campaign against "instant rationality". That he was – at least to some extent – right was acknowledged by the subsequent work on "probability kinematics" by R. Jeffrey, I. Levi, and others, where the rational justification for a theory depends on its relation to new evidence *and* to the preceding theory (or knowledge situation).

Dramatic evidence for the systematic interconnections between the two controversies was also provided by the fact that the two "enemies" of the

Popperian camp seem to have been active in allying their forces. Persuaded by the result that genuine generalizations have a zero inductive probability in his λ-continuum, Carnap adopted an instrumentalist attitude towards scientific laws as tools for prediction – and reinterpreted his inductive logic as a non-cognitive framework for rational decision making (cf. Stegmüller 1973).[3] On the other hand, some of the pragmatists, who follow Kuhn in their opposition to the correspondence theory of truth, have attempted to re-establish the strong cognitive claims of science (or at least "Peircean limit science") through an epistemic definition of truth (cf. Putnam 1981, Tuomela 1985).

Recent developments in the philosophy of science have also highlighted the intimate connection between the two controversies. Both of these disputes are related to the problem of finding adequate concepts for describing and analysing the success of science. A key issue in the Popper–Kuhn debate has been the question whether it is at all meaningful to speak about scientific progress as "approach to the truth". Various forms of instrumentalism and methodological antirealism have been proposed in order to get rid of such talk (P. Feyerabend, W. Stegmüller, L. Laudan, B. van Fraassen, R. Rorty). On the other hand, after the refutation of Popper's (1963) explication of the concept of truthlikeness or verisimilitude (D. Miller and P. Tichý in 1974), a new programme for defining truthlikeness in terms of similarity has emerged (R. Hilpinen, P. Tichý, I. Niiniluoto, R. Tuomela, G. Oddie). (See Oddie 1986 and Niiniluoto 1987). As I have argued elsewhere (Niiniluoto 1984), this work gives us a tool for defending a realist view of scientific progress as increasing truthlikeness, and thereby allows us to answer the Kuhnian challenge.[4]

The main thesis of this paper is that the theory of verisimilitude allows us to re-evaluate the Popper–Carnap controversy as well. Popper himself suggested that the degree of corroboration of a theory (i.e., the degree to which the theory has been severely tested at a given time) is an indicator of its degree of verisimilitude – and, hence, can be used as a guide to the preference between rival theories "with respect to their apparent approximation to truth' (Popper 1972, p. 103). Lakatos acknowledged that Popper's theory of verisimilitude makes it "possible, for the first time, to define *progress* even for a sequence of false theories". But, in his famous "plea to Popper for a whiff of 'inductivism'", Lakatos argued that this is not enough in order to distinguish fallibilism from scepticism:

We have to *recognize* progress. This can be done easily by an inductive principle which connects realist metaphysics with methodological appraisals, verisimilitude and corroboration, which reinterprets the values of the "scientific game" and a – conjectural – theory about the *signs* of the *growth of knowledge*, that is, about the signs of *growing verisimilitude of our scientific theories*. (Lakatos 1974, p. 255.)

In other words, in order to achieve a victory in the battle with Kuhn, a fallibilist needs some sort of a "synthetic inductive principle" for estimating progress and verisimilitude. It is clear, however, that Popper's own concept of corroboration – as well as other explicates of corroboration (cf. Hintikka 1968, Good 1975) – fails to fulfil this promise.

Lakatos thought that "Popper achieved a complete victory in his campaign against inductive logic" (Lakatos 1974, p. 259). This is a mistake, I believe.[5] Hintikka showed already in 1964 that it is perfectly possible to construct systems of inductive measures that assign non-zero probabilities to generalizations (cf. Hintikka and Suppes 1966, Niiniluoto 1977a, Kuipers 1978, Jeffrey 1980).[6] Levi (1967) showed how truth and information can be regarded as the "epistemic utilities" within a probabilistic Bayesian framework (see also Bogdan 1976). Niiniluoto and Tuomela (1973) developed a "non-inductivist logic of induction" which rejects what Lakatos (1968) called the "atheoretical thesis": inductive probabilities may depend on theories as evidence. It thus turns out that some aspects of Popper's methodology could be reconciled and even justified within a theory of induction and semantic information (cf. Hintikka 1968, Niiniluoto and Tuomela 1973, Niiniluoto 1973, Jeffrey 1975, Good 1975, Grünbaum 1978).

However, any attempt to combine Bayesianism and Popperianism remains incomplete, if it is not systematically connected to the realist's problem of scientific progress. This means that some link should be established between the concepts of inductive probability, confirmation, corroboration, and verisimilitude. Since 1975 I have advocated the view that it is indeed possible to combine the ideas of induction and truthlikeness in methodologically interesting ways (Niiniluoto 1977b, 1982, 1986c). There is a way of developing a comprehensive theory of scientific inference which in a sense merges together the weakly fallibilist programme of inductive logic and the strongly fallibilist programme of verisimilitude (Niiniluoto 1986b, 1987). It also turns out that this synthesis of Carnap and Popper contains Levi's (1967) cognitive decision theory as a special case.

This theory will be sketched in general terms below in Sections 2 and 3.

But I also mention some other, previously unexplored ways of combining probability and verisimilitude.[7]

2. DEFINITION OF TRUTHLIKENESS

Let us say that a non-empty set $B = \{h_i | i \in I\}$ of statements is a P-set relative to background assumption b, if

$$b \vdash \bigvee_{i \in I} h_i$$

$$b \nvdash \sim h_i \qquad \text{for all} \quad i \in I$$

$$b \vdash \sim (h_i \,\&\, h_j) \qquad \text{for all} \quad i \neq j; \qquad i, j \in I.$$

Given the presupposition b, and a semantically determinate language L where the elements of B are stated, there is one and only one element h_* of B which is *true* (in the Tarskian model-theoretical sense) in the fragment of the actual world W_L relative to L.

If h_* is unknown, P-set B defines a *cognitive problem* with the "target" h_*: which of the elements of B is true? The statements $h_i \in B$ are the *complete potential answers* to this problem. The *partial* potential answers are represented by the disjunctive closure $D(B)$ of B:

$$D(B) = \{ \bigvee_{i \in J} h_i | \varnothing \neq J \subseteq I\}.$$

Let $\Delta: B \times B \to R$ be a real-valued function expressing the *distance* $\Delta(h_i, h_j) = \Delta_{ij}$ between pairs of elements of B. Here $0 \leqslant \Delta_{ij} \leqslant 1$ and $\Delta_{ij} = 0$ iff $i = j$. The distance function Δ on B can be based on the underlying metric structure of B (e.g. B may be a subspace of real numbers R in an estimation problem) or on the syntactical similarity of the elements of B (e.g., B is the set of the Carnapian state descriptions or the Hintikkian constituents of a first-order language).

Let us next see how Δ can be extended to a function $B \times D(B) \to R$, so that $\Delta(h_i, g)$ expresses the distance of a partial answer $g \in D(B)$ from a given element $h_i \in B$. Let $g \in D(B)$ be a potential answer with

$$\vdash g \equiv \bigvee_{i \in I_g} h_i, \quad \text{where } I_g \subseteq I.$$

Define

$$\Delta_{\min}(h_i, g) = \min_{j \in I_g} \Delta_{ij}$$

$$\Delta_{\text{sum}}(h_i, g) = \sum_{j \in I_g} \Delta_{ij} / \sum_{j \in I} \Delta_{ij}$$

$$\Delta_{\text{ms}}^{\gamma\gamma'}(h_i, g) = \gamma \Delta_{\min}(h_i, g) + \gamma' \Delta_{\text{sum}}(h_i, g) \; (0 < \gamma \leqslant 1, 0 < \gamma' \leqslant 1).$$ [8]

Here the parameters γ and γ' indicate the weights given to our cognitive aims of hitting close to the truth (Δ_{min}) and getting rid of falsity (Δ_{sum}). Then the *degree of truthlikeness* $\mathrm{Tr}(g, h_*)$ of $g \in D(B)$ (relative to the target h_* in B) is defined by

$$\mathrm{Tr}(g, h_*) = 1 - \Delta_{ms}^{\gamma\gamma'}(h_*, g) \quad \text{(see Niiniluoto 1986b).}$$

Further, g' is *more truthlike* than g if

$$\Delta_{ms}^{\gamma\gamma'}(h_*, g) > \Delta_{ms}^{\gamma\gamma'}(h_*, g').$$

Let us say that an answer $g \in D(B)$ is *misleading* if it is less truthlike than a tautology, i.e., worse than the trivial answer 'I don't know'. Misleading answers are in a clear sense not worth having. Since $\Delta_{ms}^{\gamma\gamma'}(h_*, g) = \gamma'$ for a tautology, a complete answer h_i is misleading if $\Delta_{*i} > \gamma'/\gamma$. This gives us an "operational" characterization of the parameters γ and γ': the ratio γ'/γ gives (approximately) the radius of the circle around h_* which includes the non-misleading complete answers.

The properties of the *min-sum measure* Tr include the following:

(1) $0 \leqslant \mathrm{Tr}(g, h_*) \leqslant 1$.

(2) $\mathrm{Tr}(g, h_*) = 1$ iff $g = h_*$.

(3) Among true statements, but not among false statements, truthlikeness covaries with logical strength.

(4) Assume $j \notin I_g$. Then $\mathrm{Tr}(g \vee h_j, h_*) > \mathrm{Tr}(g, h_*)$ iff $\Delta_{*j} < \Delta_{min}(h_*, g)$.

(5) If g is false, then $Tr(h_* \vee g, h_*) > \mathrm{Tr}(g, h_*)$.

These conditions hold for a non-trivial distance function Δ. If Δ on B is trivial, i.e., $\Delta_{ij} = 1$ for all $i \neq j$, then $\mathrm{Tr}(g, h_*)$ reduces as a special case to Levi's definition of epistemic utility in *Gambling with Truth* (1967).

The concept of truthlikeness should be distinguished from the notion of 'approximate truth'. This concept can be defined by the minimum distance function: $g \in D(B)$ is *approximately true* if $\Delta_{min}(h_*, g)$ is sufficiently small. Hence, in particular, every true statement is also approximately true (but not necessarily highly truthlike).

3. ESTIMATION OF VERISIMILITUDE

If the target h_* is unknown, the degree $\mathrm{Tr}(g, h_*)$ remains unknown as well. However, it is possible to *estimate* this degree, if there is an inductive probability measure P defined on B.

If B is a discrete set, then P is a function with

$$P(h_i) \geqslant 0 \quad \text{for all} \quad i \in I$$

$$\sum_{i \in I} P(h_i) = 1,$$

where $P(h_i/e)$ is the rational degree of belief in the truth of h_i given evidence e. Then the *expected degree of verisimilitude* of $g \in D(B)$ given evidence e is defined by

$$\mathrm{ver}(g/e) = \sum_{i \in I} P(h_i/e)\mathrm{Tr}(g, h_i).$$

(Niiniluoto 1977b). If B is a continuous space, P is replaced by a probabilistic density function p on B, and

$$\mathrm{ver}(g/e) = \int_B p(x/e)\mathrm{Tr}(g, x)\mathrm{d}x.$$

(Niiniluoto 1982). Function ver gives us a comparative concept of apparent verisimilitude: g' *seems more truthlike* than g on evidence e, if and only if $\mathrm{ver}(g/e) < \mathrm{ver}(g'/e)$.

Estimated degrees of truthlikeness are also revisable by new evidence, if we allow the probabilities to change through conditionalization. When new evidence e' is obtained, the value of $\mathrm{ver}(g/e)$ is changed to

$$\mathrm{ver}(g/e \,\&\, e') = \sum_{i \in I} P(h_i/e \,\&\, e')\mathrm{Tr}(g, h_i).$$

Example. Let $\theta_* \in R$ be the unknown value of a real-valued parameter. Then maximizing the expected verisimilitude of a point estimate θ_0 is equivalent to the "Bayes rule" of the Bayesian statisticians: minimize the posterior loss

$$\int_R p(\theta/e)|\theta - \theta_0|\mathrm{d}\theta.$$

(Niiniluoto 1982.) This idea can be extended to the case of interval estimation (Niiniluoto 1986c.)

The properties of function ver (for finite B) include the following:

(a) If $P(h_i/e) = 1/|I|$ for all $i \in I$, then $\mathrm{ver}(h_i/e)$ is a constant for all $i \in I$.

(b) If $P(h_j/e) \approx 1$ for some $j \in I$, so that $P(h_i/e) \approx 0$ for $i \neq j$, $i \in I$, then ver$(g/e) \approx \text{Tr}(g, h_j)$.
(c) ver$(g/e) = 1$ iff, for some $j \in I$, $g = h_j$ and $P(h_j/e) = 1$.
(d) High posterior probability is not generally sufficient for high estimated verisimilitude. For example, if g is a tautology, $P(g/e) = 1$ but ver$(g/e) = 1 - \gamma'$.
(e) It is possible that ver$(g/e) \approx 1$ but $P(g/e) = 0$.
(f) Assume that $P(h_j/e) = 1$, $j \in I_{g_1}$, and $j \in I_{g_2}$. Then *ver*(g_1/e) $>$ ver(g_2/e) if $g_1 \nvdash g_2$, $g_2 \nvdash g_1$.
(g) It is possible that g_1 and g_2 contradict each other, but ver(g_1/e) and ver(g_2/e) are both high.

Result (a) shows that non-trivial comparisons of estimated verisimilitude presuppose a non-even distribution of inductive probabilities on B (cf. also Niiniluoto 1982). Result (b) tells that, if we are completely certain on e of the location h_j of truth h_* in B, then *ver*(g/e) reduces to the value of Tr for g with h_j as the target. Result (c) shows that ver(g/e) can take its maximum value 1 if and only if g is a completely certainly true complete answer.

Results (d) and (e) indicate dramatically the difference between posterior probability and estimated verisimilitude. A logically weak, non-informative hypothesis cannot be appraised as highly verisimilar in spite of its high probability. Further, it may happen that evidence e refutes hypothesis g, but still g is judged to be highly truthlike on e. This very natural condition is violated by Popper's concept of corroboration which receives its minimum value whenever e falsifies g.[9] Similarly, of two hypotheses g_1 and g_2 that are both refuted by evidence e, one may be estimated to be more truthlike than the other – a result that Popper's concept of corroboration could not deliver. These observations show that Popper was wrong in thinking of his degrees of corroboration as indicators of verisimilitude. However, it leaves open the possibility that some measure of the success of a hypothesis g, when g is consistent with evidence e, is directly related to the estimated verisimilitude ver(g/e) of g on e. Indeed, it turns out that in some cases Hintikka's (1968) measure of corroboration corr(g/e) agrees and covaries with the value ver(g/e) (see Niiniluoto 1977).

Result (f) shows that at least in some cases estimated verisimilitude covaries with logical strength. This will be the case if we have strong reasons for thinking that the rival hypotheses g_1 and g_2 are true. On the other hand, this result will not hold generally for rivals that are believed to be false, since the "true" degrees of truthlikeness do not generally covary with logical strength (cf. result (3) above). Hence, comparison of estimated verisimilitude does not as a rule reduce to the problem of content comparison (cf. Watkins 1984).

Finally, result (g) shows that two mutually incompatible theories g_1 and g_2 may both be appraised to be highly truthlike on the same evidence e. (This could not happen with posterior probabilities, as Popper 1972, p. 102, observed.) If g_1 and g_2 are here compatible with empirical evidence e, we have a case where the "empirical underdetermination" of theories does not preclude a rational estimation of their truthlikeness.[10] It is also possible that e refutes g_1 but not g_2 (e.g., g_1 = Newton's theory, g_2 = Einstein's theory), or e refutes both g_1 and g_2, but both are appraised to be truthlike on e.

4. PROBABLE VERISIMILITUDE

The posterior probability $P(g/e)$ is usually interpreted as a rational degree of belief in the *truth* of g on a given evidence e. In A. Shimony's (1970) "tempered personalism", probabilities $P(g/e)$ are instead understood as rational "degrees of commitment" on e to the approximate truth of h.

Given our definition of the concept of approximate truth in Section 2, the idea of *probable approximate truth* can be explicated as follows. Let B again be a cognitive problem with the target h_*. For $g \in D(B)$, $\varepsilon > 0$, let

$$V_\varepsilon(g) = \{i \in I \mid \Delta_{\min}(h_i, g) \leqslant \varepsilon\}.$$

Then the probability given e that g is (within the degree ε) approximately true is

$$PA_{1-\varepsilon}(g/e) = P(* \in V_\varepsilon(g)/e) = \sum_{i \in V_\varepsilon(g)} P(h_i/e).^{11}$$

It is possible that $PA_{1-\varepsilon}(g/e) = 1$ even if $P(g/e) = 0$. However, $PA_{1-\varepsilon}$, just as ordinary probability P, and unlike ver, always covaries with logical weakness.

Another related concept is *probable verisimilitude*. Let $g \in D(B)$, $\varepsilon > 0$, and

$$U_\varepsilon(g) = \{i \in I \mid \Delta^{\gamma\gamma'}_{\mathrm{ms}}(h_i, g) \leqslant \varepsilon\}.$$

Then the probability that the degree of verisimilitude of g is at least $1 - \varepsilon$ given evidence e is

$$PTr_{1-\varepsilon}(g/e) = P(* \in U_\varepsilon(g)/e) = \sum_{i \in U_\varepsilon(g)} P(h_i/e).$$

Then it is certainly the case that "the probability that a universal theory has a positive degree of verisimilitude need *not* be zero", even granting Popper's assumption that its probability is zero (Koertge 1978, p. 276). Indeed, it is possible that $PTr_{1-\varepsilon}(g/e) = 1$ even if $P(g/e) = 0$. Moreover, $PTr_{1-\varepsilon}$ resembles ver in that it does not generally covary with logical weakness. However, it is possible that more than one complete answer $h_i \in B$ has the maximum $PTr_{1-\varepsilon}$-value one on the same evidence, which never happens with ver.

Popper (1972, p. 103), suggests that there is an argument from "an accidentally very improbable agreement between a theory and a fact" to the "(comparatively) high verisimilitude" of the theory. The valid core of this idea is explicated by our function ver. But there is another similar passage where Popper instead speaks of probable verisimilitude. This footnote 165b to 'Replies to my Critics' is worth quoting in full, since here Popper in so many words accepts Lakatos's plea for a whiff of inductivism:

Truthlikeness or verisimilitude is very important. For there is a probabilistic though typically noninductivist argument which is invalid if it is used to establish the probability of a theory's being true, but which becomes valid (though essentially nonnumerical) if we replace truth by verisimilitude. The argument can be used only by realists who not only assume that there is a real world but also that this world is by and large more similar to the way modern theories describe it than to the way superseded theories describe it. On this basis we can argue that it would be a highly improbable coincidence if a theory like Einstein's could correctly predict very precise measurements not predicted by its predecessors unless there is "some truth" in it. This must not be interpreted to mean that it is improbable that the theory is not true (and hence probable that it is true). But it can be interpreted to mean that it is probable that the theory has both a high truth content and a high degree of verisimilitude; which means here only, "a higher degree of verisimilitude *than those of its competitors* which led to predictions that were less successful, and which are thus less well corroborated".

The argument is typically noninductive because in contradistinction to inductive arguments such as Carnap's the probability that the theory in question has a high degree of verisimilitude is (like degree of corroboration) inverse to the initial probability of the theory, prior to testing. Moreover, it only establishes a probability of verisimilitude relative to its competitors (and especially to its predecessors). In spite of this, there may be a "whiff" of inductivism here. It enters with the vague realist assumption that reality, though unknown, is in some respects similar to what science tells us or, in other words, with the assumption that science can progress towards greater verisimilitude." (Popper 1974, pp. 1192–1193.)

I am not familiar with any attempt to reconstruct Popper's line of thought here. In any case, he has not done it himself. Instead of venturing to this task, I wish to formulate a related Bayesian argument, taken from the classical theory of errors (Niiniluoto 1987). It shows how improbable success of a point hypothesis (with probability zero) leads asymptotically to very high appraisals of its probable approximate truth[12] and its estimated verisimilitude.

Let x be a random variable which takes as its values the results of the measurement of an unknown real-valued quantity $\theta \in R$. Assume that the error of x is distributed normally $N(\theta, \sigma^2)$, i.e.,

$$f(x/\theta) = \frac{1}{\sigma\sqrt{2\pi}} e^{-(x-\theta)^2/2\sigma^2}$$

Let y be the average of n observations. Then $f(y/\theta)$ is $N(\theta, \sigma^2/n)$. If the prior distribution $g(\theta)$ is normal $N(\mu, \sigma_0^2)$, but sufficiently flat, then the posterior distribution $g(\theta/y) \propto g(\theta)f(y/\theta)$ is approximately normal $N(y, \sigma^2/n)$. Then the method of estimating θ by the observed average y has the following properties:

$$P(\theta = y/y) = 0 \quad \text{for all} \quad n$$

$$PA_{1-\varepsilon}(\theta = y/y) \to 1, \quad \text{when} \quad n \to \infty$$

$$\text{ver}(\theta = y/y) \to 1, \quad \text{when} \quad n \to \infty.$$

5. FALSE PRESUPPOSITIONS

We have seen above that a fallibilist has rational methods for evaluating the "success" of a hypothesis that is known to be false. However, our discussion has so far presupposed that the relevant cognitive problem B contains true elements, and that the most informative of these truths is chosen as the target h_* in problem B.

To conclude our discussion in this paper, assume that the cognitive problem $B_b = \{h_i | i \in I\}$ is defined relative to a presupposition b (e.g., an idealizing assumption) that is known to be false. Then all the elements of B_b are false, and our cognitive goal is to find the *least false* element of B_b. Formally, the target h_b^* in B_b may be defined as that element of B_b which would be true if b were true.

As, relative to our total evidence e_0, each hypothesis in B_b is known to be false, we have $P(h_i/e_0) = 0$ for all $i \in I$, and hence ver$(g/e_0) = 0$ for all $g \in D(B_b)$.

To apply our method of estimating verisimilitude in this situation, evidence e_0 should be replaced by $e \& b$, where e tells what our actual evidence would have been if the counterfactual condition b had obtained. This proposal is applicable whenever we have reasonably reliable experimental or theoretical methods for calculating e from our actual evidence.

Another approach is to assume (inductively) that the evidence is sufficiently representative, so that it can be, as it were, chosen as the target. Then the least false element of B_b is taken to be the one that is the closest to evidence e. In other words, first we define the distance $D(h_i, e)$ between a hypothesis $h_i \in B$ and evidence e (cf. Niiniluoto 1986a). Then we evaluate on e that $h_i \in B$ is more truthlike than $h_j \in B$ if $D(h_i, e) < D(h_j, e)$.

Example. Let h be a hypothesis that expresses a functional connection between k real-valued quantities $f_i: E \rightarrow R (i = 1, \ldots, k)$:

$$f_k(x) = g(f_1(x), \ldots, f_{k-1}(x)).$$

Let evidence e consist of a finite set of measurements $\{\langle f_1(a_j), \ldots, f_k(a_j)\rangle \mid j = 1, \ldots, n\}$ for n objects $a_1, \ldots, a_n$. Then we may define a Minkowski-type distance between h and e:

$$D(h, e) = \left(\sum_{n=1}^{n} |g(f_1(a_i), \ldots, f_{k-1}(a_i)) - f_k(a_i)|^p \right)^{1/p}$$

where $p \geqslant 1$.

For $p = 2$, $D(h, e)$ is the traditional method of LSD (Least Square Difference).

Examples of this sort are important for the reason that they show the relevance of the idea of truthlikeness to the actual practice of science. Verisimilitude – one of the fruitful notions developed by philosophers of science after the 1965 London Colloquium – is not only an artificial philosopher's concept, but rather a methodological tool for analysing the real-life success of scientific inquiry.

NOTES

[1] Suszko's paper was not included in the Lakatos–Musgrave volume, however. It is a pioneering attempt to apply formal logical tools to the analysis of scientific change. This topic has later become one of the main trends within contemporary philosophy of science. See Stegmüller (1976), Niiniluoto and Tuomela (1979), Niiniluoto (1984), Pearce (1987).

[2] Cf. the title of Michalos (1971).
[3] Popper also accepted the claim that the logical probability of genuine generalizations is zero – but for bad reasons (cf. Niiniluoto and Tuomela 1973; Niiniluoto 1983).
[4] It should be noted, however, that the theory of verisimilitude alone does not give a solution to the problem of incommensurability. To compare two theories T_1 and T_2 for truthlikeness, we need to translate them to a common linguistic framework (Niiniluoto 1984, 1987). For a lucid discussion of the problem of translation, see Pearce (1987).
[5] For reviews of the programme of inductive logic, see Hintikka and Suppes (1966), Carnap and Jeffrey (1971), Cohen and Hesse (1980), Jeffrey (1980), and Niiniluoto (1983).
[6] The recent argument of Watkins (1984) reduces to the old Carnapian position: the only way of guaranteeing that $\alpha > n$ holds in Hintikka's two-dimensional continuum, where the size n of evidence could be *any* natural number, is to choose $\alpha = \infty$. This would lead us back to Carnap's λ-continuum, where the prior probability of all genuine generalizations is zero.
[7] For more details, see Niiniluoto (1987).
[8] Here we assume for simplicity that B is finite, i.e., $|I| < \infty$. For the case of infinite cognitive problems, see Niiniluoto (1986c, 1987).
[9] This was observed already by Lakatos (1974, p. 270). Cf. also Cohen's (1977) concept of support.
[10] The definition of ver(g/e) does not presuppose that evidence e is observational. Niiniluoto and Tuomela (1973) study probabilities of the form $P(g/e \,\&\, T)$, where e is a singular observation statement and T is a theory in a language enriched with new theoretical concepts.
[11] If B is a continuous space, the sum has to be replaced by an integral sign in this definition.
[12] As point hypotheses are maximally informative answers to an estimation problem, this entails also a high appraisal of probable verisimilitude.

BIBLIOGRAPHY

Bogdan, R. J. (ed.), *Local Induction*. Dordrecht: D. Reidel, 1976.

Carnap, R. and Jeffrey, R. C. (eds.), *Studies in Inductive Logic and Probability*, Vol. I. Berkeley: University of California Press, 1971.

Cohen, L. J., *The Probable and the Provable*. Oxford: Oxford University Press, 1977.

Cohen, L. J. and Hesse, M. (eds.), *Applications of Inductive Logic*. Oxford: Oxford University Press, 1980.

Good, I. J., 'Explicativity, Corroboration, and the Relative Odds of Hypotheses', *Synthese,* **30** (1975) 39–73.

Grünbaum, A., 'Popper vs. Inductivism', in Radnitzky and Anderson (1978), pp. 117–142.

Hintikka, J., 'Towards a Theory of Inductive Generalization', in Y. Bar-Hillel (ed.), *Proceedings of the 1964 Congress for Logic, Methodology, and Philosophy of Science*. Amsterdam: North-Holland, 1965, pp. 274–288.

Hintikka, J., 'Induction by Enumeration and Induction by Elimination', in Lakatos (1968), pp. 191–216.

Hintikka, J. and Suppes, P. (eds.), *Aspects of Inductive Logic*. Amsterdam: North-Holland, 1966.

Jeffrey, R. C., 'Probability and Falsification: Critique of the Popper Program', *Synthese,* **30** (1975) 95–117.

Jeffrey, R. (ed.), *Studies in Inductive Logic and Probability*, Vol. II. Berkeley: University of California Press, 1980.

Koertge, N., 'Towards a New Theory of Scientific Inquiry', in Radnitzky and Anderson (1978), pp. 253–278.

Kuipers, T., *Studies in Inductive Probability and Rational Expectation*. Dordrecht: D. Reidel, 1978.

Lakatos, I., 'Changes in the Problem of Inductive Logic', in Lakatos (1968), pp. 315–417.

Lakatos, I. (ed.), *The Problem of Inductive Logic*. Amsterdam: North-Holland, 1968.

Lakatos, I., 'Popper on Demarcation and Induction', in Schilpp (1974), pp. 241–273.

Lakatos, I. and Musgrave, A. (eds.) *Criticism and the Growth of Knowledge*. Cambridge: Cambridge University Press, 1970.

Levi, I., *Gambling With Truth*. New York: Alfred A. Knopf, 1967.

Michalos, A., *The Popper–Carnap Controversy*. The Hague: Martinus Nijhoff, 1971.

Niiniluoto, I., 'Review of A. Michalos, *The Popper–Carnap Controversy*', *Synthese*, **25** (1973) 417–436.

Niiniluoto, I., 'On a *K*-dimensional System of Inductive Logic', in F. Suppe and P. D. Asquith (eds.): *PSA 1976*, Vol. 2. East Lansing: Philosophy of Science Association, 1977a, pp. 425–447.

Niiniluoto, I., 'On the Truthlikeness of Generalizations', in R. E. Butts, and K. J. Hintikka, (eds.), *Basic Problems in Methodology and Linguistics. Part Three of the Proceedings of the Fifth International Congress of Logic, Methodology and Philosophy of Science, London, Ontario, 1975*. Dordrecht: D. Reidel, 1977b, pp. 121–147.

Niiniluoto, I., 'What Shall We Do With Verisimilitude?', *Philosophy of Science*, **49** (1982) 181–197.

Niiniluoto, I., 'Inductive Logic as a Methodological Research Programme', *Scientia: Logic in the 20th Century*. Milano, 1983, pp. 77–100.

Niiniluoto, I., *Is Science Progressive?* Dordrecht: D. Reidel, 1984.

Niiniluoto, I., 'Theories, Approximations, Idealizations', in R. Barcan Marcus, G. J. W. Dorn, and P. Weingartner, (eds.): *Logic, Methodology, and Philosophy of Science VII*. Amsterdam: North-Holland, 1986a, pp. 255–289.

Niiniluoto, I., 'The Significance of Verisimilitude', in P. D. Asquith and P. Kitcher (eds.): *PSA 1984*, Vol. 2. East Lansing: Philosophy of Science Association, 1986b, pp. 531–613.

Niiniluoto, I., 'Truthlikeness and Bayesian Estimation', *Synthese*, **67** (1986c) 321–346.

Niiniluoto, I., *Truthlikeness*. Dordrecht: D. Reidel, 1987.

Niiniluoto, I., and Tuomela, R. *Theoretical Concepts and Hypothetico-Inductive Inference*. Dordrecht: D. Reidel, 1973.

Niiniluoto, I., and Tuomela, R. (eds.), *The Logic and Epistemology of Scientific Change*. Proceedings of a Philosophical Colloquium, Helsinki, December 12–14, 1977. (Acta Philosophical Fennica 30.) Amsterdam: North-Holland, 1979.

Oddie, G., *Likeness to Truth*. Dordrecht: D. Reidel, 1986.

Pearce, D., *Roads to Commensurability*. Dordrecht: D. Reidel, 1987.

Popper, K. R., *Conjectures and Refutations: The Growth of Scientific Knowledge*. London: Routledge and Kegan Paul, 1963.

Popper, K. R., *Objective Knowledge*. Oxford: Oxford University Press, 1972 (2nd ed. 1979.)

Popper, K. R., 'Replies to My Critics', in Schilpp (1974), pp. 961–1197.

Putnam, H., *Reason, Truth, and History*. Cambridge: Cambridge University Press, 1981.

Radnitzky, G. and Anderson, G., *Progress and Rationality in Science*. Dordrecht: D. Reidel, 1978.

Schilpp, P. A. (ed.), *The Philosophy of Karl Popper*. (The Library of Living Philosophers, Vol. XIV, Books I–II.) La Salle: Open Court, 1974.

Shimony, A., 'Scientific Inference', in R. C. Colodny, (ed.): *The Nature and Function of Scientific Theories*. Pittsburgh: The University of Pittsburgh Press, 1970, pp. 79–172.

Stegmüller, W., 'Carnap's Normative Theory of Inductive Probability', in P. Suppes *et al*. (eds.): *Logic, Methodology and Philosophy of Science IV*. Amsterdam: North-Holland, 1973, pp. 501–513.

Stegmüller, W., *The Structure and Dynamics of Theories*. New York, Heidelberg, Berlin: Springer-Verlag, 1976.

Tuomela, R., *Science, Action, and Reality*. Dordrecht: D. Reidel, 1985.

Watkins, J. *Science and Scepticism*. Princeton: Princeton University Press, 1984.

University of Helsinki

JOSEPH D. SNEED

MACHINE MODELS FOR THE GROWTH OF KNOWLEDGE: THEORY NETS IN PROLOG

I. INTRODUCTION

In this paper I shall sketch one way that scientific knowledge or information might be stored in a digital computer and used to model the problem solving activity of empirical scientists. Among the types of problem solving activity that might be modeled in this way is that which is a part of some systematically informed "research program" associated with a scientific community. Thus, what I am proposing is, in part at least, a way of making a computer model the kind of scientific activity first described by Imre Lakatos.[13] This, I hope, makes the paper an appropriate contribution to a conference in his honor.

The theoretical base for my approach to representing scientific knowledge is provided by the structuralist, or model-theoretic, approach to axiomatizing scientific theories.[1,2,3,15,16,19,20]. The implementation of these ideas is envisioned to be in the programming language PROLOG.[6] Regrettably, space limitations force me to assume some familiarity with both structuralist methods and PROLOG. About the latter, it must suffice to say here to those completely unaquainted with PROLOG that it is an attempt to implement a significant fragment of the first order predicate calculus as a programming language. Thus, the PROLOG sentences encountered in this paper may be, as a good approximation, regarded as definitions of set-theoretic predicates in first order logic. A summary of the relevant structuralist methodology appears below.

II. PROBLEMS IN EMPIRICAL SCIENCE

Structuralist representation of scientific knowledge suggests a taxonomy of scientific problems as well as a formal representation of these different kinds of problems. It is instructive to begin with an informal sketch of these ideas. A taxonomy of problems will be considered first. This taxonomy has some intrinsic, intuitive plausibility – apart from the context of structuralist representation. Next, an informal description of how the central

K. Gavroglu, Y. Goudaroulis, P. Nicolacopoulos (eds.), Imre Lakatos and Theories of Scientific Change. 245–268.

problem type is represented with model-theoretic methods will be offered and some indication of its relation to other problem types suggested. In the next section, model-theoretic reconstruction of large fragments of scientific knowledge will be informally sketched and its application to problem representation made explicit.

II.1. *Kinds of Problems in Empirical Science*

Let us consider problem solving in empirical science from the perspective of constructing a computer program to reproduce the full spectrum of expertise possessed by the most sophisticated practitioners of the theory. Such computer programs – expert systems – have been successfully constructed for a variety of knowledge domains.[7,8] For the sake of concreteness, let us consider an expert problem solver for a specific empirical theory T. For example, T might be classical thermodynamics. Let us call our expert system 'T-CONSULTANT'. We may think of T-CONSULTANT either as replicating the expertise of a single individual who knows how to use theory T or as replicating the expertise of the scientific community committed to the practice of T. We may make the envisioned operation of T-CONSULTANT more explicit by categorizing problem solving activities in the following way.

(1) CONCEPTUALIZATION – translating descriptions of specific situations into descriptions in the vocabulary of a specific theory
(2) LAW APPLICATION – applying laws within theories to specific situations described in the vocabulary of a specific theory
(3) LAW DISCOVERY – formulating laws in the vocabulary of a given theory to account for given data
(4) CONCEPTUAL INNOVATION – discovering emendations of the vocabulary of a given theory required to formulate laws to account for given data.

We may think of a CLIENT presenting problems to T-CONSULTANT described perhaps in some vocabulary different from that associated with T. T-CONSULTANT's first task may well be to translate CLIENT's problem into the vocabulary of T. This makes the problem amenable to activity (2) – law application. Law application may plausibly be regarded as the core activity for T-CONSULTANT in the sense that all other activities are means to more effective execution of law application. Clearly, activity (1) – conceptualization – is pursued only as a means to (2). Though it may not be immediately evident, activity (3) – law discovery – is pursued, as we shall shortly see, only for the purpose of making (2) easier. Activities (1), (2) and (3) are similar to what Lakatos called 'a research programme'[13] and what Kuhn[11,12,19] later called 'normal science'. Activity (4) is similar to Kuhn's revolutionary science.

II.2. *Law Application*

Because of its core role, it is expedient to focus on activity (2) – law application. For simplicity, let us suppose task (1) to have been accomplished by CLIENT before it hands the problem to T-CONSULTANT. Specifically, CLIENT presents T-CONSULTANT with a problem described in some sub-set of T-CONSULTANT's own vocabulary. Problems in category (2) that CLIENT hands to T-CONSULTANT have a kind of "canonical form". Roughly, the form is this. Several partially described "applications" of the theory are presented. The "problem" is to complete the description of one or more of the applications. The method of "solution" is roughly this:

(1) identify some "measurement applications" among those given;
(2) use the laws of theory T in measurement applications to "calculate back" to complete descriptions of these applications;
(3) use consistency requirements to transfer features of these complete descriptions to other non-measurement applications.
(4) use the "laws" again, together with transferred features of measurement applications to "calculate forward" to remaining unknown features of these applications.

Measurement Applications

Measurement applications are just those in which the laws of the theory "uniquely" determine the way the given partial description may be completed. What counts as "unique" is generally dependent on the theory. In many cases, it will be possible to identify so-called "T-theoretical" parts of T-CONSULTANT's vocabulary. CLIENT knows nothing of T-theoretical terms and its descriptions of applications *never* contain them. However, they play a role in T-CONSULTANT's laws and are used in its calculations in (2) and (4). In these cases, we may think of (2) as measuring "values" in T-theoretical terms, (3) as transferring these values and (4) as using these transferred values to calculate additional values of non-theoretical terms.

Measurement Identification

This problem solving process can fail in two distinct ways:

(A) No acceptable complete descriptions are found.
(B) Acceptable complete descriptions are not unique.

In case (A), theory T is demonstrated not to be applicable to the given set of "putative" applications. In rather atypical cases where CLIENT insists that T should be applicable to JUST these putative applications, CLIENT would discard T – or fire T-CONSULTANT. More typically, CLIENT might try to adjust the given applications. In situation (B) where the procedure sketched in (1)–(4) does not suffice to uniquely determine complete descriptions of all given applications, we might envision T-CONSULTANT doing the following:

(5) suggest partial descriptions of additional measurement applications whose full descriptions would lead to unique complete descriptions for all of the emended set of applications.

Intuitively, T-CONSULTANT is recommending experiments that CLIENT might do which would provide information sufficient to solve his initial problem.

Text-Book Problems as Law Application

In the special case where T-CONSULTANT is presented with only *ONE* partially described application and the laws of T suffice to uniquely determine the remainder of the description, we obtain one paradigm of a "text-book" problem. Slightly more complex situations in which a small number of applications are *given* and the student is expected to use the laws of the theory in some applications to obtain information, via consistency requirements, he/she needs to make calculations in others occur less commonly as "text-book" problems.

Research Programs and Measurement Identification

The more general case of this type of problem, which rarely appears in text-books, is that in which the *given* information is not sufficient to permit the inference or calculation of complete descriptions with the laws and consistency requirements T. Here, the essence of the problem is determining a "research program" – a sequence of "experiments" designed to yield the information needed to "fill-out" the given incomplete descriptions. These problems become more interestng when a kind of "cost function" for experiments is given and the problem becomes one of resource allocation to research programs. Substantial and complex historical examples of "problems" in basic science with this same logical structure include: determination of planetary masses in Newtonian mechanics; determination of atomic weights and molecular structures using Daltonian stoichiometry; determination of properties of elementary particles and the structure of nuclei with theories of high energy physics; calculation of energy band structure of specific "doped" solids using theories of solid state physics.

II.3. *Law Discovery*

Special Laws and Measurement

Up to this point, our sketch of problem solving assumes that the same laws apply to all applications of theory T. This assumption is not generally plausible. Many theories, such as classical particle mechanics, have several laws associated with them. Applications of these theories are "sorted" according to which laws are claimed to hold for them. The laws in such theories may be ordered by a "specialization relation". For example, in classical particle mechanics the law of gravitation is a specialization of Newton's third law in that all gravitational forces are central, action–reaction pairs, but not conversely.

Reconsider the problem solving process just sketched. Call the laws of T 'L'. The laws L apply to all applications. Suppose the situation is changed by adding laws L' that are a specialization of L and apply just to some applications. If some of the given applications are among the applications of L' we may well have an easier problem to solve than the initial one. The additional requirements L' may just happen to turn some given applications into measurement applications that were not such before. This suggests a way that law discovery may be regarded as a means to making law applications easier. Roughly, the more and/or stronger laws you have, the easier it is to use them to fill out partial descriptions. From a somewhat utilitarian perspective on the growth of scientific knowledge, the activity of law discovery is a kind of "overhead" for the activity of law application.

II.4. *Conceptual Innovation*

A somewhat similar point can be made about conceptual innovation. In the face of repeated failures or significant difficulties in solving CLIENT's problems with T, we might envision T-CONSULTANT experimenting with incremental and even radical modifications in its vocabulary. One kind of incremental vocabulary modification is the addition of new theoretical terms. For example, T might be a theory about billiard ball collisions using only the concept of velocity. Such a theory could be successful, but quite labor intensive to use. A rather sophisticated T-CONSULTANT might be able, at some level of "dissatisfaction", to invoke a procedure to search for a new theoretical term. In this case, this procedure might be expected to come up with collision mechanical mass. It should be noted here that conceptual innovation and law discovery are, in actual historical examples, often difficult to distinguish. This is most often the case in "revolutionary" theory change in which new concepts and laws formulated with them appear simultaneously.

III. THEORY NETS

The taxonomy of problems just described follows naturally from a model-theoretic reconstruction of the "global" structure of scientific knowledge. Globally, scientific theories are a "net" of "local" theories – called 'model elements' connected by "intertheoretical links".[2,3,21] This global picture can be exploited to provide an account of the "intended applications" of model elements as well as a formulation of the "empirical claim" made about these intended applications using the model element.

III.1. *Model Elements*

A model element consists of two categories – potential models and models. The potential models characterize the conceptual apparatus, or vocabulary associated with a specific part of empirical science. The models – a sub-category of the potential models – characterize the laws associated with this part. Model elements are functionally analogous to "frames" in some formal representations of knowledge common to artificial intelligence work.[18] They differ from frames chiefly in that they are constructions in a traditional set-theoretic or category-theoretic apparatus. Both potential models and models are needed because we will frequently consider situations in which the same conceptual apparatus is used to formulate different laws. In the present discussion, the morphisms associated with these categories will largely be ignored and they may be viewed simply as model classes. A fully adequate treatment would require consideration of the morphisms.

III.2. *Intertheoretical Links*

These local theories or model elements are connected by intertheoretical links. Intuitively, intertheoretical links are relations between theories – analogous to what some have called 'bridge laws' and 'correspondence rules' – that correlate the values of components in potential models of different model elements. For example, simple equilibrium thermodynamics is linked to Daltonian stoichiometry via the mole number function, to classical hydrodynamics via the pressure function and to physical geometry via the volume function. Dynamical theories like Newtonian particle mechanics are linked to "underlying" kinematical theories that specify acceptable means for measuring kinematic functions like time and distance. Classical population genetics is linked to molecular genetics by "identifying" genetic properties with kinds of molecular structures. Several types of inter-theoretical links such as specializing links, reducing links and bi-lateral reducing, or equivalence, links have been considered which correspond to well-known relations among theories.

External and Internal Links

The concept of intertheoretical link is general enough to describe these and other examples of a model element's "external links" with other model elements, as well as "internal links" among the model element's own potential models. Internal links describe the ways in which some components of the element's models have related values in other applications. These relations are an "essential" feature of the element's conceptual apparatus, on a par with the empirical laws. For example, the fact that mass in classical collision mechanics is regarded as an "intrinsic" property of particles may be expressed as an internal link. These internal links represent the "consistency requirements" among different models of the same theory mentioned above and provide an alternative way of viewing what were called 'constraints' in other contexts.[16] For present purposes, it is convenient to regard internal links or constraints as an essential part of a model element.

Model Elements in PROLOG

The following is a sketch of how model elements might be described in Clocksin–Melish PROLOG.[6] Here and in subsequent discussions of PROLOG implementations, the PROLOG procedures mentioned are "sketches" in the sense that sub-procedures appearing in them have not actually been written. The entire discussion is a "sketch" in the sense that implementations of the simplest ideas only are described.

The classes of models which are objects of the categories appearing in model elements K may be represented by PROLOG procedures of the form:

```
potential_model(K,Mp)   :— . . .  .
model(K,M)   :— . . .  .
constraint(K,C)   :— . . .  .
```

where K will be instantiated to the *name* of the model element – e.g. "classical_collision_mechanics"–and the procedures will define set-theoretic structures of the appropriate kind.

For example, we might consider a model element "equivalence" whose potential models were binary relational structures and whose models were equivalence structures.

```
potential_model (equivalence,P)   :—
                    relstr2(_,P).
```

where relstr2(M,L) if L is a binary relation structure over set M:

```
relstr2([M,S])   :—
      genset(M),
      cross(M,M,Z),
      gensub(Z,S).
```

"genset" generates sets of individuals from some set of "urelementen", "cross" generates the Cartesian product of two sets and "gensub" generates sub-sets.

Models for "equivalence" would then be characterized in the following way:

```
model(equivalence,M)   :—
       equivalence(_,M).
```

```
eqv(M,R)   :—
            relstr2([M,R]),
            ref_ax(M,R),
            sym_ax(M,R),
            trn_ax(M,R).

ref_ax(M,R)   :—
            not(no_ref(M,R).

no_ref(M,R)   :—
            mem(X,M),
            not(mem([X,X],R)).

sym_ax(M,R)   :—
            not(no_sym(M,R).

no_sym(M,R)   :—
            mem(X,M),
            mem(Y,M),
            not(mem([X,Y],R)).

trn_ax(M,R)   :—
            not(no_trn(M,R)).

no_trn(M,R)   :—
            mem([X,Y],R),
            mem([Y,Z],R),
            not(mem([X,Z),R).
```

In most interesting cases, it will be relatively straightforward to produce procedures that will *test* candidates for the set-theoretic structures in question. It will be somewhat more difficult to *generate* such structures in a useful way. One problem is simply that these structures usually involve real valued functions and their model classes will have uncountably many members. Any implementation strategy must use generation of models with restraint and/or insight. These procedures may be viewed as "intentional descriptions" or definitions of "set theoretic predicates" characterizing the model classes we have used in other contexts.[16]

It is convenient to have "constraint" procedures work in the following way.

```
constraint(K,X1,X)   :— . . .  .
```

will succeed only if list X satisfies the constraint and individual $X1$ can be added to list X yielding a list that also satisfies the constraint. That is, the list $[X1{:}X]$ satisfies the constraint. Thus, "constraint" procedures will be defined recursively. Intuitively, they will be defined as principles for "constructing" bigger lists of potential models from smaller ones. This has some intuitive appeal in representing the growth of applications of theory elements. That all "real life" constraints can be formulated this way is not entirely obvious.

This formulation of model elements takes the basic individuals to be individual set-theoretic structures. An attractive alternative would take isomorphism equivalence classes to be the basic individuals. Thus potential models and models would be classes of "empirically equivalent" structures. Intuitively, they would be things like "physical systems" in a somewhat more robust sense. Technical details of this approach are discussed in Sneed[17].

Theory Nets PROLOG

The simplest conception of theory nets is that they are simply graphs. This is consistent with our earlier treatment of nets[1,15] where we suppose that all intuitive links between $K1$ and $K2$ have been rolled into one. These graphs might be represented in a PROLOG data base by statements of the following form:

```
link(K1,K2).
```

where $K1$ and $K2$ are instantiated to names of model elements whose components – potential models, models and constraints – are defined in the same PROLOG data base by procedures of the form described above. Initially, we don't need to say anything about the specific properties of these graphs. However, some crucial special cases – interpretation nets – will be directed graphs. That is, if link($K1$,$K2$) is in the data base then link($K2$,$K1$) is not.

A more useful picture of a theory net would be provided by a PROLOG data base with statements of the following form:

```
link(K1,K2,L12).
```

where $L12$ is instantiated to the *name* of an extensional description of a link between $K1$ and $K2$. That is, there is some procedure:

```
L12(X1,X2) :–
    potential_model(K1,X1),
    potential_model(K2,X2).
```

that characterizes ordered pairs of $K1$ and $K2$ potential models in the specific link.

Note that, on our present conception of nets, intuitively unlinked model elements are linked by the vacuous link that contains all possible ordered pairs of potential models. Thus, for intuitively unlinked $K1$,$K2$ we need to have: link($K1$,$K2$,vacuous12).

```
vacuous12(X1,X2) :–
    potential_model(K1,X1),
    potential_model(K2,X2).
```

in the PROLOG data base.

Content of Theory Nets

The global structure of empirical science is characterized as a "theory net" – a set of model elements together with certain configurations of inter-theoretical links connecting them. The "content" of a theory net contains configurations of structures that are compatible with both the laws and the links in the net. Intuitively, the model classes tell us what potential models are empirically possible in the absence of links. Links tell us what combinations of potential models are empirically possible. Together, they tell us what combinations of models are empirically possible. Very roughly, the content of a theory net is the "extension" of the net viewed as something like a predicate used to make a claim about "the world". Formally, the members of the content are just classes of models of model elements in the net which exhibit the relational structure of the links in the net.

III.3. *Interpretation Nets*

It remains to say how the formalism just sketched can be used to say something about what the world is like. The fundamental idea is that certain kinds of links – interpreting links – between the mathematical building blocks – model elements – may provide empirical interpretation for those model elements. This interpretation characterizes the intended applications of the model elements – what the model element talks about.

Interpreting Links

Generally, intertheoretical links serve to carry information about the values of relations and functions from the applications of one theory to those of another or across different applications of the same theory. Some of a theory element's external links serve to "import" information from other theories that limit or determine the values of at least some of the functions and relations in the element's intended applications. Intuitively, an interpreting link is a formal description of a situation in which one kind of system – a measuring device – is used to determine the values of some quantities appearing in the laws of a theory. For example, clocks and measuring rods – kinematic measuring devices – are used to determine the values of the kinematic functions appearing in the laws of mechanical theories. More formally, an interpreting link correlates the values of some components in an interpreting model element – whose models are intuitively measuring devices – with the values of some components in an interpreted model element – whose potential models contain components whose values are measured by the measuring devices that are models for. Having some interpreting external links of this sort appears to be a necessary condition for a model element to be associated with an empirical theory. Such links provide a kind of "physical semantics" for a model element that make it more than "just a piece of mathematics".

Interpreting links illuminate and generalize earlier treatments[16] of intended applications and the distinction between theoretical and non-theoretical components in model elements. General necessary conditions for intertheoretical links to be amenable to this kind of intuitive interpretation can be provided. However, which links actually appear in empirical science as interpreting links between model elements – just as the question of which model elements actually appear – is a question about the practice of science that must elude completely formal characterization. How interpreting links appear together in interpretation nets – a special case of theory nets – can be described. The essential role of interpretation in empirical science may then be described by telling how interpretation nets must appear as sub-nets in any model element net that represents a significant fragment of empirical science at a given time.

Local Empirical Claims

The local empirical claim of a model element may be described by characterizing the "intended applications" of a model element K in an interpretation net N. The fundamental idea is that the intended applications of K are models for model elements that interpret K which satisfy all the requirements of the net N that do not, in some way, "presuppose" the laws of K. This conception of intended application is holistic in that we must look beyond just laws in the model elements that "directly" interpret K to determine what is demanded of intended applications for K. Indeed, we must, in principle at least, consider all the rest of empirical science – excluding only those parts in which the laws of K are required to hold.

The local empirical claim of a model element K is roughly that K's intended applications in all its interpreting model elements are all together linked to models for K. More intuitively, the data gleaned from measuring values of all components in K's potential models whose values may be determined using measuring instruments that are not required to satisfy K's laws, is consistent with the laws of K.

Generally, the model elements that interpret K will only provide information about some proper subset of the components in K. The other components are "uninterpreted" and perhaps "K-theoretical". Roughly, the uninterpreted components in K function in the laws of K as, perhaps essential, conceptual tools in formulating K's claim even though the claim is not "about" them. However, the uninterpreted components in K generally have a larger role than this in that their values may in turn be correlated with the values of components in model elements that K interprets. Thus, the distinction between interpreted and uninterpreted components is relative to a single model element. In the global picture of empirical science it becomes blurred to the point that all empirical concepts appear to have the same epistemological and ontological status.

Local Content in Two Member Theory Nets in PROLOG

Consider first the case of a two member net in which the link is intuitively a "theoretizing interpreting link" connecting the "non-theoretical" $K1$ with the theoretical $K2$. For example, $K1$ might be the kinematic part of a mechanical theory and $K2$ the "full" theory including the dynamical components. This net is described simply by a statement of the form:

```
link(K1,K2,L12).
```

in a PROLOG data base that contains procedures of the kind described above for potential models, models and constraints for $K1$ and $K2$ as well as for the link$L12$.

Intuitively, the "local content" of $K1$ in this net consists of lists of models of $K1$, satisfying the constraints of $K1$ – an intended application of $K2$ – that must be "filled out" with theoretical components yielding a list of models of $K2$, satisfying the constraints of $K2$. The following procedure generates members of the local content of $K1$.

```
content(K1,[X1:X])  :-
        content(K1,X),
        model(K1,X1),
        constraint(K1,X1,X),
        L12_list([X1:X],[Y1:Y]),
        model(K2,Y1),
        constraint(K2,Y1,Y).
```

where $L12$ linking of lists "$L12$_list" is defined in the obvious way in terms of $L12$ linking of individual potential models in the obvious way:

```
L12_list([X1:X],[Y1:Y])  :-
        L12(X1,Y1),
        L12_list(X,Y).
L12_list([ ],[ ]).
```

Note however that adding the following "net terminating conditions" to the data base:

```
link(K2,[ ],vacuous2[ ]).
vacuous2[ ],(_,[ ]).
content([ ],[ ]).
```

to indicate that $K2$ is not forward linked to anything else allows us to write the "content" procedure in a recursive form as:

```
content(K1,[X1:X])  :-
        content(K1,X),
        model(K1,X1),
        constraint(K1,X1,X),
        link(K1,K2,L12),
        L12_list([X1:X],[Y1:Y]),
        content(K2,[Y1:Y]).
```

When this procedure calls "content($K2$,[$Y1$:Y])", the only sub-goals that matter are "model($K2, Y1$)" and "constraint($K2, Y1, Y$)". The remainder automatically succeed because of the net terminating conditions.

Local Content of Three Member Theory Nets in PROLOG

Consider the three member theory net described by:

```
link(K0,K1,L01).
link(K1,K2,L10).
```

and appropriate procedures defining structures and links.

In the case that these are intuitively interpreting links, we may see this as a situation in which the intended applications of $K2$ are $K1$-models that are also "backward linked" to $K0$-models. For example, $K0$-models provide acceptable means of measuring or determining values of components in $K1$-potential-models. With this intuitive understanding of the interpretation net, a plausible procedure for the local content of $K1$ is – assuming appropriate "net terminating conditions":

```
content(K1,[X1:X])  :—
        content(K1,X),
        model(K1,X1),
        constraint(K1,X1,X),
        link(K0,K1,L01),
        L01([Z1:Z],[X1:X]),
        content(K0,[Z1,Z]),
        link(K1,K2,L12).
        L12_list([X1:X],[Y1:Y]),
        content(K2,[Y1,Y]).
```

Intuitively, lists in the $K1$-content must be linked "back" to lists in the $K0$-content and "forward" to lists in the $K2$-content.

Local Content in General Theory Nets in PROLOG

The basic intuitive idea here is that one may use recursion to determine the local content of any model element in the net. The fact that the net is a *directed* graph assures that the recursion will terminate at the "boundaries' of the graph.

A first cut at this requires that we deal with the possibility that a single model element K may be linked – both forward and backward – to multiple model elements. Intuitively, we want to require that a list X in the K-content is linked to elements in the content of all the model elements that are linked to K. To do this universal quantification in PROLOG we must resort to the – "not (there exists K' not . . .)" locution.

```
content(K,[X1:X])   :-
        content(K,X),
        model(K,X1),
        constraint(K,X1,X),
        all_good_link(K1,K,[X1:X]),
        all_good_link(K,K1,[X1:X]).

all_good_link(K1,K,[X1:X])   :-
        not(bad_link(K1,K,[X1:X]).

bad_link(K1,K,[X1:X])   :-
        link(K1,K,L),
        not(L([Z1:Z],[X1:X]),content(K1,[Z1:Z])).

all_good_link(K,K1,[X1:X])   :-
        not(bad_link(K,K1,[X1:X])).
   bad_link(K,K1,[X1:X])   :-
        link(K,K1,L),
        not(L([X1:X],[Z1:Z]),content(K1,[Z1:Z])).
```

Note that this formulation obviates the need for net terminating conditions since "all_good_link" will succeed when there are no links.

A finer cut at this requires that we have some way of deleting the forward cone of *K* from the net pertinent to the "input elements" *K*. One way to do this is to make the "content" procedure explicitly relative to specific nets. Then the content of interpreting nets would be evaluated relative to the net obtained from the original net by deleting the forward cone of *K*. This appears to require that our data base, at least temporarily, keep track of more than one net. While this is not impossible to implement, a more elegant solution would be desirable.

IV. PROBLEMS REPRESENTED IN THEORY NETS

Very roughly, the *theoretical* knowledge that a T-CONSULTANT – person or machine – has about an area of empirical science at a specific time may be represented as a sub-net *Np* of some theory net *N* representing "full theoretical knowledge" in this area. Generally, the sub-net *Np* will contain several theory elements together with interpreting links, specialization links, etc. In addition to this theoretical knowledge, T-CONSULTANT typically has some *pragmatic* knowledge representable as search heuristics in the theory net. Finally, T—CONSULTANT may have (or be given by its CLIENT) full and partial descriptions of some specific models for theory elements in N. T-CONSULTANTS's job is to use its theoretical knowledge to extend the partial descriptions to full descriptions. In doing this T-CONSULTANT will naturally find itself doing each of the kinds of things described intuitively as problem types in Section II.1. The more formal representation of these problem types in terms of theory nets suggests one way to design an expert system for solving them.

IV.1. *Conceptualization*

Conceptualization may be modeled in the theory net formalism provided one assumes that T-CONSULTANT's net Np includes theory elements that represent the vocabulary used by CLIENT in describing his problems. That is, we must assume that "everyday" "ordinary language" ways of describing these problems can be formally represented as a theory element E – known to T-CONSULTANT. In addition, T-CONSULTANT knows about links between E and other theory elements with richer structure representing "scientific" theories. These are links that "translate" CLIENT's ordinary language description of his/her problem into some vocabulary that makes it amenable to "law-application". There is more to be said about a model theoretic formulation of conceptualization. However, these matters are not the primary focus of this paper.

IV.2. *Law Application*

Model theoretic formulation of law application is straightforward. Here one is typically dealing with a single theory element, its specializations and the internal links among their models. Interpreting connections to other theory elements in the net may be ignored. However, reduction and equivalence links may play a role in some cases.

Problems here are simply partial descriptions of putative models for a single theory element and its specializations. Easier problems just require using the laws in the specialization net, together with the consistency requirements, to fill out the given partial descriptions in the manner of (1)–(4) Section II.2. In many cases, problems of this sort may have an algorithmic solution (Section II.4). However, this algorithm may be difficult to formulate generally so that it can be applied mechanically to all problems of this type arising within a given specialization net. That is, different algorithms may work depending on the configuration of special laws required for a specific problem. Even when an algorithmic solution is feasible, it may not correspond to problem solving processes used by human experts. For pedagogic purposes, and perhaps for others as well, heuristics may be more useful than algorithms in describing T-CONSULTANT's knowledge about this kind of problem solving.

More difficult problems here require identifying measurement experiments and planning research programs (Section II.5). Viewed model theoretically, this amounts to looking for models of the theory element in question and its specializations in which information available from interpreting links, together with the laws of the theory element suffice to determine uniquely the values of components not available from interpreting links – the theoretical components in the theory element. Values determined in this way are then "imported" into the given problem via consistency requirements that operate across different models of the same theory element. The fact that consistency requirements are "inherited" in specialization nets makes this generally possible. In some cases, consistency requirements that operate over entire specialization net in a non-hierarchical way are required to depict the connection that actually obtain between different parts of the net. In some cases, reduction in equivalence links may also afford measurement possibilities.

T-CONSULTANT's knowledge in his area consists of:

(A) algorithms or heuristics for identifying information required to solve the given problem;
(B) algorithms or heuristics for identifying theory elements and partial descriptions of some or their models that afford measurement methods for the required information.

In many cases question (A) will not have a unique answer. That is, several different sets of additional information will suffice – together with laws and consistency requirements – to fill out the given descriptions. Heuristic principles for choosing among these appear to be the most plausible way of representing T-CONSULTANT's knowledge here. In some cases, the search space for (B) will be characterized quite sharply. For example, mathematical characterizations of models and data configurations that afford uniqueness may be available. But even here (B) will not have a unique solution, except perhaps in some trivial cases. Indeed, some additional constraints, for example "cost" may be required to make the search space for (B) interesting – not to say tractable. In other cases, it may be impossible even to characterize the search space sharply. Some kind of heuristically guided "generate and test" may be the best that T-CONSULTANT can do in these cases.

Law Application in PROLOG

Law application can be represented at a rather general level in a PROLOG data base describing a theory net. To see how this works, consider the PROLOG description of a two member theory net described above (Section III.2). Suppose that, in addition to descriptions of the model elements and links, the data base keeps track of things that users have found to satisfy both

content(K1,X) and content(K2,Y).

More precisely, suppose list

$x = [xn, \ldots, x1]$

is intuitively the successful applications of $K1$ discovered up to now and list

$y = [yn, \ldots, y1]$

contains some list of theoretical emendations that is linked to x. Intuitively, y contains the theoretical emendation of x that suffice to show that x is in the content of $K1$. In the simplest cases, there will be just one content statement about x while there may be multiple statements about theoretical emendations of x. When x and y are conceived as sets of isomorphism equivalence classes the multiple y's are alternative values of theoretical concepts consistent with the data x.

Suppose assertions of the form:

```
content(K1,x)  :—  !.
content(K2,y)  :—  !.
```

appear at the "top" of the PROLOG data base so that when "content" is called, they are checked before the rule for content. Intuitively, these clauses embody the users current knowledge about the "range of applications" of $K1$ and $K2$. A full implementation of these ideas would contain procedures for updating these records as new successful applications of model elements were discovered. Most of the search procedures suggested above in Section III will only be feasible in an environment where these records are rather rich and many branches terminate in "off the shelf" knowledge of previous successful applications.

Queries of this PROLOG data base correspond to various types of law application. Consider first the case of checking whether a new application may be added to already existing applications.

(1) Testing to see if a specific putative non-theoretical application x_n+1 can be added to x is done with the query:

?-content(K1,[x_n+1:_]).

This query will first find that x is in the content of $K1$, check to see if x_n+1 is a $K1$-model, then whether $[x\_n+1{:}x]$ satisfies $K1$-constraints. Then it will look for some y_n+1 that is linked to x_n+1, check to see if y_n+1 is a $K2$-model and finally whether $[y\_n+1{:}y]$ satisfies the $K2$-constraints. Suppose, as in the most common case, x_n+1 is a $K1$-model and $[x\_n+1{:}x]$ satisfies the $K1$-constraints. As things are set up here, PROLOG will just keep generating $K2$ potential-models until the query succeeds – which, of course, may be never. More subtle treatments would incorporate some kind of terminating condition that would fail the query at some point.

Note that, even without the "cut" after "content($K1$,x)" this would never try other values besides x for the anonymous variable slot in $[x\_n+1{:}\_]$. Without the "cut" after "content($K2$,y)", it will generally try other values besides y to satisfy the anonymous variable slot in $[y\_n+1{:}\_]$. For any reasonable examples, trying something other than x without first trying something other than y looks dumb. In interesting cases, they will be interpretable as more precise information about the values of theoretical components – revealed by data from x_n+1. However, it is not completely clear that – in the face of repeated failure – one should not try something besides x. Intuitively, this form of the query models a very conservative growth policy for applications – "once established, never give up an application". How to model less conservative growth policies is an open question.

Consider next the case of searching for new applications to be added to the stock of already existing applications.

(2) Looking for new non-theoretical applications.

?-content(K1,[X1:X]).

Again this query will first instantiate X to the "known $K1$-applications" x. Then it will start generating $K1$ models, proceeding as the above query for each one generated until it succeeds – or the user turns it off. This is clearly a pretty mindless way to run what might be called the "internally generated research program" associated with a model element. Something besides the convenience of the designer of the model generating procedure should be guiding what is tried. An obvious suggestion is that properties of the known applications x should "somehow" suggest what to try next. The same suggestion has plausibility at the theoretical level as well. Choice of linked $K2$ potential models should be guided by existing theoretical emendations. Minimally, constraint satisfaction can probably be built into the generating procedure.

A special case of this kind of query is one in which $X1$ is a "partially described" $K1$-model. That is, some components or parts of components have been instantiated to "known values" while others remain variables. This is a somewhat more general formulation of the "text book" and "research" planning problems discussed above. Simply, generating $K1$-models by instantiating the variables and then testing to see whether they can be "emended" to $K2$-models satisfying $K2$-constraints is clearly not very satisfactory. However, it is not clear that satisfactory procedure for such problems can be provided at this level of generality. The research proposed here may be conceived as addressing this problem for a specific, very simple example.

IV.3. *Law Discovery*

Model theoretically, law discovery amounts to enlarging a given theory net by adding specializations of existing theory elements. This must be done in such a way that theory elements representing special laws appear as well as theory elements that interpret these. That is, the emended theory net tells us about special laws *and* to what they are intended to apply.

At this point, the model theoretic formulation of law discovery does not appear to yield any specific new insights into processes T-CONSULTANT might have for discovering laws. However, it does suggest that any account of this kind of "expertise" that does not address the question of identifying the range of application of the laws discovered cannot be fully adequate. Simon *et al*'s work with BACON.5[4,5,14] is at least suggestive of how one might design PROLOG processes to search systematically for specializations of a given model element once the appropriate specialization its interpreting model elements has been provided. This, in turn, suggests that law discovery, like law application, should be modeled as a recursive process that propagates through a theory net.

From this perspective, the problem of special-law discovery involves two components: searching for a sub-class of applications to which the law is to apply; and searching for the law. Simon *et al.*'s BACON attacks the second component, essentially assuming the first component has been solved.[14] The first component appears to be equally important and lacking an obvious solution. For example, from data about momentum conserving collisions including both elastic and inelastic collisions, how might one identify the elastic ones as a prelude to the search for the special-law – energy conservation – that applies to them? A number of plausible search procedures suggest themselves in this example. More generally, however the "utilitarian" perspective on special-laws suggests that, whatever procedures T-CONSULTANT might have for law discovery, they should be triggered by some measure of "work load" in law application.

IV.4. *Conceptual Innovation*

Model theoretically, conceptual innovation amounts to enlarging a given theory net by adding new theory elements that contain conceptual elements that do not appear in old theory elements. These new conceptual components may be linked to components in the old theory elements, but they need not be. Of course, these new elements must be linked in some way to the old, but the links may be quite tenuous – as in the case of "revolutionary" conceptual innovation.

Again, the model theoretic formulation offers no specific new insights into specific processes that might search for conceptually innovative theory elements. BACON.5[14] shows how conceptual innovation may be coupled with law discovery in a restricted class of situations. Essentially, the restriction is that the theoretical concept discovered be explicitly definable in all models. That is, BACON.5 discovers the mass concept in two particle collision models. It would apparently also discover this concept in other models in which mass was uniquely determined by velocities. It would not discover this concept by looking at data about collisions in which mass was not uniquely determined by velocity. Obviously, BACON.5 needs to know how to generalize what it has learned about simple collisions to more complicated collisions and check whether the generalization holds. This emendation appears to be feasible.

Whether it serves as the basis for some more general procedure that searches both for specializations and conceptually innovative extensions of a theory net remains to be explored.

IV.5. *Implementation in Concurrent PROLOG*

Concurrent PROLOG[9,22] is designed to allow certain designated procedures to be run concurrently by separate "parallel" processors. Intuitively, this appears to be a very natural way to implement a model element net representing connected parts of empirical science. Roughly, for each different theory element K, "content(K,X)" would be processed concurrently with some kind of control procedure assigning priorities to calls received from other linked processes. This appears to provide a quite natural model scientific activity in which various parts of science – model elements – pursue their own research agenda – generate additional successful applications – in an environment in which they depend on results from neighboring parts of science and, in turn, provide results "on demand' to other parts of science. A somewhat similar picture of the practice of empirical science as a "metaphorical" model for control in parallel processing is provided by Kornfeld and Hewitt.[10] The suggestion here is to turn the metaphor around and model scientific practice by specific control procedures in a parallel processing situation. To make this more than a metaphor would require explicit processes for the "internal research program" associated with each model element as well as processes for setting priorities among internally and externally generated calls.

BIBLIOGRAPHY

1. Balzer, W., and Sneed, J. D., 'Generalized Net Structures of Empirical Theories, I and II', *Studia Logica* **XXX** (3), 1977, 195–212; and *Studia Logica* **XXXVII** (2), 1978, 168–194.
2. Balzer, W., Moulines, C.-U. and Sneed, J. D., 'The Structure of Empirical Science: Local and Global', to appear in *Proceedings of the 7th International Congress of Logic, Methodology and Philosophy of Science, 1983*. North Holland, Amsterdam.
3. Balzer, W., Moulines, C.-U. and Sneed, J. D., *An Architectonic for Science*. Reidel, Dordrecht, 1987.
4. Bradshaw, G., Langley, P., Simon, H., 'Studying Scientific Discovery by Computer Simulation', *Science*, 2 December 1983, **222**, No. 4627, 917–975.
5. Bradshaw, G., Langley, P., Simon, H., 'Rediscovering Chemistry with the BACON

System', in R. Michalski, J. Carbonell, and T. Mitchell (eds.) *Machine Learning: An Artificial Intelligence Approach*. Tioga, Palo Alto, CA., 1983.

6. Clocksin, W. and Mellish, C., *Programming in PROLOG*. 2nd ed., Springer, Berlin, 1985.
7. Duda, R., and Shortliff, E., 'Expert Systems Research", *Science*, 15 April 1983, **220**, No. 4594, 261–268.
8. Gasching, J., 'Application of the PROSPECTOR System to Geological Exploration Problems', *Machine Intelligence*, **10**, 1988, 301–323.
9. Hogger, C., 'Concurrent Logic Programming', in K. Clark and S.-A. Taernlund (eds.) *Logic Programming*. Academic Press, New York, 1982.
10. Kornfeld, W. and Hewitt, C., 'The Scientific Community Metaphor', *IEEE Transactions on Systems, Man and Cybernetics*. SMC-11, 1, 1981.
11. Kuhn, T. S., 'Theory-Change as Structure-Change: Comments on the Sneed Formalism', *Erkenntnis,* **10**, 1976, 179–199.
12. Kuhn, T. S., *The Structure of Scientific Revolutions*. University of Chicago Press, Chicago, 1962.
13. Lakatos, I., 'Falsification and the Methodology of Scientific Research Programmes', in I. Lakatos and A. Musgrave (eds.), *Criticism and the Growth of Knowledge*. Cambridge, 1970, pp. 91–195.
14. Langley, P., Bradshaw, G., and Simon, H., *BACON.5: The Discovery of Conservation Laws*. CIP Working Paper No. 430, Department of Psychology, Carnegie Mellon University, Pittsburgh, PA, 1981.
15. Moulines, C.-U., 'Theory-Nets and the Evolution of Theories: The Example of Newtonian Mechanics', *Synthese,* **4**, 1979, 417.
16. Sneed, J. D., *The Logical Structure of Mathematical Physics*, revised edition. Reidel, Dordrecht, 1979.
17. Sneed, J. D., 'Invariance Principles and Theoretization', in I. Niiniluoto and R. Tuomela (eds.), *The Logic and Epistemology of Scientific Change*. North Holland, Amsterdam, 1979.
18. Sowa, J. F., *Conceptual Structures: Information Processing in Mind and Machine*. Addison-Wesley, Reading, MA., 1984.
19. Stegmueller, W., *Theorienstrukturen und Theoriendynamik*. Springer, Berlin, 1973.
20. Stegmueller, W., *Theorie und Erfahrung*. Springer, Berlin, 1970. English translation by W. Wohlhutter, *The Structure and Dynamics of Theories*. 1976.
21. Stegmueller, W., *Theorie und Erfahrung: Band II, Dritter Teilband; Die Entwicklung des neuen Strukturalismus seit 1973*. Springer, Berlin, 1986.
22. van Emden, M., and de Lucena Filho, G., 'Predicate Logic as a Language for Parallel Programming', in K. Clark and S.-A. Taernlund (eds.), *Logic Programming*. Academic Press, New York, 1982.

Colorado School of Mines

ARISTIDES BALTAS

LOUIS ALTHUSSER AND JOSEPH D. SNEED: A STRANGE ENCOUNTER IN PHILOSOPHY OF SCIENCE?

INTRODUCTION

Given the title of the present paper, I feel singularly privileged by having to speak here today just after Professor Sneed and just before Professor Balzer. This can be well understood and I leave it at that. However, despite (or, rather, because of) this privilege, my feelings are not unmixed – to say the least. The question-mark in my title – implying that my conclusions are not definite (and hence that my presumed Marxism is probably not of the dogmatic variety) – as well as the adjective "strange" there – intending to predispose my audience favourably by inciting it to keep an open mind towards what it is going to hear – are very poor defenses indeed against the risks involved in what I am about to propose, namely that the encounter in question – the question-mark notwithstanding – *is* effective.

Let me rapidly rehearse those risks in guise of an introduction.

First of all, who is Louis Althusser? In the English-speaking world, Althusser's work is usually taken to be the "Marxist component" of what is called there the "structuralist/post-structuralist movement", a movement which apparently has nothing to do with science but concerns itself only – or so they say – with literary criticism. Of course, after taking even a superficial look at Althusser's work, one may grant that, in its idiosyncratic way, this work is somehow preoccupied with science. To the extent that this is done, one cannot fail to notice that the philosophy of science of Louis Althusser constitutes an embittered battle with no concessions against positivism and empiricism in their different variants and versions and in their fundamental philosophical tenets.

On the other hand, no stretching of the notion is required in order to call Joseph Sneed a "philosopher of science'. But Sneed's philosophy seems to be at the antipodes of any philosophy of science that can be distilled from Althusser's work. The work of Sneed (as well as, among others, the work of Balzer, Moulines and Stegmüller done in conjunction with his) is generally considered as today's surrogate of logical positivism.

K. Gavroglu, Y. Goudaroulis, P. Nicolacopoulos (eds.), Imre Lakatos and Theories of Scientific Change. 269–286.

And accordingly, within the "mood" that the historicist turn created, it is discarded – mostly in a cavalier manner – as a retrograde move that tries to resuscitate the well-forgotten efforts of the logical positivists to formalize philosophy of science. Curiously enough, the critics of the Sneed approach also maintain that this approach constitutes a formal reconstruction of Kuhn's views. This claim is not only historically wrong (although in effect, as Kuhn himself (1977) has expressly recognized, the Sneed approach does provide such a reconstruction, it was not devised to that end), but is also clearly at odds with the previous appraisal. But in any case and however one may judge it, it is certain that the wildest imagination would be reluctant to relate this approach to any kind of Marxism in general and to Althusser's idiosyncratic antipositivist and antiempiricist Marxism in particular. If the pragmatic elements and the vocabulary introduced by Kuhn may hint at some Marxist influence, the heavy formalism of the Sneed approach and the logical purity it is striving for certainly suffice to do away with even the slightest suspicion at such an influence.

Given this situation, the risks I am taking are obvious. On the one hand, the Marxists in general and the Althusserians in particular will justifiably cry anathema accusing me that, following the worst traditions of Marxist revisionism, I am trying to reconcile the Althusserian view with its explicit arch-enemy: positivism. On the other hand, the Sneedians in particular and the Anglo-Saxon philosophers of science in general will raise an eyebrow with hardly anything short of contempt. Marxism does not enjoy any high prestige in Anglo-Saxon philosophy. Its only brand which is relatively respected is that related to Marx's early works; that is, the brand which Althusser expressly repudiates as non-Marxist. Not only am I unjustifiably bringing Marxism into the picture but I am bringing in its worst variety.

Thus, as the encounter I propose is at least unwelcome to all interested parties, my flanks are uncovered on all sides. Speaking in an audience that includes Professors Sneed, Balzer and Moulines is not an unmitigated privilege. I run the danger of turning into my philosophical enemies those whom, above many, I would like to make my philosophical friends. To be perfectly frank, it was my becoming aware of the risks involved in this hazardous enterprise which finally induced me to change the title I had first considered for the present paper. That title was: 'Is Sneed a Marxist?' – with a question mark again, of course. Realizing that, after all, this title would probably not be appreciated as a humorous and provocative way to open the discussion but that it would sound more like a provocation, I finally declined to use it.

Now that I have stated my case in all its dangers let me briefly state how I propose to defend it.

As should be evident from what precedes, within the confines of a short paper, my case is, of course, indefensible. Nothing short of a longish book or an extended series of essays can hope to overturn the received view on *both* the Althusserian and the Sneedian approaches in a way which, first, can show that there really does exist a space in which my question makes sense and which, second, can locate the particular areas where the two approaches effectively meet (as well as those where they diverge). Accordingly, the present paper cannot pretend to be anything more than the barest outline of an argument which only tries to render somewhat plausible the encounter I am suggesting.

The paper will be divided into four parts. In the first part, I will try to sketch, very roughly, the Althusserian approach. In the second part, I will present a brutal summary of the Sneed approach. In the third part, I will outline what I take the *synthesis* of the two approaches to amount to. This synthesis, however, will not be considered at the abstract philosophical level. It will be directly "applied" to the particular case of the structure of Physics as a science. In the fourth part of the paper I will make some comments on how I view the relations holding between my proposed synthesis and the Sneed approach. Given that the formulation of this synthesis will be more "Althusser biased" than "Sneed biased" I think that these comments will be necessary for the clarification of the interpretative conditions under which the encounter I am suggesting becomes effective.

1. A SKETCH OF ALTHUSSER'S APPROACH TO SCIENCE

For reasons related to the particular political and theoretical conjuncture in which he set to work – and the details of which need not concern us here – Althusser defined his own project as an effort to assess the theoretical status of Marxism. Althusser subscribes to the tradition that considers Marxism as being made up from a new science (Historical Materialism) and a new philosophy (Dialectical Materialism) but esteems that both these "parts" as well as the relations holding between the two are in need of a drastic reappraisal. Accordingly, Althusser's project involves the following interrelated facets: (1) Establish the "scientificity" of Historical

Materialism by showing that this discipline, as founded by Marx's *magnum opus The Capital*, opens for the first time to scientific investigation everything that makes up the social and the historical. For Althusser, Historical Materialism thus constitutes the science of History (or, rather, in his terminology, the scientific "continent" of History potentially consisting of many different "regional sciences".[1] (2) Mark out and pinpoint the specificity of Marxist philosophy and, in particular, of the epistemological aspect of this philosophy and of the way it views science. (3) In the light of the above, clarify the relations holding between science in general and philosophy in general. Althusser's project is thus an expressly epistemological project involving a particular "philosophy of science".

Althusser's philosophy of science places itself in radical and explicit opposition both to political (as well as any kind of) relativism and to "dialectical materialist" (as well as any kind of) empiricism. Moreover and – in respect to the Marxist tradition – even more radically, Althusser maintains that a specifically Marxist philosophy of science (and a specifically Marxist philosophy in general) has not been expressly worked out anywhere. This philosophy exists only in the "practical state" and is scattered throughout the Marxist *corpus* most parts of which speak of quite different matters. It is only through the "symptomatic reading"[2] of this *corpus* that this philosophy can finally become explicitly articulated. Althusser embarks on such a reading whereby he extracts the following fundamental thesis: A science enjoys a "relative autonomy" in respect to the social relations within which it is engendered and develops further. The autonomy of a science disqualifies relativism while the relative character of this autonomy disqualifies empiricist (as well as idealist) foundationalism. Marxist philosophy of science will become theoretically articulate to the extent that it succeeds in spelling out with rigour the particulars of this autonomy. This spelling out can be carried out only *in conjunction* with what Historical Materialism (*the science* as Althusser reconstructs it, of whatever constitutes the social and the historical) establishes to that effect.

On the basis of this thesis, Althusser articulates his conception of science roughly as follows:

Every science is born through a particular process, the process of an epistemological break. The epistemological break is a break with the ways in which ideology (and more specifically the interplay between what Althusser (1974) calls "practical" and "theoretical" ideologies) carves out reality into different domains, constitutes these domains as objects of our experience and accounts for them. The epistemological break is epistemological in the sense that a new, non-ideological, mode for the cognitive appropriation of a particular domain of reality is thereby instituted. But this particular domain of reality is different from all those that ideology had, up to then, carved out. It is a *new* reality domain specific to the science that the epistemological break brings about. The epistemological break institutes a new mode for the cognitive appropriation of a particular domain of reality by *constructing itself* this domain. The construction of this domain (of the *object* of the science in question) is effectuated by the production of a *system of concepts* which both delimits this object and accounts for it and by the establishment of the particular procedures by which object and conceptual system check and control each other (and which we may call *experimental procedures*). The object, the conceptual system and the experimental procedures of a science are thus not produced independently from one another; they are congenitally interrelated and in effect interdefined within the unique process (the process of the epistemological break) which composes all three of them and assigns to each its particular epistemic function. The cognitive effects of the science in question result *exclusively* from these interdefining relations.

The fact that the object, the conceptual system and the experimental procedures of a given science are defined *only* through their interrelations (the fact that a science forms, in this sense, a structure) renders the knowledge that this science produces *independent* from the ideological (and thence social)[3] conditions within which the process of the corresponding epistemological break is effectuated. In other words, although this knowledge results from a particular social process and from the particular social practice that operates this process, it is independent from the social determinations of this process and of that practice. But this independence can never be absolute. This knowledge cannot function as knowledge if it cannot be related to – if it is not always already tied to – our ideologically constituted experience. Thus ideology can never be completely done away with and within scientific knowledge the ideological element remains always present. The growth of this knowledge, as it comes about by the further development of the given science, encounters the ideological element in the form of particular epistemological obstacles that have to be overcome. Therefore the autonomy that every science enjoys in respect to the social relations that bring it about and develop it further is only *relative autonomy*.

The (relative) independence of scientific knowledge and the (relative) autonomy of the corresponding science's development entail that the acceptance or the rejection of any particular result of this development is an internal affair of the science in question. The conditions, criteria and norms of this acceptance or rejection – whatever constitutes truth and objectivity for that science – cannot be determined at the general philosophical level and cannot be imposed on this science from the outside. It is the interrelations defining the object, the conceptual system and the experimental procedures of this science which determine by themselves these conditions, criteria and norms and which define the rigour specific to that science.[4] Thus the established results of a given science are not vulnerable to any extrascientific criticism but *only* to that science's further development. For this reason, it is only these established results which can tell if and to what extent *past* results are or are not true or are or are not scientific.

2. A PRECIS OF THE SNEED APPROACH[5]

Professor Sneed's project can be defined as the effort to understand as adequately as possible what the logical structure of theories of Mathematical Physics in fact amounts to, where this "in fact" should be taken seriously. This is to say that Sneed is not interested in ideal and abstract reconstructions but only in the logical structure of what physicists *actually* come up with. Sneed presents his view as a set of *empirical claims* concerning the science of Physics, that is as a set of claims which should be put to the test of what is really happening in that science.

The unit of investigation of the Sneed approach is a "mature" physical theory T (for example Classical Particle Mechanics – CPM for short), that is a theory which can be axiomatized in Suppes's sense (Stegmüller 1976). But as such theories are always theories *of* something (of natural phenomena), Sneed's project cannot become empirically meaningful in the above sense unless this "*of* something" is somehow taken care of. Accordingly, Sneed develops a complex set-theoretic formalism which encompasses both the mathematical structure of the theory T and the set of physical systems to which this theory effectively applies and for which it accounts. This set constitutes the "universe of discourse" of the theory in question.

Sneed maintains that the logical structure of the theories of Mathematical Physics does not allow us to conceive of this "universe of discourse" as one unique universal application. The world does not present itself to a theory *T* as an undifferentiated whole but only as a set of particular individuated physical systems. For CPM, the Earth–Moon system constitutes a different application from the Sun–Earth system. Moreover, each such physical system does not present itself to the theory *T* as it "really exists" in the world out there. To be considered as a potential application of *T*, the physical system must present itself as a set of individuals already "dressed" by a set of numerical functions. For example, the Sun–Earth system presents itself to CPM as a set of two individual particles ("representing" the Sun and the Earth) *together with* the set of functions which give the position, the velocity and the acceleration of the Earth in respect to the Sun for each instant of time. These functions partake of the mathematical structure of the theory *T* but *are not specific to T*. The values of these functions are determined by a theory different from *T*. (In our case by Euclidean Geometry and Classical Particle Kinematics).

A physical system thus given becomes effectively subsumed under *T* (cognitively accounted for by *T*) if the quantities *specific to T*[6] (mass and force for CPM) can be ascribed to the individuals of that system in such a way that the following three conditions are simultaneously satisfied: (1) The *T*-specific quantities can be given numerical values which always satisfy the relations (the general laws) that implicitly define these quantities within the mathematical structure of *T*. (In the case of CPM these relations are Newton's three laws). (2) The numerical values of the *T*-specific quantities are consistent with the values of the functions not specific to *T*. (3) The numerical values of the *T*-specific quantities also satisfy one or more particular relations (special law(s)) *specific to the physical system* in question. Such particular relation(s) involve the *T*-specific quantities and are consistent with the relations implicitly defining these quantities. (In the case of the Sun–Earth system the "special law" is the inverse square gravitation law).

From the above it follows that the mathematical structure of *T* always contains two kinds of quantities (functions) – those specific to *T* and those not specific to *T* – and two kinds of relations – those that implicitly define the *T*-specific quantities and which have to be satisfied in all applications and those which involve the *T*-specific quantities but may be different for different applications. And here Sneed comes up with a remark (or rather a *discovery*) which, to my mind at least, has revolutionary consequences: The logical relations holding between the mathematical structure of *T* and the applications of T are such that the effort to assign numerical values to a T-specific quantity in a particular application (through the experimental measuring of this quantity) cannot be successful unless (and therefore *necessarily presupposes* that) such values have *already* been *successfully* assigned in *another* application. Or, to put it in other words, the experimental test in a given application of a claim involving a T-specific quantity *always presupposes* that the *same* claim *already holds* in another application. Thus a theory T cannot be applied *seriatim* to one physical system after the other. The applications of T always come in *sets* of "mutually supportive" members: *Each* member of the set is an application only because *the other* members of the set guarantee it. The more general "structural" function of the T-specific quantities is to tie together within a coherent structured whole the mathematical structure of T, the *set* of the applications of T and the experimental procedures necessary for assigning numerical values to the different quantities of T and for testing the claims formulated by it. The Sneed approach is also called the *structuralist approach* because it highlights the existence of such structural relations.

The structural relations holding among the mathematical structure of a physical theory T, the set of applications of T and the relevant experimental procedures are not exhausted by the function of the T-specific quantities. There should also exist structural relations assuring that the quantity expressing a given physical property of a given physical object is assigned *the same* value in *different* applications. (E.g. the object "Earth" must be assigned the same mass value in both the Sun–Earth and the Earth–Moon systems). The Sneed approach takes account of such relations and calls them "constraints". Moreover, there should exist structural relations between *two different physical theories* T and T' assuring that the quantities which appear in *both* these theories (e.g. velocity in Classical Particle Kinematics and in CPM) can be treated as *one* physical quantity. Such relations have recently drawn the attention of the structuralist approach (Moulines 1984) and are called "intertheoretic links". Space does not allow me to enter into any details here.

As a conclusion to this section, it should be added here that the Sneed approach casts all the above ideas in a formalism which is both rigorous and powerful. Within this formalism, many of the "standard" relations among physical theories which the practice of Physics takes for granted (e.g. the "equivalence" of the Newtonian and Lagrangian formulation of CPM making these two formulations different formulations of one and the same physical theory, the "reduction of Rigid Body Mechanics to CPM, etc.) can be proved and some essential features of a theory's dynamics can be accounted for.

3. THE ABSTRACT OF A SYNTHESIS OF THE ALTHUSSERIAN AND THE SNEED APPROACHES

The above presentation of the Althusserian and of the Sneed approaches, despite its extreme roughness, should have rendered, I hope, at least somewhat plausible the thesis of the present paper: strangely enough and quite unexpectedly, these two approaches bear enough similarities to justify an attempt at their synthesis. As I have tried to develop this synthesis elsewhere (Baltas 1987, 1988), I will present here only its "abstract". As I said in the introduction, this synthesis will be directly "applied" to the case of the science of Physics.

Physics is a structured whole made up from three constitutive elements that mutually define each other: The conceptual system of Physics, the object of Physics (the phenomena Physics cognitively accounts for) and the experimental procedures specific to Physics.

The conceptual system of Physics constitutes a complex structure of concepts and of relations among concepts. For the "mature" parts of this system (physical theories axiomatizable in Sneed's sense), these relations are mathematical relations independently formulable within the realm of pure Mathematics. The concepts of the system acquire their identity from two sources of meaning. On the one hand, *the position* of each within the system determines the *systemic component* of its meaning. On the other hand, the fact that this system *congenitally talks about the phenomena it is established to account for* makes it an *always already interpreted* system. The concepts of the system are therefore always already endowed with the *empirical component* of meaning which this interpretation confers to them. The interpretation of the conceptual system of Physics ties this system to our overall experience at the time and makes it understood on the basis of this experience. Therefore, as this experience is always ideologically determined, the interpretation of the conceptual system of Physics cannot but harbour the ideological element within itself. The ideological element residing in the interpretation of the conceptual system of Physics is responsible for the difficulties and obstacles that the further development of Physics will have to overcome.[7]

The fact that the conceptual system of Physics congenitally talks about the phenomena it is established to account for is an expression of the fact that, *in the very process of their own production* the concepts and relations making up this system become *immediately "attributed"* to these phenomena conceptually transforming them. It is convenient to distinguish the attribution of concepts from the attribution of the relations.

The process of attribution of the concepts, first, engages, in general, single phenomena by themselves and results in a particular conceptual transformation of each such phenomenon. This process picks up and singles out one particular aspect of the phenomenon which is idiosyncratic to the conceptual system that is being simultaneously established, it transforms conceptually that aspect and it identifies the whole phenomenon with just this singled out and transformed aspect. For example, the concepts of Classical Mechanics are attributed to the phenomenon of a falling apple thereby transforming it to the phenomenon describable as "a point mass moves attracted by a gravitational force". The thus transformed phenomenon constitutes a *physical phenomenon* as opposed to a natural phenomenon. The set of physical phenomena make up the *reference domain* of Physics.

The process of attribution of the relations making up Physics' conceptual system, second, always involves *a set* of phenomena and works as follows: *Let us for the moment suppose* that, somehow, these relations *already hold* for a given physical phenomenon. For example, let us suppose that we somehow already know that "the small oscillations of a pendulum are isochronous". Then this phenomenon can be "frozen" into the form of an *experimental device or apparatus* which can be used as a gauge to measure if the relations of the system hold for *another* phenomenon, "empirically independent" from the "frozen" one. (I call "empirically independent phenomena" phenomena such that the experience of the one does not involve the experience of the others and which, accordingly, were considered as having nothing to do with each other before Physics decreed otherwise). In our example, the pendulum can be used as *a clock* that measures time intervals in "free fall" thereby checking that the relations that the conceptual system of Physics stipulates for "free fall" do indeed hold. *The experimental procedures specific to Physics* consist in such "correlations" among phenomena already constituted as physical.[8] Physical phenomena for which such correlations have proved to hold make up the *domain of Physics' effective applications*.

But on this basis, there is no way that any *one* physical phenomena can be singled out and turned by itself into *the first* of Physics' effective applications. *The transformation of a physical phenomenon into an effective application of Physics always presupposes the completion of this transformation for another physical phenomenon.*[9] Therefore the experimental procedures that assure this transformation have necessarily to take in their stride *a set* of physical phenomena and make them *collectively* establish their "correlability". The set of physical phenomena which have first proved their correlability constitute the *founding members* of the domain of Physics' effective applications. The *object* of Physics consists of that science's reference domain *together with* the domain of its effective applications.

On the basis of the above we can see in what sense the three constitutive elements of Physics mutually define each other thereby constituting Physics into a structure. Each one of these elements is composed and is assigned its particular epistemic function *only through the relations* that this element entertains with the others. Thus the *abstract* concepts of the conceptual system, by being *ab ovo* attributed to the phenomena they are produced to account for, are *themselves always already* the *empirical* concepts involved in the treatment of these phenomena. These concepts become empirically meaningful by transforming a natural into a physical phenomenon. In other words, these concepts acquire their physical (in contradistinction to their mathematical) identity *only through the relations* that tie them to Physics' object. On the other hand, the experimental procedures which transform physical phenomena into effective application of Physics thereby structuring Physics' object do not constitute a dialogue between two heterogeneous parties which are essentially external to each other – an abstract conceptual system on the one hand and a phenomenon given to our experience independently of that system on the other. Instead, they consist in a test of mutual "correlability" between *two phenomena both of which* have been "constructed" as physical *by the conceptual system of Physics alone*. Physics forms a structure because its three constitutive elements are defined *only* through their interrelations. And the cognitive effects of Physics depend *exclusively* on these interrelations.[10]

It should be added here that the interrelations defining the three constitutive elements of Physics do not exhaust their role and function in just composing these elements and putting them into place. They also institute the rules of the incessant interplay among these elements, interplay which allows Physics to develop further. However, the analysis of this development (as well as the analysis of the process through which Physics was born and which, following Althusser, we could call process of an epistemological break) lies outside the scope of the present paper. Suffice it to say that such an analysis would again rely on elements drawn from both the Althusserian and the Sneed approaches.

4. SOME COMMMENTS ON THE RELATIONS BETWEEN THE SNEED APPROACH AND MY PROPOSED SYNTHESIS

The above synthesis may well be faithful to what I take the overall spirit of the Sneed approach to be. Nevertheless it should be evident that at some crucial points it departs significantly from it. The following comments are intended to locate some of these points and to give my main reasons for such a departure.

Sneed chooses as his unit of investigation a particular "mature", physical theory T while I speak from the outset of the conceptual system of *Physics* in its entirety. This system is considered as including the conceptual system of *all presently accepted* physical theories or quasi-theories, "mature" or "immature", from "protophysics" up to the tentative models (in a physicist's sense) proposed in the different research fronts. Also this system is considered as forming a structure in the strict, technical, meaning of the term: each element of the system (each concept), is what it is only in virtue of the relations it entertains with the other elements.

Now, the reasons why I consider that the conceptual system of Physics forms a structure are those that the Sneed approach itself has advanced: (1) Each "mature part" T of the system contains T-specific concepts (Sneed speaks of T-theoretical quantities or functions – I will return to this below) which, as far as what I call the systemic component of their meaning is concerned, are defined *implicitly*, that is only by the fundamental relations (the fundamental laws) that confer to T it's identity. (2) The different "parts" of the conceptual system are all tied together by "inter-theoretic links".

On the other hand, I am well aware, of course, that such a huge and heterogeneous thing as the conceptual system of Physics in its entirety is extremely awkward to handle in a logically precise manner; accordingly, I fully understand why Sneed chooses his unit of investigation the way he does. But here I am not interested in a precise logical analysis. I am much more concerned to highlight a general philosophical point which the Sneed approach, as it stands, waters down: The world is not given to

Physics (or, for that matter, to any science) in the raw as most versions of scientific realism and empiricist foundationalism would have it. The part of reality for which Physics cognitively accounts is not allotted to Physics prior to and independently from that science's constitution. The object and the conceptual system of Physics *constitute each other* (with the help of the corresponding experimental procedures) in the sense that Physics *always* speaks *only* of things which are *exclusively* determined in a *Physics specific way*. Sneed would probably agree with this but his formulations do not underscore the point as clearly as one would wish: A theory T speaks of things which are *not* determined in a T-specific way; the values of the quantities not specific to T are determined by different theories T'. We have to take into account the fact that, for Sneed, *theories* are involved here (a physical system is given by a set of individuals *together with* a set of functions and therefore is, in some sense, a *theoretical* entity) as well as the fact that the theory T and the theories T' are structurally related by "intertheoretic links", in order to see the Sneed approach as finally advocating the same kind of constructivism as Althusser.[11]

I have in mind the same philosophical point when I use the terminology T-specific instead of T-theoretical. Sneed's distinction between T-theoretical and T-non-theoretical terms certainly allows the empirical testing of T in a way such that most if not all of the puzzles created by the theoretical/observational dichotomy, as postulated by the empiricist conception, become solved or dissolved. But on a deeper philosophical level this terminology is misleading: Both T-theoretical and T-non-theoretical terms (T-specific and T-non-specific concepts) are *theoretical terms of* T; the former are determined by T itself, the latter by theories T' different from T but structurally related to it. Therefore not only the non-theoretical is not equivalent to the observational – as a superficial reading of the Sneed approach would have it – but, within this approach, no room at all is left for *purely* observational terms without this giving rise to any relativism. Despite appearances, the Sneed approach is *not* a refinement of logical empiricism. The terminology T-theoretical, T-non-theoretical helps to perpetuate the myth that the Sneed approach constitutes only such a refinement and correspondingly blurs its radical novelty and its antiempiricist, constructivist thrust.

The above may also help to relieve some tensions in Sneed's writings. For example, in his (1971), Sneed "confess(es) to have almost nothing to say about what counts as a physical system" (p. 250). And then (p. 290) he goes on to say that "a (physical) theory determines its own range of applications; it is intended to apply to just exactly what it, in fact applies to – nothing more nothing less" adding that he knows "of no philosopher of science who explicitly holds the view that this is the correct way to regard the characteristic range of intended applications in theories of mathematical physics" (*ibid.*). Now, if the burden of the whole paper is to show that one finds in Althusser a philosopher of precisely those specifications, the burden of the preceding remarks is, among other things, to show that there exist no grounds to entertain the guilt that this confession presupposes. Physical systems are *just* the systems that Physics as a science accounts for. The structural relations that the Sneed approach itself has marked out show quite clearly why, despite appearances, this formulation has nothing to do with a tautology.

The second important point where the Sneed approach and my proposed synthesis depart from each other in the following: Sneed speaks of functions and their values while I speak of concepts and their meaning (with its "systemic" and "empirical" component). Again, I can very well understand the reasons for Sneed's choice. Functions and their values are eminently suited to a logical treatment while concepts and meanings are notoriously elusive kinds of things. So let me rapidly sketch what I consider as the relative advantages of my choice as regards, in particular, the empirical component of meaning. The systemic component can be readily accommodated within the Sneed approach as it stands.

As I have said a few pages before, for me, the conceptual system of Physics is always already interpreted. It is this interpretation which ties the system to our overall experience thereby making it understood and which confers to its concepts the empirical component of their meaning. This interpretation and the empirical component of meaning that ensues from there assure what Sneed, in his (1971, p. 299) calls "connection with intuitively familiar notions", a connection that constitutes "a crucial feature of theories of mathematical Physics that "generates" genuinely explanatory claims" (*ibid*., p. 298). My introducing the empirical component of meaning at the very beginning of my account builds in the explanatory function of the corresponding theory from the outset. Moreover, such an account remains closer to what practising physicists intuitively take the theories they work on to be without sacrificing anything essential from what the Sneed approach has taught us.

Second, it is the empirical component of meaning which assures that the concepts of the system "dress" *real*, *material* phenomena (what I called *natural* phenomena) and transform them into *physical* phenomena, that is turn them into those particular conceptual-material kinds of things (what Bachelard calls "abstract-concrete objects") for which Physics accounts.[12] It is this transformation which permits the "correlation" of *empirically independent* phenomena in the way that defines a particular experimental procedure. In two words, it is the empirical component of meaning which conceptually supports and thereby permits us to make physical sense of those *practical material moves and procedures* (building of devices, connections of wires, readings of dials etc.) in which a physical experiment consists. By confining his discussion to functions and their values, Sneed renders himself unable to encompass physical experiments in his account otherwise than as essentially unglossed procedures which just assign particular values to the functions. In this way he becomes unable to exploit in full *his own* fundamental insight on the interdependence between the different applications of a given physical theory T. My account of experiments in Physics does no more than spell out this insight on the basis of Bachelard's (and Althusser's) conception of physical objects (physical phenomena).

Third, my talking of the conceptual system of Physics in its interconstitutive relation (through the empirical component of meaning) with the phenomena making up Physic's object allows a description of *the mechanism* through which Physics was born. This mechanism (in Althusserian terminology, the mechanism of the corresponding "epistemological break") may be conceived as simultaneously operating on three interconnected levels: (1) It transforms pretheoretic notions into the concepts making up the conceptual system of Physics. (2) It works on the natural phenomena themselves and transforms them into physical phenomena by attributing to them the concepts of the system which is being simultaneously produced. (3) It performs "correlations" among the phenomena that are being constituted as physical thereby arriving to single out the founding members of the domain if Physics' effective applications. The *historical* process which led to Galileo's achievements and from then on to Newton's synthesis can be seen as this mechanism at work. It is up to the historian of Physics to assess if the overall characteristics of this process can be accounted for by this mechanism as well as to examine how this process has been actually carried through.

Fourth, the way I construe the empirical component of meaning allows a discussion of some fundamental *dynamical* aspects of Physics' development that remains close to the questions which the historian (or the sociologist) of Physics has actually to face. For example, as I have already noted, the interpretation of the conceptual system of Physics and the empirical component of meaning that ensues from there inevitably harbour the ideological element. The development of Physics encounters this element in the form of obstacles that it has to overcome. In particular, this element is present in the form of self-evident, scientifically unwarranted, "assumptions" and presuppositions to the unwarranted character of which or even to the existence of which Physics remains blind. A *revolution* in Physics consists precisely in the *disclosure* of a sufficiently central such "assumption" and in the *reinterpretation* of the conceptual system that this disclosure brings about.[13] Such a construal of what a revolution in Physics consists in does not contradict, I take it, Sneed's (or Kuhn's) approach to the question. But to my mind at least, it leads to an account of the actual "flow" of Physics that does not seek refuge too early in such "not definable (at least within the present metatheory)" pragmatic concepts as "historical intervals", "scientific communities", "epistemic relations" etc. (Moulines 1979). Within the "metatheory" I am trying to work on, the "pragmatics" of Physics should not constitute an independent realm, externally appended, as it were, to the "more fundamental" parts of this "metatheory". The account of the *structure* of Physics has itself to be such that what we use to call the pragmatics of that science follows naturally and rigorously from it. As I see it, the empirical component of meaning opens precisely this door: With its help, both the "internal" and the "external" aspects of Physics structure and development can be tackled *together* by a *unified* conceptual system in a way that remains as close as possible to what is really happening in that science.[14]

NOTES

[1] Althusser uses this idea in many of his writings. See in particular his (1971b).

[2] See Althusser and Balibar (1970) where Althusser on the one hand explains what this kind of reading amounts to and, on the other hand, applies it to Marx's *The Capital*.

[3] For Althusser, the "ideological instance" constitutes – along with the economic and the political – one of the three fundamental "instances" or "levels" of the social whole performing an essential function in the reproduction of every society. Ideas are always elements of particular "ideological formations" and thereby socially determined. See Althusser (1971c).

[4] The traditional Marxist (or rather "Marxist") criterion of practice, barely distinguishable from its empiricist analogue, becomes thus radically transformed by being relativized to each particular science. In his (1970) p. 79 Althusser speaks of "l'interiorité radicale du critère de la pratique à la pratique scientifique" thereby creating a proper scandal for traditional Marxism.

[5] This approach is expounded in full in Sneed (1971) and in Stegmuller (1976) and more succinctly in Stegmuller (1979).

[6] Sneed calls these quantities *T*-theoretical. In the next section I will explain why I do not use this terminology.

[7] For more details see Baltas (1987).

[8] This conception of the experimental devices employed in Physics is based on Bachelard's ideas. For him (Bachelard 1933, p. 140), "un instrument dans la science moderne est veritablement un theorème réifié". It is in this sense that the object with which Physics is involved are "abstract-concrete".

[9] This, at least according to me, constitutes Sneed's fundamental insight which this author derives from his account of what he calls "theoretical terms".

[10] I underline "exclusively" here in order to emphasize that, according to me, nothing outside these interrrelations can contribute to those cognitive effects. However, this does not imply that these interrelations are in themselves scientifically "pure", uncontaminated by anything "extrascientific". For details see Baltas (1987).

[11] If we identify the conceptual system *of Physics* with the network of the conceptual systems of *all* theories related by "intertheoretic links" we have a starting point from where we can try to clarify in what precise way Physics is related to and distinguished from *the rest of the natural sciences*. (E.g. how Physics is related to and distinguished from Chemistry). And from there we can go on to examine the relations holding or not holding between *the set* of the natural sciences and those social and human disciplines which purport to be scientific. In Althusserian terminology this amounts to the relations holding or not holding among scientific "continents".

[12] In this way, I do not seem to need for my account Sneed's notion of "constraint". The transformation of the natural object "Earth" into a physical object the sole "inherent" characteristic of which is mass seems to imply that this object must "possess" the same mass in all the applications that it appears. It is possible, however, that a finer grain analysis would show that the notion of "constraint" is indispensable.

[13] For a development of this idea see Baltas (1987).

[14] For more details see again Baltas (1987).

REFERENCES

Althusser, L. and Balibar, E., *Lire le Capital*, Maspéro, 1970.

Althusser, L., 'Lenin and Philosophy' in his *Lenin and Philosophy and Other Essays*, NLB, 1971a.

Althusser, L., 'Ideology and Ideological State Apparatuses', in *op. cit.*

Althusser, L., *Philosophie et Philosophie Spontanée des Savants*, Maspéro, 1974.

Bachelard, G., *Les Intuitions Atomistiques*, Boivin, 1933.

Baltas, A., 'Ideological "Assumptions" in Physics: Social Determinations of Internal Structures', in A. Fine and P. Machamer (eds.), *PSA 1986*, Vol. 2, East Lansing, Philosophy of Science Association, 1987.

Baltas, A., 'The Structure of Physics as a Science', in D. Batens and J. P. van Bendegens (eds.) *Theory and Experiment*, Dordrecht, Reidel, 1988.

Kuhn, T. S., 'Theory Change as Structure Change: Comments on the Sneed Formalism' in

R. E. Butts and J. Hintikka (eds), *Historical and Methodological Dimensions of Logic Methodology and Philosophy of Science*, 1977. Part IV of the Proceedings of the 5th International Congress of Logic Methodology and Philosophy of Science, London, Ontario, Canada, 1975.

Moulines, C. U., 'Theory Nets and the Evolution of Theories: the Example of Newtonian Mechanics', *Synthese* **41** (1979) 417–439.

Moulines, C. U., 'Links, Loops and the Global Structure of Science', *Philosophia Naturalis*, Band 21, Heft 204, 1984.

Sneed, J., *The Logical Structure of Mathematical Physics*, D. Reidel, 1971.

Stegmuller, W., *The Structure and Dynamics of Theories*, Springer-Verlag, 1976.

Stegmuller, W., *The Structuralist View of Theories*, Springer-Verlag, 1979.

National Technical University of Athens

W. BALZER

ON INCOMMENSURABILITY*

It is my aim in this paper to introduce a precise core for a theory of incommensurability – a "core" as contrasted to a pure definition or an explication in the sense of logical empiricism. The picture of a theory I have in mind here is that of structuralist meta-theory[1] according to which a theory consists of a basic core which covers the phenomena common to *all* applications of the theory, and which is empirically rather empty, together with various specializations of this core which are valid only in special subsets of intended applications, and which usually have respect able empirical content. It turns out that the basic requirements characterizing this core are not thrillingly new. In fact, they were stated in some form quite some time ago by Paul Feyerabend[2,3] when he described incommensurability as logical inconsistency. The way we are led to these requirements, however, starts from a basic intuition which may be traced back at least to Thomas Kuhn's PSA paper[4] where incommensurability in one place is characterized as overall structural change in the light of a stable taxonomy.

Since the reasoning towards the conditions for incommensurability here is quite independent of former writings the coincidence of these conditions may be regarded as a piece of confirmatory evidence for them. Moreover, by stressing the character of a theory, we insist on the one hand on taking seriously the concrete historical examples. On the other hand we are free in adjusting the conceptual model to fit these phenomena. Finally, this theory contributes to the study of intertheoretic relations and therefore also to that of scientific progress and rationality.

I. THE BASIC PHENOMENA

The phenomena on which discussions about incommensurability are based are given in the form of periods in the historical development of a science in which one theory is replaced by another one such that there is much controversy about whether the new theory is better than the old one and should be accepted instead of the old one, and about whether the new theory really can reproduce all the achievements of the old one. Usually,

K. Gavroglu, Y. Goudaroulis, P. Nicolacopoulos (eds.), Imre Lakatos and Theories of Scientific Change. 287–304.

both theories are about the same phenomena, and the controversies are possible only because, essentially, both the defenders of the old and the new theory use the same language. The languages of the adherents of both theories are the same only "essentially", not strictly. That is, there are some few – but important – terms in which both languages differ: these terms either are simply different or they are identical but are used in different ways (have different meaning) in both theories.

In addition, such periods may be distinguished from other types of developments by means of psychological and sociological features of the behaviour of the individuals and groups adhering to the two theories, as described in particular by Thomas Kuhn. I will deliberately neglect these features here and concentrate on those conceptual issues which can be stated in an extensional language. It is not entirely clear whether the psycho-sociological features are essential or necessary to the phenomenon of incommensurability: In the absence of any group-fighting would we say that we are confronted with incommensurable theories? Very likely we wouldn't. So the phenomenon would have an essentially intentional character, and my neglecting this would render my whole account inadequate. To this I have two replies. First, we all know that essentiality comes in degrees, and with respect to incommensurability the purely conceptual aspects are much more essential than are the intentional ones. So my account of incommensurability may miss some admittedly important features to be added by psychologists and sociologists. But – secondly – if I were given the choice between really clarifying some phenomenon in *some* essential aspects or between making only some very broad and vague statements in order to be sure not to commit any slight inadequacy, I would tend to prefer the first alternative.

More concrete examples of the kind indicated above are given in the following list which by now may be called standard: Aristotelian theory of motion versus (pre-Cartesian) kinematics; Ptolemean versus Copernican theory of planetary motions; impetus theory versus Newtonian mechanics; phlogiston theory versus stoichiometry; phenomenological thermodynamics versus statistical mechanics; classical versus special relativistic mechanics. If we study these examples on the conceptual level, i.e. if we try to state the respective theories with a sufficient degree of precision and comprehension, we detect the following general pattern. Whenever we try to *match* the concepts of the two theories one by one we will reach a stage in which not all concepts are as yet matched up but in which it is also impossible to continue the process of matching without getting into conflict with what both theories require of their concepts. More precisely, the situation will be this. We start making a manual for matching each concept of the old theory with a corresponding concept of the new theory,[5] and we succeed in including most of the concepts in this manual. However, at a certain point in this process the following problem arises. The concepts of the old theory, T, are strongly interrelated in T. That is, T contains many assumptions, laws and hypotheses, by which the concepts of T get linked to each other and get determined by each other. A concept of T cannot be explained or learned in isolation but only "in the context of T", that is, in one process simultaneously with many other concepts of T. This may be expressed by saying that T's concepts *fit into the structure of* T. Now the problem arising in the endeavour to match both theories' concepts is that, at a certain point, we cannot continue matching without destroying the way in which the concepts matched fit into the structures of their respective theories. In other words, if we succeed in matching the concepts $c_1, \ldots, c_n$ of T with concepts $c'_1, \ldots, c'_n$ of the new theory T', such that $c_1, \ldots, c_n$ fit into the structure of T and $c'_1, \ldots, c'_n$ fit into the structure of T', it will not be possible to include another concept c_{n+1} *of* T. That is, for given c_{n+1}, *whatever concept* c'_{n+1} of T' we choose as a candidate to match with c_{n+1}, *the chosen concept* c'_{n+1} will not fit into the structure of T'. Expressed still differently: the way both theories' concepts fit into the respective structures does not allow for matching the concepts one by one. (The assumption of a "one by one" match is not really essential here, and is used only for reasons of simplicity.)

The basic phenomenon of incommensurability at the conceptual level thus consists of a *tension* between a one by one match of the concepts and the way these fit into the structures of both theories. It is intuitively clear that this tension is a feature entirely internal to the two theories and their conceptual correlation. No intentional aspects are involved and reference to meaning ("meaning variance") and translation is completely avoided. This latter feature deserves special attention, for most of the discussions usually ended up with questions of meaning or meaning variance. In particular, this holds for discussions referring to the notion of translation, for it does not seem possible to say what a translation is without being able to say what the meaning is of some suitable chunks of a language. But there is no good theory of meaning that would help here, and so the people engaged in such discussions usually end up "with mud on their head', as Paul Feyerabend put it.[6]

Before I proceed to make this basic phenomenon more precise, a picture may serve to further clarify the intuition. Suppose the concepts of our two theories are represented by the nodes in Fig. 1 below, and the way the concepts fit into the structure of T and T', respectively, is represented by the structure as established by the arcs drawn between some nodes. It is clear that no way of matching the concepts (nodes) in the left-hand structure with concepts in the right-hand structure can preserve all the relations between the concepts. If, for instance, we match the concepts as indicated by the dotted arrows in Fig. 1 then c_1 and c_2 will be related in T but their counterparts c_1' and c_2' in T' will not be related (there is no arc between c_1' and c_2' in T'). There is a tension between the one-by-one match of the nodes on both sides, and the ways in which the nodes are interrelated in the two different structures.

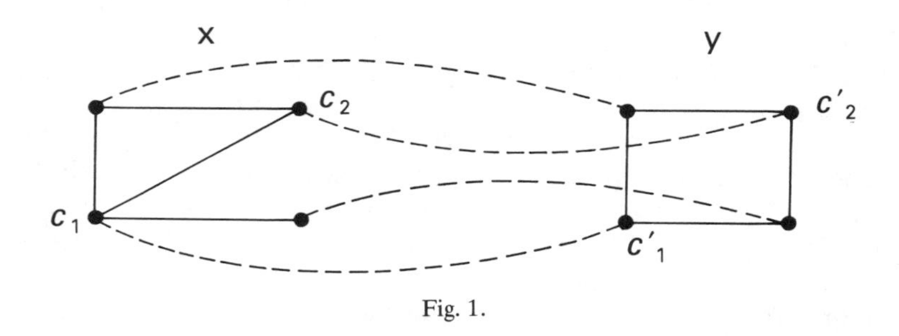

Fig. 1.

II. A PRELIMINARY APPROACH

Let us now see how the phenomenon described may be conceptually clarified. Let T and T' be two theories. If t_i denotes an arbitrary word or term from the vocabulary of T then this vocabulary may be written in the form $(t_1, \ldots, t_i, \ldots)$. Similarly, we write $(s_1, \ldots, s_j, \ldots)$ for the vocabulary of T'. The only assumption about both languages is that they be extensional, otherwise the following definition cannot be stated. Again, it might be objected that this assumption leaves out essential features, and our reply to this is the same as that concerning intentions in the preceding section.

Suppose we have some concrete system x before us, which is an intended application of theory T. Then all of T's terms can be interpreted in x. That is, for each term t_i of T we can point out some entity which is a "part" of the system x and which in the context of T typically is denoted by

t_i. This entity is called the *interpretation* or the *denotation* of t_i in x, and we will denote it by t_i^x in the following. There may be actual problems in precisely determining t_i^x but it is commonly assumed that this entity at least exists, however limited our access to it. If we also accept the (rather trivial) additional assumption that t_i^x may be conceptualized as a (possibly very complicated) *set* then we have come to the point where model theory begins. It is not necessary here to go into the technicalities of model theory. All we need is some agreement that each term t_i of T has an interpretation t_i^x in any concrete system x. If we forget about those features of x which are irrelevant from the point of view of T we may identify the system x with the collection of interpretations of T's terms ("in x"), and write $x = (t_1^x, \ldots, t_n^x, \ldots)$ whereby some assignment of interpretations to terms is assumed as given (for instance by the indices: t_i^x is the interpretation of term t_i). Furthermore, it is convenient to consider not only real systems but abstract ones as well. If an abstract entity is given by a collection $(\bar{t}_1, \ldots, \bar{t}_n, \ldots)$ of sets such that each set $\bar{t}_i$ has the right type to be an interpretation of the term t_i we may treat such an entity as a system as well, and extend the notion of an interpretation to such abstract systems. We say that T's terms have interpretations in all abstract systems $x = (\bar{t}_1, \ldots, \bar{t}_i, \ldots)$ which are of the right type, and the interpretation t_i^x of t_i in $x = (\bar{t}_1, \ldots, \bar{t}_i, \ldots)$ is just $\bar{t}_i$.

Now a *model of T* is a (real or abstract) system $x = (t_1^i, \ldots, t_n^i, \ldots)$ such that each t_i^x is an interpretation of term t_i (in the system), and such that all t_i^x are connected with each other precisely in the way expressed by the statements of T by means of the terms t_i. This is just an intuitive version of the well-known definition form model theory.[7] These notions can of course be applied to T' and we obtain models of T' in the form $y = (s_1^y, \ldots, s_j^y, \ldots)$.

A crucial term in Section I was that of a match between the terms of two theories. This notion becomes accessible if we transfer it to the level of models. If $x = (t_1^x, \ldots, t_i^x, \ldots)$ and $y = (s_1^y, \ldots, s_j^y, \ldots)$ are models of T and T', respectively, we say that x and y *match up* iff for all t_i^x there is precisely one s_j^y such that t_i^x matches with s_j^y and conversely, or equivalently, if there is some permutation π of the indices such that for all i: t_i^x matches with $s_{\pi(i)}^y$. This is of course a purely formal move as long as the meaning of "match of interpretations" is not explained (more on this below). For the moment let's stick with the vague expression which seems to be quite intuitive in concrete cases. If t_i^x, for instance, is the mass function in a model x of classical mechanics, and if s_j^y is the mass function in a model y of special relativistic mechanics such that both x and y "describe" the same, real, moving particles it is clear that t_i^x matches with s_j^y. If s_j^y, on the other hand, is the global force function of the relativistic model then, clearly, t_i^x does not not match with s_j^y.

In order to state our definition we have to extend the notion of "match" to cover cases of "partial match", too. If S is a subset of the vocabulary of T we say that models x and y of T' *match in* S iff, for all w in S, the interpretation w^x of w in x matches with some member w^y of y and conversely. If, for instance, S contains only kinematical terms, and x, y are, as before, classical and relativistic models of "the same system" then x and y intuitively match in S: there is a natural correspondence between, say, the particles, the instants, and the position functions in both models, namely identity. If, on the other hand, y is a model of the motion of quite different particles there will be doubts as to whether x and y match in S.

Now we can state a preliminary definition of incommensurability which will be emended further below. Suppose we have two theories T and T' before us such that our notational conventions apply. Both T and T' are about the same phenomena, so their joint vocabulary, i.e. those terms common to T and to T', will be rather large relative to all the terms of T and T'. In particular, the joint vocabulary will be non-empty. Let x and y be given models of T and T', respectively. (It will be helpful to think of x and y as "describing" the "same" real system from different points of view, maybe by using different terms as provided by T and T'). The idea developed in Section I was to start matching the terms of both theories and see where this clashes with how they fit into the structures of both theories. Quite naturally, one will start matching terms from the joint vocabulary first, among those one will concentrate on the unproblematic ones, i.e. on those for which there is some straightforward correspondence like identity or identity of meaning. Suppose in this way we succeeded in matching the terms of a relatively large subset S of the joint vocabulary. But if S is sufficiently large ("maximal" in this respect) we cannot extend the match to further terms without running into problems, that is, without contradicting the basic laws of both theories. In terms of models this means that we cannot extend the match between two structures x, y beyond S and still insist that both x and y are models (of T and T', respectively). In particular, the match cannot be extended to the *full* joint vocabulary. In the following we will concentrate on the latter special case, first, because it facilitates the formulation, and second, because it still covers the intended examples. In summary, we obtain the following simple condition. For some suitable subset S of the joint vocabulary of T and T', and for any two models x of T and y of T': if x and y match in S then it is impossible to extend this match to the full joint vocabulary of T and T'. An equivalent formulation stressing the clash with the assumptions of both theories is this. There is no suitable subset S of the joint vocabulary such that for all structures x, y: if x and y match in S then it cannot be the case that x is a model of T, y is a model of T' and x and y match in the full joint vocabulary. By adding the requirement of the joint vocabulary being non-empty we obtain the following preliminary characterization of incommensurability$_p$ (with "p" for "preliminary").

Theories T and T' are *incommensurable*$_p$ iff

(1) the joint vocabulary of T and T' is non-empty
(2) there is some non-empty subset S of the joint vocabulary such that for all x, y: if x is a model of T and y is a model of T' and x and y match in S then x and y do not match in the full joint vocabulary.

Note that condition (1) follows from (2). This account immediately raises the following question: which subset S of comparable terms should be chosen to do the job? Thomas Kuhn in a private discussion at once pointed out that it is not easy to justify some particular choice of terms as those for which comparison ("match") is unproblematic. For this amounts to drawing a distinction among the terms of a theory such that the meaning of one subclass of terms is independent of the meaning of the other subclass. A holistic picture about science and about meanings of terms in a theory throws some doubt on the possibility of such a distinction. In addition, if we had criteria for the choice of the "right" class of comparable terms we would also have a criteria for distinguishing some sound empirical or observational basis of theories which is not affected by "turbulences" on the theoretical "Ueberbau" – and thus criteria for continuity and progress. Of course, this does not completely demolish the idea of choosing some correct class of comparable terms, for we are not without *any* clue to "the right" choice. It may be pointed out that terms which have "lived through" a long scientific development without major modifications (like Euclidean distance or mass (=inertial mass = rest mass)) have achieved some dignity which gives them enough independence to serve as a basis of comparison. Also, it may be pointed out that terms get more standing in connection with their multiple referents, that is, with multiple established and important links from other theories (like "mass" in mechanics which is linked to the weight-function in stoichiometry, to energy in thermodynamics, to the stress-energy tensor in general relativity and so on). These possibilities notwithstanding it seems that we are far away from having interesting and applicable criteria in order to choose "the correct" set of comparable, common terms (if there is such a thing).

There is a simple way out of this problem: just strengthen the above condition (2) to hold for *all* proper subsets of the joint vocabulary. So T and T' would be incommensurable$_p$ iff condition (2) above is replaced by

(2′) for all proper subsets S of the joint vocabulary and for all x, y: if x is a model of T and y is a model of T' and x and y match in S then x and y do not match in all of the joint vocabulary.

Though such a condition may be cumbersome to check in concrete cases it seems a move in the right direction. If commensurability has holistic roots insofar as it does not allow one to break both theories' structures apart for the sake of one by one comparison then there is no reason why such one by one comparison should be justified for just one distinguished subclass of terms. Rather what should be expected is a kind of symmetry: whatever subclass we take as a basis of comparison we will end up in difficulties. So let's accept this move as an emendation. The transition from (2) to (2′) only apparently complicates the issue. On closer inspection it turns out to simplify things considerably. For now reference to a subset S of common terms in the "if"-clause of condition (2′) becomes redundant. In fact, (2′) is obviously equivalent to the following

(2*) for all x, y: if x is a model of T and y a model of T' then x and y do not match in all of the joint vocabulary.

III. REQUIREMENTS FOR COMMENSURABILITY

Still there remains another problem, namely to make precise the meaning of "to match", and by "solving" this problem we arrive at the desired characterization of incommensurability. The problem may be stated as a problem of choosing the correct solution from a whole spectrum of possible solutions. This spectrum ranges from taking "to match" simply as "to be identical" on the one extreme to defining "to match" by means of some specific set theoretic function or relation at the other extreme. If we take "to match" as "to be identical" then two models x, y match in S (where S is a subset of the joint vocabulary) iff for any term t in S its interpretations t^x *in* x and t^y *in* y are identical. Since we agreed on the set theoretic nature of all interpretations this amounts to stating that the two sets t^x and t^y – however complicated – are identical, and set theory provides a clear and easy criterion here. On the other hand the weakest definition of "match" would be to require that the interpretations of a term in two models of both theories be merely related by some set theoretic relation. The term "state", for instance, in a model of thermodynamics is interpreted by unspecified, basic entities whereas in a model of statistical mechanics it is interpreted by sequences of function values for positions and momenta. The classes of these different entities may well be related with each other by some set theoretic relation.

In between these two extremes there are many other possibilities, like replacing set theoretic relations by relations definable in various different ways in formal calculi of different logical properties and strength.[8] Intuitively, the closer we get to "identity" as defining "match" the weaker our concept of incommensurability. For the preliminary definition says that a match in a given subset S can *not* be extended to a full match. And the more we require for "match" the less we have to show in a proof that terms can *not* be matched. On the other hand, the weaker we choose our definition of "match" the stronger the resulting concept of incommensurability.

A brief reflection shows that the extremely weak definition of "match" as a mere set theoretic relation yields a concept of incommensurability so strong that it hardly will have any real instances in the history of science. For in this case we usually will be able to establish set theoretic relations – contrived ones or plausible ones – among the interpretations of *all* the common terms. (Roughly, between two classes of entities there always exist *some* set theoretic relation.) On these grounds we may dismiss this first possibility.

Next, as concerns the intermediate cases, it seems that to adopt one of these will make the notion of incommensurability dependent on a particular logical system, its strength, or on some special syntactic features of definitions. This also is reason enough to dismiss such possibilities for it would be strange that an important meta-scientific and philosophical concept should depend on the logical subtleties mentioned. I admit that this is a rash conclusion, and I will remain open for intermediate cases which do not depend on such subtleties. For the time being, however, and in the absence of "interesting" intermediate proposals (which also have to save the phenomena) I think that the remaining extreme case is the most plausible and adequate one. It is plausible because it is simple and easily applicable; it is adequate because it mirrors actual talk in scientific dispute and because it can successfully deal with the examples mentioned in

Section I (see also Section V below) while not turning other real-life examples of non-incommensurable theories into incommensurable ones. I therefore adopt "identity" as the correct solution for the definition of "match". Consequently I will say that, for some subset S of the joint vocabulary of T and T', and for two models x of T and y of T', x and y *match in S in identity* iff for all terms in S the interpretations t^x and t^y of t in x and y are identical. We than have the following final requirement for incommensurability.

Theories T and T' are *incommensurable* iff

(1) the joint vocabulary of T and T' is non-empty
(2) for all x, y: if x is a model of T and y is a model of T' then x and y do not match in the joint vocabulary in identity.

Some remarks may be added. First, it has to be stressed that condition 1) is only a rather poor expression of what we intend to cover, namely that the joint vocabulary of both theories is rather large in comparison with the union of both vocabularies. In general, it would be too much to require that both theories have the same vocabulary. But as soon as we admit differences it becomes very difficult to say that what subset of joint terms is "relatively large". In the absence of any reasonable criterion we retreat to the much weaker condition (1) on the purely formal side. In real-life examples the worst deviation from full identity of the languages will be cases where one theory (or both) contains very few (one, two or three) "theoretical" concepts not available in the other theory.

Second, it has to be stressed that these requirements only provide a basic core for a theory of incommensurability. Various specializations due to the particular circumstances in particular real examples are to be expected (see Section V).

Third, the logician will be eager to point out that these requirements are equivalent to the definition of two theories' being inconsistent. I agree.

IV. POSSIBLE OBJECTIONS

A first objection to this approach is that it does not refer to meaning and translation. Usually, the subject is discussed in terms of meaning variance or untranslatability. As already mentioned I have some reservations about such discussions. Not that I want to say they are useless. But it seems to me that the phenomena at hand can be conceptualized and clarified in less ambitious terminology. To refer to translation and meaning is to rely on deep philosophical issues which are far from being clearly understood up to now in order to deal with phenomena that do not reach quite as deep into overall philosophical themes. (Still, incommensurability is interesting enough and certainly not a surface-phenomenon.)

There was some discussion with David Pearce[9] recently who suggested a two level picture with the level of meaning represented by some kind of set theoretic relation among models (a "reduction relation") and the level of language represented by a "translation", i.e. a mapping of the sentences of both languages satisfying certain additional properties. It was clear from the beginning that a set theoretic relation among models does not cover all aspects of meaning relevant for the comparison of theories. What *became* clear in the course of the discussion is, I believe, that the same holds for translation. We are far away from being able to compress all the aspects of the notion of translation into a mapping of the sentences satisfying certain additional, precise requirements. These brief considerations confirm what was said in the last paragraph. To summarize the point: it seems to me that the absence of meaning and translation does not raise any objection against my account. On the contrary, it has to be counted as one of its positive features.

A second objection focusses on my central use of identities: considerable identity of the vocabularies, and identities of the interpretations of the common terms. From a logician's point of view it may be said that identity of the vocabulary cannot be too important because of the possibility of replacing ("renaming") the established terms by new, say, artificial ones without changing the theory. Identity of interpretation may be found unsatisfactory because in first-order logic predicates can only be characterized up to isomorphism. The first part of the objection concerning identity of terms in fact shows that the present definition is still too limited with respect to the possibility of "equivalent" reformulations of theories. If T and T' are incommensurable in the sense above it still may be the case that we find *real-life* reformulations T_1 of T and T'_1 of T' such that T_1 and T'_1 are no longer incommensurable (because, for instance, they have no terms at all in common). To this objection two things can be said. First, we may extend the definition to cover reformulations in the following way. We say that T and T' are *invariantly incommensurable* iff there exist equivalent versions T_1 of T and T'_1 of T' such that T_1 and T'_1 are incommensurable in the sense defined above. The second observation to the point is critical with respect to the actual importance of such reformulations. Of course, we know the standard examples of supposedly equivalent formulations, like Newtonian and Lagrangian mechanics, or matrix- and wave-formulations of early quantum mechanics. We do not know, however, of any detailed analysis of such examples that demonstrates equivalence, nor do we know general concepts of equivalence that were successful when applied to concrete examples of the kind mentioned. Our own recent attempts[10] tend to support this negative picture. So there is some doubt, to put it mildly, as to whether the idea of equivalent formulations is an important one as far as real cases are concerned.

The second part of the objection, namely the one concerned with identity of interpretations, from my point of view only demonstrates once more the limitations of first-order logic, and I would hesitate to accept these restrictions just on the basis of the "beauty" of completeness theorems, and compactness and Loewenheim-Skolem properties. Moreover, it seems possible to provide a formal treatment of my account that takes interpretations as basic objects (of a many-sorted first-order language), and within such a treatment the identities under discussion may be stated without any problems.

A third objection is that my characterization includes inconsistent pairs of theories, or even more sharply, that it consists simply of the requirements for inconsistency. The natural way to define inconsistency of two theories is by requiring that they have no joint model, that is, no structure for the union of all terms which satisfies the axioms of both theories. This is, in fact, equivalent to the above requirements. But, so the objection, inconsistency is a case of commensurability because inconsistent theories can easily be compared (the negative outcome of the comparison does not count against its being some comparison). Again, there are two points in reply. First, I have some difficulties in relating the terms "incommensurability", "commensurability" and "comparability" with each other. Even the relation between the first two is not one of simple negation because this would make theories commensurable which have no terms in common at all. It seems that condition (1) above should apply also in cases of commensurability, so that the relation is this:

incommensurability = condition (1) and condition (2)
commensurability = condition (1) and condition (not-2)

With respect to comparability I believe that the central adherents of incommensurability – Feyerabend and Kuhn – did not claim that incommensurability implies incomparability. On the contrary. Incommensurable theories can be and are compared with each other – even if such comparison usually takes place much later than the original quarrels. It may well be that at the moment we are not very smart in comparing incommensurable theories. But structuralist, holistic accounts of comparison (like "reduction" relations) seem to be possible even in the light of incommensurability. Therefore I would like to deny that comparability implies commensurability. But then incommensurability does not imply incomparability, and the objection fails.

Secondly, as already mentioned, my requirements "boil down" to those of "mere" inconsistency. As Feyerabend held a similar view long ago, what's new? I think the value of the present paper is not in its result (=requirements for incommensurability) but rather in the "derivation" of the result. The way in which the result was reached is independent of the writings and explications of other authors. If, in such a situation, the same result occurs, the better for the result. It is confirmed (according to the bootstrap view of confirmation). I take this as an essential contribution of this paper: to confirm that incommensurability is just inconsistency.

A last objection is that my account is not operational. In order to check whether two theories are incommensurable we have to find out identities about their respective interpretations. That is, we have to find out whether two entities (objects, relations, functions) denoted by the same term in two models of the respective theories are identical. This affords criteria for identity independent of both theories, and thus a kind of "Archimedian point" for the comparison. But such an independent point of view is feasible only for the metaphysical realist (it is given by "reality"). Therefore, so the objection, this account leads to metaphysical realism. The short reply which is dictated here by reasons of space (and which perhaps will not be compelling for many readers) refers to the distinction between form and content. There are certain things and relations which we have learned to accept by purely formal means; from this fact logics derives its right to exist. The comparison of two given, axiomatized theories according to my definition may be regarded as a purely formal set theoretic exercise, and the claim associated with my definition is that this exercise in the real-life examples mentioned above (once they are axiomatized) yields positive (i.e. incommensurable) examples.

Still, one may feel uncomfortable with this and wonder how the realist's problem of checking "cross-theory" or "cross-world" identities is circumvented or neutralized by purely formal means. Roughly, this happens as follows. From a realist, non-formal perspective, condition (2) which involves the identities under discussion may be checked for "real" cases, that is, interpretations occurring in "real" systems. T and T' would be incommensurable if no match is possible in this domain of real systems. But a formal view includes many more, "abstract" systems (models) as well. A priori there might be cases where there is no match in the domain of real systems but where there exist two abstract models that can be matched. Such a case would come out as non-incommensurable from the formal point of view adopted here, but it might be claimed to constitute an example of incommensurability by the realist. This shows that the formal perspective yields a much stronger concept of incommensurability. In view of the real cases to be covered it simply happens that this more narrow concept nicely applies. I do not want to belittle the problem of cross-world identities, I just want to say that a theory of incommensurability can go along without paying too much attention to this problem.

V. A BRIEF LOOK AT EXAMPLES

To conclude, let us really look at the phenomena as given by the examples mentioned in Section I, and see how the "new theory" applies to these cases. Each example on its own certainly needs an extensive treatment: first reconstruct the two theories involved, and then check whether the above requirements apply. It is clear that this is a program for a whole book. So I have to be very brief and sketchy (as everybody is in this context when it comes to examples).

The vocabularies of Aristotelian theory of motion and pre-cartesian kinematics do not have special terms associated with them. So both vocabularies by and large can be taken to be identical. The crucial tension may be located around the term "motion" which in Aristotle is much more comprehensive, including changes of features different from position. If we identify the interpretations in two models of the other terms strongly linked with "motion" we see that the interpretation of "motion" in an Aristotelian model is different from that in a kinematical model.[11]

Similarly, the vocabularies of the Ptolemean and Copernican theories of planetary motion are essentially the same. Apart from ordinary language the technical terms also seem to coincide.[12] But clearly there are terms the interpretations of which in the (unique) two models are different, for instance "centre of the sun's path" which is the centre of the earth in Ptolemy's and a point inside the sun in Copernicus' theory.

In phlogiston theory there is the term "phlogiston" in addition to Lavoisier's terminology, and conversely, "oxygen" does not occur in phlogiston theory. As Thomas Kuhn argued in his PSA paper there is a group of terms to which condition (2) above nicely applies,[13] and he convincingly worked out the tension between the theoretical structures on both sides and between one by one identification of those terms.

In impetus theory and Newtonian mechanics the vocabularies are roughly the same: the technical terms are present on both sides. A term creating problems of comparison is, for example, the term "natural motion". In models of impetus theory this term denotes the path of a particle at rest relative to the surface of the earth. In Newtonian mechanics it denotes the path of a particle at rest or in uniform motion relative to an "inertial system". Since the surface of the earth is not an inertial system the term has different interpretations in any two models of the respective theories. It has to be noted that this informal sketch leaves implicit a crucial point, namely the characterization of inertial systems in Newton's theory. In my opinion, if we reject recourse to metaphysics, inertial systems can only be characterized by second-order sentences talking about sets of models of the theory. If this is so then we have here a first example which for its proper treatment requires some specialization of the general core. The "specialization" (on the meta-level of incommensurability) here will have to include the full range of specializations ("special laws" at the level of mechanics) of both theories involved, and therefore will yield a notion of incommensurability for whole theory-nets.[1] This is, of course, only a hint and will have to be worked out in detail.

The last example I want to consider is that of classical and (special) relativistic mechanics.[14] Here, the only difference in the vocabularies is that "velocity of light" in the relativistic theory has acquired the status of a technical term. At a first intuitive glance it seems that condition (2) fails because in models with zero-forces the masses on both sides will coincide, and therefore there are joint models of both theories.[15] There is, however, a little problem with this intuitive account for it neglects that, properly speaking, the term "mass" is not the same in both theories. Sure, we use the same word in ordinary talk but we also admit from the beginning that mass in the relativistic theory depends on velocity whereas it does not depend on velocity in the classical theory. So, properly speaking, the term has different type in both theories and it therefore would be dumb to try checking whether it has identical interpretations in two models: it cannot have, for reasons of typification. If this is accepted, another problem arises. If "mass" is not a common term then it need not be considered in the evaluation of condition (2). But then the two theories become "commensurable" (more precisely: non-incommensurable), for the other terms may be interpreted identically on both sides at least in some models. Here we have another case which requires specialization of the basic picture. Intuitively, the adequate treatment will be to include the term "mass" in the attempt of matching but to relax the condition of identity for its interpretations. Formally, the corresponding specialization may be stated by extending condition (2) to certain well specified cases of terms which are not common to both theories but which have some strong, say, syntactic similarity (to be precisely specified).

Still the case is not settled, it really is a borderline case. Things will now depend on precisely how we relax the identity requirement for interpretations. The most natural way for the present case (which may be representative for many similar cases in which just a new argument of a well-known function is "discovered", and the theory adjusted accordingly) is to replace identity by some relation of "being part of". The classical mass function in the crucial case of zero-forces is just a "part of" the relativistic mass function, namely in the sense that each pair (particle p, mass of p) is a component of the corresponding triple: (particle p, velocity of p, velocity-dependent mass of p). Thus condition (2) becomes: for any two models x, y it is not the case that (all their interpretations in the joint vocabulary are identical *and* their interpretations of "similar" terms are related by the "part of" relation). In the example before us this condition fails because there are models in which the classical mass function, in fact, *is* contained in the relativistic one in the sense just defined. The conclusion then is that classical and special relativistic mechanics are commensurable. This result is in accordance with judgements of physicists which have taken seriously the idea of incommensurability (the majority, for which this qualification does not apply, will of course simply deny any phenomena of incommensurability in physics).

It has to be stressed that this result depends on a particular view of special relativistic mechanics,[15] namely as "being built" on individual frames of reference which have a *classical* space-time structure. I used this approach here for reasons of simplicity, but I do not at all want to defend it. I would conjecture that the inclusion of the level of space-time (at which the problem of comparison properly has to be considered) in the reconstruction of the dynamical theories will yield a case of incommensurability,[16] just as will the study of the relation of classical and general relativistic space-time and mechanics.[17]

NOTES

* This paper was written under DFG-project Ba 678/3-1. I am indebted to George Berger for critical suggestions on an earlier draft as well as for correcting my English.

[1] See, for instance, Balzer–Moulines–Sneed (1987), in particular Chap. IV.
[2] See, for instance, Feyerabend (1965, 1970).
[3] Compare also Sheibe (1976) on this point.
[4] Kuhn (1983).
[5] The term "translation" is avoided on purpose. See Section IV.
[6] In Feyerabend (1977), p. 363.
[7] See, for instance, Shoenfield (1964).
[8] One such intermediate possibility is found in Graham Oddie's contribution to this volume. Alas, its effect is to turn all the examples from Sec. I into commensurable ones. Another possibility is based on a frame used by David Pearce. See, for instance, Pearce (1982) and Balzer (1985a, 1985b).
[9] See the references in note 8 and also Pearce (1986).
[10] See Balzer–Moulines–Sneed (1987), Chap. VI.
[11] See Kuhn (1981) from which the example is drawn, for further details.
[12] Compare Heidelberger (1976).
[13] See Kuhn (1983).
[14] I have nothing to say about thermodynamics and statistical mechanics, mainly because of the lack of attempts at working out the structure of statistical mechanics. Obviously, mere identity of interpretations of common terms (like "state") will not do. Some other specialization – typical for "micro-reductions" in general, perhaps – will be needed.
[15] I here refer to the reconstructions provided by McKinsey *et al.* (1953) and Rubin and Suppes (1954).
[16] See Balzer (1984) for an axiomatic attempt at comparison at the level of space-time.
[17] See, however, Ehlers (1986) and Malament (1979) for different opinions.

REFERENCES

Balzer, W., 'On the Comparison of Classical and Special Relativistic Space-Time', in W. Balzer *et al.* (eds.), *Reduction in Science*. D. Reidel, 1984, 331–357.

Balzer, W., 'Incommensurability, Reduction, and Translation', *Erkenntnis* **23** (1985a) 255–267.

Balzer, W., 'Was ist Inkommensurabilitaet?', *Kant-Studien* **76** (1985b) 196–213.

Balzer, W., Moulines, C. U. and Sneed, J. D., *An Architectonic for Science*. D. Reidel, 1987.

Ehlers, J., 'On Limit Relations between, and Approximative Explanations of, Physical Theories', in R. Barcan Marcus *et al.* (eds.): *Logic, Methodology and Philosophy of Science VII*. North-Holland, 1986, pp. 387–403.

Feyerabend, P., 'Problems of Empiricism', in: R. G. Colodny (ed.). *Beyond the Edge of Certainty*. Prentice-Hall, 1965, pp. 145–260.

Feyerabend, P., 'Against Method', in: M. Radner *et al.* (eds.): *Analyses of Theories and*

Methods of Physics and Psychology (Minnesota Studies, Vol. IV). University of Minnesota Press, 1970, pp. 17–130.
Feyerabend, P., 'Changing Patterns of Reconstruction', *Brit. J. Phil. Sci.* **28** (1977) 351–369.
Heidelberger, M., 'Some Intertheoretic Relations Between Ptolemean and Copernican Astronomy', *Erkenntnis* **10** (1976) 323–336.
Kuhn, T., 'What are Scientific Revolutions?', *Occasional Paper* # 18. Center for Cognitive Science, M.I.T., 1981.
Kuhn, T., 'Commensurability, Comparability, Communicability', in: P. Asquith *et al.* (eds.): *PSA 1982*, pp. 669–688. East-Lansing, 1983.
Malament, D. B., 'Newtonian Gravity, Limits, and the Geometry of Space', to appear in: *Pittsburgh Studies in the Philosophy of Science*.
McKinsey, J. C. C., Sugar, A. C. and Suppes, P., 'Axiomatic Foundations of Classical Particle Mechanics', *Journal of Rational Mechanics and Analysis* **2** (1953) 253–272.
Pearce, D., 'Logical Properties of the Structuralist Concept of Reduction', *Erkenntnis* **18** (1982) 307–333.
Pearce, D., *Roads to Commensurability*, Habilitationsschrift. Freie Universität Berlin, 1986.
Rubin, H. and Suppes, P., 'Transformations of Systems of Relativistic Particle Mechanics', *Pacific Journal of Mathematics* **4** (1954) 563–601.
Scheibe, E., 'Conditions of Progress and the Comparisons of Theories', in R. S. Cohen *et al.* (eds.): *Essays in Memory of Imre Lakatos*. D. Reidel, 1976, pp. 547–567.
Shoenfield, J. R., *Mathematical Logic*. Addison-Wesley, 1964.

University of Munich

GRAHAM ODDIE

PARTIAL INTERPRETATION, MEANING VARIANCE, AND INCOMMENSURABILITY

According to philosophical folklore the thesis that theories partially define their own terms entails the thesis of meaning-variance. Meaning variance in turn is supposed to lead to the thesis of incommensurability, or at least pose severe problems for commensurability. And the incommensurability thesis goes hand-in-glove with anti-realism. The aim of this paper is to show that these purported links do not exist. But the mere contemplation of incommensurability has tempted so many more able philosophers into a swamp of confusion that it might be prudent to chart the logical geography of the territory before setting out. It has to be admitted that the map I offer has some novel features, and swamp-lovers may well object that those very features make the going too easy. I thus recommend its use only to lazy travellers who value firm ground and a well-marked path.

1. REALISM

A fundamental tenet of realism, call it the *truth doctrine*, is that the aim of an inquiry is the truth about the structure of the world. An inquirer need not seek the whole truth about *every* aspect of the world. Each inquiry delimits an *aspect* of the whole truth (for example, the truth about the number of planets, or about the laws governing planets, or about the immortality of the soul, or whatever) but the aim of the inquiry, *qua* inquiry, is to get at the whole truth about that particular aspect of the world's structure.

Another tenet of robust realism is that an inquiry may seek an aspect of the truth which is, strictly speaking, beyond complete empirical decidability, even in principle. Empirically adequate theories may nevertheless be false. Call this the *verification transcendence doctrine*.[1]

A third tenet of realism, not widely appreciated as such, is that it is at least possible for an inquiry to make (non-trivial) progress in its aim, without actually achieving it completely. That is to say, of two theories one may be a better approximation to the truth than the other, despite the fact

K. Gavroglu, Y. Goudaroulis, P. Nicolacopoulos (eds.), Imre Lakatos and Theories of Scientific Change. 305–322.

that both fall short of the whole truth of the matter in question. Call this the *possibility of progress* doctrine.

There may be more to realism than these three doctrines, but one view which it is implausible to impute to realism is the view that our current inquiries, scientific or otherwise, have already attained their goal, or even that they represent progress over previous stages of the inquiry. Realism will not have been refuted, or even damaged, if current theories turn out to be wildly wrong, or more wrong than their predecessors. Realism is a philosophical view about the nature of inquiry generally, and the relation of an inquiry to a world which is in some sense inquiry-independent. It is not a contingent thesis about the successes or failures of actual inquiries, and it is entirely compatible with robust fallibilism.

While there may be more to realism than these three doctrines, any view which eschews one or any combination of these is either a severely impoverished realism, hardly worthy of the title, or not realism at all. A brief look at the possibilities may make this claim plausible. There is no logical difficulty in espousing doctrines one and three and eschewing two; but the result usually goes under the name of *internal realism*, to mark its distance from robust, full-bodied, metaphysical, or external realism. Again, doctrines one and two may be espoused while doctrine three is rejected, but the resulting position would be profoundly depressing for most realists. If the only theory which is any good is the one which fulfils the aim of inquiry completely then every miss is as good as a mile. There can be no gradual progress in an inquiry (*qua* inquiry). Either the whole truth is hit upon, or the enterprise is an utter failure. Even a *truth* which gives only part of the truth of some matter would have to be considered worthless, because it would not realise the aim completely. A realism which eschewed the possibility of progress would be forced to accept an extreme pessimism not only about the course of actual inquiries, but about all possible inquiries, excluding only the most trivial.[2] It is only by accepting the possibility of progress that the realist's aim can seem anything other than hopelessly utopian. Both the second and third doctrines presuppose the first, and a rejection of the truth doctrine would entail a rejection of all three. Needless to say, any such rejection would be totally alien to any kind of realism.

It is worth noting here that the doctrines of truth and of the possibility of progress demand a notion of closeness to truth (or truthlikeness, verisimilitude). For if progress towards the truth is to be possible then it must make sense to say of two theories both of which fall short of the whole truth, that one is closer to the truth (or more truthlike) than the other. Popper was, of course, the first philosopher to stress the importance of the concept of truthlikeness, but having a coherent account of the concept is not merely a requirement of a Popperian account of knowledge. It is a general requirement of realism. The reason Popper was the first to stress its importance is simply that Popper was the first philosopher to take seriously two theses: firstly, that our current theories, like their predecessors, are (almost certainly) false; secondly, that our current theories *may* nevertheless realise the aim of inquiry better than did their predecessors. Popper's concern for truthlikeness thus sprang out of his combination of fallibilism and optimism. But strictly speaking, all that is required to generate the concern are the two rather modest doctrines of truth and the possibility of progress.

2. INCOMMENSURABILITY

The thesis of incommensurability is vaguer than that or realism, but if we take, more or less at random, three different characterizations given in the literature, then a few core tenets emerge.

Devitt (in 'Against Incommensurability') writes that the incommensurability thesis is:

> ... the thesis that different theories in one area may be radically incomparable because of meaning changes.[3]

Musgrave (in 'How to Avoid Incommensurability'):

> The incommensurability thesis says that the successive major scientific theories, or paradigms, or world-views are incommensurable because the meanings of terms occurring in them are different.[4]

English (in 'Partial Interpretation and Meaning Change'):

> Because the meanings of their terms change, we are told, competing theories are expressed in different, not readily intertranslatable, languages. Rival theorists tend to 'talk past' each other: their arguments 'fail to make contact'; it is as though they were members of different language-culture communities. It is in this sense that rival theories may be said to be 'incommensurable'.[5]

Thus incommensurability seems to amount to incomparability due to changes in meaning of crucial terms. But comparability is always relative to some feature or aim, and it is never made very explicit exactly what feature or aim is at issue here. Discussions almost automatically slide into the controversy over meaning changes before it is settled what kind of incomparability such meaning changes ensure. But if, as is usually assumed, incommensurability is a problem for the realist, and we take as our characterisation of realism the doctrines outlined in Section 1, then it is easy to pinpoint the aim or feature with respect to which theories are claimed not to be comparable: incommensurability is incomparability with respect to the aim of inquiry. That is to say, two theories are incommensurable if they cannot be compared for the degree to which they succeed in capturing the truth. Or what amounts to the same thing, they cannot be compared for truthlikeness.

There are two distinct, but connected, senses of 'comparable' here: a purely logical sense, and an epistemological sense, and there are two distinct, but connected, problems of truthlikeness. The *logical* problem of truthlikeness is to specify what it takes for one theory to be closer to the truth than another. The *epistemological* problem is to specify what would constitute evidence for a judgement of relative truthlikeness. The logical problem has priority, in that a solution to the epistemological problem presupposes a solution to the logical problem. Consequently there are two kinds of incommensurability. Two theories are epistemologically (or weakly) incommensurable) if there are no evidential criteria for judgements of the relative truthlikeness of the two theories. They are logically (or strongly) incommensurable if judgements of their relative truthlikeness simply do not make sense. As logical incommensurability entails epistemological incommensurability, and it seems to be what most proponents of incommensurability have had in mind, it is the kind of incommensurability discussed here.

Other aspects of incomparability have been canvassed but it is not implausible to regard these as important only because of their indirect links, or purported links, with truthlikeness. Consider just two such aspects: the lack of apparent logical conflict between incommensurable theories, and incomparability for content. One simple but appealing kind of theory of truthlikeness is based on the idea that aggregation of truths and removal of falsehoods constitutes progress towards the truth. Without too much reflection on the issue this may well be the most natural kind of theory for a philosopher to start working on. Although Popper was the first to try to articulate this theory in a precise form, it is nevertheless probably true that the theory is lurking in the back of most people's minds.[6] If this is right then it is clear why, for example, logical conflict is important for truthlikeness. One important way forward would be to root out a falsehood and replace it with its negation. But if apparently rival theories 'talk past each other' and apparently contradictory statements are not really so, then this simple model cannot be applied.

Or consider content. One way of articulating the crude theory adverted to above is to compare the truth content and falsity content of two theories. The truth content is the set of truths a theory implies, the falsity content the set of falsehoods it implies. Popper's first theory of truthlikeness, and it is very tempting, claims that either a set-theoretic increase in truth content, or a set-theoretic decrease in falsity content, guarantees a step towards the truth. But if the contents of theories cannot be compared (because the same sentences have very different meanings) then obviously this idea cannot be applied.[7]

We have then a definition of incommensurability. The thesis of incommensurability can be split into two parts. The first is a purely conceptual claim to the effect that meaning-variance guarantees incommensurability. The second is a factual claim to the effect that apparent rivals in the history of inquiries are incommensurable because of such meaning-variance. Most of the analysis that ensues will be concerned with the conceptual claim, but the conclusion will have relevance for the factual claim as well.

3. ANTI-REALISM AND COMMENSURABILITY

For our starting point we may take a brief, but characteristically bold, article by Alan Musgrave: 'How to Avoid Incommensurability'. In it Musgrave points to the slippery slope from incommensurability to anti-realism, and then argues that there is no good reason to get onto the slippery slope in the first place. This is because there is only one decent argument for the incommensurability thesis, and the major premise of that argument is false. This master argument begins with the doctrine that a theory partially defines its own theoretical terms (the doctrine of partial interpretation). Granted this doctrine it seems that the meanings of theoretical terms will be theory dependent. Substantial differences in theory will ensure substantial differences in meaning, and incommensurability seems to follow. Musgrave argues that the doctrine of partial interpretation is fundamentally mistaken, and that this emerges from the failure of the various attempts to articulate it rigorously.

Musgrave is right, I think, to stress that the main inspiration for meaning-variance theories is the idea that theories partially define or interpret their own theoretical terms. But there is available one very attractive account of partial interpretation which is not susceptible to Musgrave's criticisms: namely, Carnap's last proposal.[8] What is not widely realised, and will be argued below, is that Carnap's proposal does not generally entail meaning-variance for rival theories.[9] Moreover, the proposal is compatible with a very large measure of commensurability amongst theories, even in those cases in which meaning-variance does arise. Thus a consideration of Carnap's proposal shows not only that the arguments from partial interpretation to meaning-variance, and from meaning-variance to incommensurability, are invalid. Because it is, in at least one important sense, anti-realist, it also furnishes an example of one brand of anti-realism which violates the dogma of the supposed link between anti-realism and incommensurability.

According to the positivist tradition, out of which Carnap's proposal grew, the meaning of theoretical terms is problematic. The positivist programme was to show how such terms could be invested with meaning by being connected with observational terms. Without such a connection statements incorporating the terms would be metaphysics, or meaningless, or both. Carnap in particular was deeply concerned with this problem of the meaning of the theoretical terms, and made a series of proposals each of which was more holistic and less restrictive than its predecessor. In his early work he demanded that theoretical terms be explicitly definable in observational terms. His last proposal incorporates Quine's insights to the effect that theoretical terms gain their significance only by virtue of their role in a theory, or whole cluster of theories, a cluster which only confronts experience at the borders; and that even then the meanings of such terms are underdetermined. However, Carnap rejected Quine's pessimism as to the possibility of seperating off the stipulative from the factual content of a theory. And his last proposal demonstrates the compatibility of Quinean holism with the thesis that analytic and synthetic components of a theory are distinct.

As is well known, Ramsey gave a general method for separating off the observational content of any (finitely axiomatisable) theory or system of theories.[10] Let V_O be a collection of observational terms (however these are to be characterised) and V_T a collection, $\{T_1, \ldots, T_m\}$ of non-observational terms, and let A be a finite axiomatisation of some theory, along with whatever else a Quinean holist might want to include (low-level observational theories, correspondence rules connecting theoretical and observational terms, and so on). Let $A\{t_1 T_1, \ldots t_m/T_m\}$ be the open sentence obtained from A by replacing each theoretical term T_i with a variable t_i ranging over the sort of entity T_i is supposed to denote. Then the Ramsey sentence of A, A^R, is the existential closure of $A\{t_1/T_1, \ldots, t_m/T_m\}$:

$$A^R \qquad (\exists t_1) \ldots (\exists \tau_\mu) A\{t_1/T_1, \ldots, t_m/T_m\}.$$

Ramsey proved that for any sentence S_O which contains only observational terms, S_O follows from A if and only if S_O follows from A^R.

$$\text{(Ramsey)} \qquad A \models S_O \quad \text{if and only if} \quad A^R \models S_O.$$

Thus A^R gives us all the observational content of A. To say that A is observationally adequate (or empirically adequate) is just to say that every observational consequence of A is true. And Ramsey showed that this is tantamount to the truth of A^R.

Being a positivist, Carnap was happy to identify the factual (synthetic) component of A with its observational content, A^R. Thus Ramsey had done almost half the work in splitting up factual and stipulative components of A. The problem remained of isolating the analytic or stipulative component of A. Two desiderata immediately suggest themselves. Firstly, the analytic component must not entail any non-trivial observational consequence: it must be *observationally uncreative*. Secondly, the analytic and synthetic components together must be formally tantamount to the original theory. These two desiderata are satisfied by what is sometimes called the *Carnap sentence* of A, A^C.

$$A^C \qquad A^R \supset A.$$

Carnap's proposal is that A^C be taken to be the analytic component of A, and A^R the synthetic component. There are other sentences satisfying the two desiderata. For example, if B is any sentence over V_T (a purely theoretical sentence) which A entails then the conjunction of A^C and B also satisfies the desiderata. In fact A^C is, formally, the 'weakest' statement to satisfy the desiderata, and Carnap does not give any convincing justification for preferring it to other stronger statements. Fortunately, Winnie has supplied just such a justification. Winnie shows that the Carnap sentence is the strongest sentence of the theory satisfying the condition of *observational vacuity*.[11] A sentence satisfies the condition just in case it plays an inessential role in generating observational sentences from the theory, even in combination with other sentences of the theory. Observational vacuity entails the first desideratum, but it is stronger, and in conjunction with the second desideratum, yields A^C as the only candidate for the analytic component.

Note that Carnap's proposal does not suffer from the defect Musgrave criticises in other accounts of partial interpretation: namely, that the theoretical postulates are rendered true by terminological fiat.[12] In fact *none* of the purely theoretical non-logical, postulates of A turns out to be analytic on this account.

The semantic justification for Carnap's proposal is interesting. It is an admirably clear version of Putnam's much discussed model-theoretic argument against realism, and it was effectively available a decade before the latter. What can be shown, almost immediately, is that if A is observationally adequate then there is a way of interpreting the theoretical terms of A in such a way as to render A true. For suppose A is observationally adequate, that is A^R is true under the interpretation I_O of the observational vocabulary V_O: Since A^R is

$$(\exists t_1) \ldots (\exists t_m) A\{t_1/T_1, \ldots, t_m/T_m\}$$

it follows, from the usual Tarskian rules, that there is an assignment of sets $S_1, \ldots, S_m$ to the variables $t_1, \ldots\ t_m$, which satisfies the open sentence $A\{t_1/T_1, \ldots, t_m/T_m\}$. Now consider the extension I of I_O which assigns S_i to T_i, for each i. I is an interpretation of the full vocabulary V and, again by the usual Tarksian rules, A is true in I. Assuming for the moment that to assign an extension to a term is to interpret it, this shows that the observational adequacy of a theory guarantees the existence of interpretations which render the theory true. To stipulate, with Carnap, that A^C be analytic is to rule out as inadmissible other interpretations of the theoretical terms. It is to rule that whenever a theory is observationally adequate then its theoretical terms are to be understood in one of the ways which renders the theory true. Alternatively, if implication and equivalence are defined not in terms of *all* interpretations, but in terms of *admissible* interpretations, then Carnap's proposal is tantamount to taking the theory to be equivalent to its Ramsey sentence. In all admissible interpretations, A and A^R have the same truth value.

Carnap had no compelling argument that theoretical terms *must* be so interpreted. Indeed, given that the proposal is tantamount to the positivist doctrine that the factual content of a theory is identical to its observational content it is hard to see how any argument for the proposal could command support generally. Somewhere in the premises there would have to be a bias towards positivism.

It is clear in what sense a theory partially defines its theoretical terms. Although taking the Carnap sentence to be analytic restricts interpretations to those that are admissible, it will not usually single out a unique interpretation, or even a unique interpretation for each interpretation of V_O. Generally there will be a range of admissible interpretations of the theoretical terms. A partial interpretation is thus a specification of a *range* of completely determinate interpretations, each of which is equally legitimate from the point of view of the theory.

As I have set it out here Carnap's proposal is couched within an extensionalist theory of interpretation, according to which an interpretation of a vocabulary is an assignment of extensions (sets to predicates, truth values to sentences). While this kind of approach may be perfectly acceptable for mathematical languages (for which it was explicitly designed) it has obvious and well-documented shortcomings for factual languages. It is strange that Carnap should not have been sensitive to this given that he himself laid the foundations for an adequate intensionalist account of interpretation, in terms of possible worlds in his (1947). Many philosophers tend to think of worlds and (extensional) interpretations as interchangeable, and in this they have been encouraged somewhat by Carnap's early work. However, Carnap himself was extremely critical of this conflation in his later work, although he did not spell out in detail the kinds of confusions it engenders.[13] Briefly, an interpretation should assign to a (factual) predicate a property, and this property does not change its *identity* in different possible worlds, though it may well have different *extensions* in different possible worlds. A world is not an assignment of either sets or properties to syntactic items, but is a way things might be, or might have been. As such it can be explicated as an assignment of extensions to traits or properties, but these are non-syntactic items. When one considers higher-order systems and languages the distinction is absolutely crucial, because in these cases (extensional) interpretations cannot ape the features worlds have. But even in the first-order case it is important to make the distinction, in order, for example, to frame an adequate account of the structure of a state of affairs.[14] In any case, it is interesting to investigate whether or not the Ramsey–Carnap methods can be extended to an intensionalist theory of interpretation, and this is done below. It is shown that the methods can be extended but only under certain restrictive conditions. But my main purpose here is not to argue that the intensionalist approach is the correct one: rather, it is to show that whether one accepts an intensionalist or an extensionalist approach, Carnap's proposal allows a very large degree of meaning compatibility amongst rival theories. As such Musgrave's criticism of it, that it does not allow rival theories of, say, electrons, is unfounded.

Recall that according to the proposal a theory A yields a partial interpretation of the theoretical terms, and that a partial interpretation is a range of completely determinate interpretations, any one of which is equally legitimate from the point of view of the theory. It is then natural to say that A and B are *meaning compatible* just in case there is at least one way of interpreting V_T which is compatible with the partial specifications of both A and B.[15] That is to say, A and B are meaning compatible just in case they do not ensure that their common theoretical terms have different meanings. The conditions for meaning compatibility will depend on the theory of interpretation one adopts.

Firstly, consider an extensionalist account of interpretation. A cuts down the set of such interpretations to those in which A^C is true: the A-admissible interpretations. And a sentence is analytic if it is true in all A-admissible interpretations. Obviously the sentence $A \supset A^R$ is analytic, so that A is equivalent to its Ramsey sentence. A and B will be said to be *weakly* meaning compatible just in case there is one interpretation which is A-admissible and B-admissible. We will introduce a stronger notion below in response to an objection to weak meaning compatibility.

Proposition 1. *A and B fail to be weakly meaning compatible if and only if they are logically incompatible and both are observationally uncreative*. (For the details of the proof, see the appendix of Oddie, 1988.)

From this result it is clear how extraordinary meaning incompatibility is. For A to be meaning incompatible with B, A and B must contradict each other but neither must have any non-trivial observational consequences. From the positivist point of view this is an extremely desirable result. It demonstrates that two *purely* metaphysical theories which *appear* to be in conflict are really just 'talking past each other': there is no way of giving a common interpretation to their theoretical vocabulary. But as a matter of course, any two theories with some non-trivial observational content, and this includes observational rivals, will be meaning incompatible. Rival theories of electrons are possible, provided at least one of them has a non-trivial observational consequence.

The philosopher committed to meaning-variance might balk at this definition of (weak) meaning compatibility. She might insist that some of the interpretations which guarantee compatibility would be bizarre. It might assign to observational terms extensions which are not their actual 'intended' interpretations. (The same cannot be said about the theoretical terms without abandoning the idea that it is the theory alone which delimits the legitimate interpretations of these.) The objection is a difficult one for an extensionalist to press, because it quickly brings to light the implausibility of the underlying extensionalist theory. It is strange indeed to say that the actual set of green things is the *intended* interpretation of the predicate 'green' simply because in order to use and understand the term perfectly it is hardly required that the speaker know *which* set that is. And since speakers do not know which set that is, it is peculiar to insist that that is the set they *intend* to pick out. It may be tempting to defend the theory by pointing out that although the speaker does not know the exact identity of the extension of the set he intends (and since sets are extensional, he does not know which set he intends) he does intend to pick it out as the set falling under a certain description ('the set of green things'). But of course this is to cast the very same problem in a different guise, for we are now faced with rival extensional and intensional accounts of the meaning of definite descriptions.

In any case, let us grant to the objector the thesis that the intended interpretation of the observational vocabulary V_O is an assignment of extensions to the terms in V_O, say I_O. In this case it would be natural for the extensionalist to restrict admissible interpretations to those that are extensions of I_O. An interpretation of V is *strongly A-admissible* just in case it is an extension of I_O and A^C is true in it. And A and B are *strongly meaning compatible* just in case there is at least one interpretation which is both strongly A-admissible and strongly B-admissible.

PROPOSITION 2. *A and B fail to be strongly meaning compatible if and only if A and B are observationally adequate while A &B is not observationally adequate.*

(Again, the proof is in the appendix of Oddie, 1988.) Although it is more difficult to be strongly meaning compatible than it is to be weakly meaning compatible it is still rather easy. To fail to be meaning compatible both theories have to be individually observationally adequate while their conjunction fails to be so. Again observational *rivals* (that is theories which, in conjunction with known or assumed observational facts, deliver incompatible observational predictions) will be meaning compatible. Rival theories of electrons are thus possible.

In order to extend the Ramsey–Carnap methods to a full-blown intensionalist theory of interpretation it is necessary to spell out in some detail how such a theory is to be constructed. This is done in Oddie (1988) for a class of higher-order languages ideographically very similar to first-order languages. Briefly, an interpretation of a language assigns to symbols appropriate objects over a logical space, a class of possible worlds. Subsets of the logical space are propositions, and each interpretation induces a mapping of closed sentences to propositions. The usual logical notions which can be explicated within a possible-worlds ontology (necessary truth, implication, and so on) can, by virtue of the mapping, be transferred from propositions to the sentences which denote them. Thus a sentence A is *necessarily true* (relative to I) just in case I assigns to A the necessarily true proposition. *A implies B* (relative to I) just in case the proposition I assigns to B is true in every world in which the proposition I assigns to A is true. Truth is relative not only to interpretation, but to an interpretation *together* with the state of the world: A is true in W (relative to I) just in case the proposition I assigns to A contains W. A is true simpliciter (relative to I) just in case the proposition I assigns to A contains the actual world. But note that which world is actualised is no part of the *interpretation* of a language, the assignment of meanings to terms, and one cannot ascertain the truth value of a sentence from an examination of the interpretation alone. One must check to see which states of affairs obtain in fact.

The notion of *logical* implication, as usually defined in logic textbooks, gives the so-called logical constants a special status: they are the only terms which receive a fixed interpretation throughout, whereas all other terms are allowed to vary in their interpretation. Implication for sentences, as defined here, is relative to a fixed interpretation of *all* terms, not just the logical constants. Thus, while all the usual first-order implications will hold in each intensional interpretation, the converse does not hold. A may imply B under interpretation I, but fail to do so under interpretation I', which differs in its assignment of meanings to some terms. Logical implication and interpretation-relative implication thus stand at two different extremes, and we could consider various intermediate notions. Let Λ be a class of interpretations of a language. A Λ-implies B just in case A implies B relative to all interpretations in Λ. Implication relative to one particular interpretation is the special case in which Λ is a singleton. Logical implication is the special case in which Λ is the set of all interpretations. There are various intermediate notions. Consider a partition of the vocabulary $\{V_1,$

$V_2\}$, and suppose that I is an interpretation of V_1 over some logical space ω. Let Λ^I be the set of all extensions of I to the whole vocabulary. (Obviously, each such extension must use the same logical space as does I.) Implication relative to Λ^I is the case in which the members of V_1 have fixed meaning, but the members of V_2 do not. One particularly interesting case will be that generated by the partition of the vocabulary into observational and theoretical, together with a fixed interpretation I_O of the observational part. Call this O-implication.

Return to Ramsey's problem as understood within this intensionalist account. How can the observational content of a theory be extracted, given the positivist assumption that the interpretation of the theoretical terms is problematic, that is, they are not understood prior to their participation in a theory (along with the usual correspondence rules, observational theories, and so on). It is natural to say that B (over V_o) is an observational consequence of A just in case A O-implies B. In the appendix of Oddie (1988) there is a proof of the Ramsey result extended to this intensionalist account of interpretation:

PROPOSITION 3. *A^R has all and only the observational consequences of A.*

An anti-realist of the sort who denies the verification-transcendence doctrine will then be justified in taking A^R to represent the factual content of A. The question arises whether Carnap's method can be extended to extract the analytic component of the theory. As it turns out there is a simple argument to show that if A^R is true in a world W under I_O then there is an extension of I_O, I, to the whole vocabulary, under which the Carnap sentence $A^R \supset A$ is true in W. And so, for any given world W, it is possible to ensure that the theoretical terms are interpreted in such a way that the theory, if observationally adequate, is true in W. But it does not follow from this that the Carnap sentence is appropriately *analytic* under that extended interpretation. For A^C to be analytic (relative to I) it must be true in *all* worlds of the logical space under consideration. We could define weak and strong admissibility, as in the extensional case, but on the intensionalist account there seems to be no good reason to consider weak admissibility. Thus an interpretation is said to be A-admissible if it is *an extension of the intended interpretation of V_O* and A^C is analytic relative to that extension.

Although A-admissible interpretations do not always exist (see Oddie, 1988), they do exist in one large class of cases particularly attractive to positivists: namely, first-order theories. Indeed some positivists would be inclined to impose the condition of being first-order on a theory for it to be acceptable in the first place.

PROPOSITION 4. *If A is a first-order theory then there is an A-admissible interpretation*. (Proof in the appendix of Oddie 1988.)

Thus Carnap's method can be extended at least to first-order theories. To say that A^C is to be regarded as analytic is to say that the only legitimate interpretations are those which are A-admissible. Under what conditions are two theories A and B meaning compatible? Obviously, when there exists at least one interpretation which is both A-admissible and B-admissible. We have the following (intensionalist) analogue of Proposition 2:

PROPOSITION 5. *If A and B are first-order theories then A and B fail to be meaning-compatible just in case there is a world W in which A and B are each observationally adequate, but in which A &B is not observationally adequate*. (Proof in the appendix of Oddie 1988.)

It follows immediately from the theorem that observational rivals must be meaning compatible. If A and B deliver incompatible observational sentences then there can be no world in which both are observationally adequate. When it is remembered that included in A and B are all the low-level observational laws as well as correspondence rules necessary to generate specific observational predictions it is clear that meaning compatibility is rather extensive. In particular, it will be entirely possible to have rival theories of electrons. Meaning incompatibility arises not so much from *massive* disagreement at the theoretical level, as from rather subtle differences: namely, those which permit piecemeal observational adequacy together with global observational inadequacy.

So far it has been shown that the doctrine of partial interpretation permits a very large degree of meaning compatibility. It remains to show that with or without meaning compatibility the kind of anti-realism espoused by Carnap is rather hospitable to commensurability.

In order to give a reasonably rigorous demonstration of commensurability it is necessary to have to hand a reasonably rigorous account of the comparability of theories for truthlikeness. Although there are a number of theories in existence only one of these has been developed explicitly for higher-order frameworks. In Oddie (1986) it is shown how to define truthlikeness for higher-order systems by means of a thoery of higher-order normal forms (called permutative normal forms). Constituents are (possibly higher-order) sentences which normalise the informative content of theories which are maximally informative (relative to some parameters, like quantificational complexity), and all sentences can be shown to be equivalent to disjunctions of constituents. (This result holds whatever kind of equivalence, logical or interpretation relative, or something between, is considered). Thus every sentence is equivalent to one which gives its information in a highly standardized form. Furthermore, because the syntactic structure of a constituent reflects the structure of the worlds in which it is true, the distance (or likeness) between the structure of worlds can be gleaned from a measure or distance (or likeness) between constituents. A theory is the more truthlike the closer are the constituents in its normal form to the true constituent. However, it is also shown in Oddie (1986) that constituents are suitable for defining distance between structures only if the language in which they are couched is suitably interpreted. For a language to be suitably interpreted the primitive predicates must stand for certain special properties: namely, those which generate the logical space, the primary properties.[16] Each logical space is a set of distributions of extensions through some primary or basic properties, and it is only in virtue of this that the space itself has any determinate structure which enables judgements of likeness. And only if the primitive predicates stand for these primary properties will the syntactic structure of the constituent reflect the common structure of the worlds in which it is true.

What seems to lie behind the anti-realism of Carnap (and perhaps Putnam and Dummett as well) is the assumption that the space of possibilities at issue in any 'legitimate' inquiry is generated solely by observational properties and relations. That is to say, if two states of affairs are identical as far as the extensions (or values) of observables go, then they are identical *simpliciter*. Put slightly differently, everything supervenes on a base of observational properties. There can be no change in the world without some change in the distribution of observational properties. Given this assumption it is entirely natural to demand that legitimate interpretations be extensions of interpretations of the observational vocabulary. Suppose I_O interprets V_O over a space of possibilities ω_O generated entirely by observational properties and relations. Then, since the meanings of theoretical terms are not given independently of the theory, the full factual content of a theory is given by its Ramsey sentence. Consequently to get at the whole factual content of a theory it is not necessary to countenance logical spaces any more complicated or richer than ω_O itself. If V_T needs to be interpreted at all it may as well be interpreted over the space ω_O. (Note that this way of interpreting Carnapian positivism or anti-realism is available only once we have distinguished between worlds and interpretations in the right way.) Furthermore, it is also entirely natural to demand that the primitive predicates stand for the very observational properties which generate the space ω_O. That is, I_O is a suitable interpretation of V_O.

It follows that for the purposes of comparing A and B for truthlikeness it is not necessary at all to ascend to the theoretical vocabulary and interpretations of it. A^R and B^R are couched in the observational vocabulary which is, by assumption, suitably interpreted. Moreover, they give the full factual content respectively of A and B. Given that we have an adequate theory of truthlikeness for higher-order sentences it follows that A^R and B^R are comparable for truthlikeness, that is to say, commensurable. This result holds even if A and B are *not* meaning compatible. The existence or non-existence of admissible interpretations of the theoretical aspect of the theories is irrelevant to their comparability for truthlikeness. In order to compare theories for truthlikeness they have to be couched in suitably interpreted languages, and since the only legitimate space (for an anti-realist of the Carnapian kind) is generated by observables, the interpretation of theoretical terms plays no part in such comparisons. From the point of view of commensurability, V_T is redundant, and all sentences formulated over V_O are (relative to I_O) comparable for truthlikeness, whether they are first-order or (like A^R and B^R) higher-order.

In fact it should come as no surprise that Carnap's anti-realism should be hospitable to such widespread commensurability. The whole point of cutting theories down to observational size is to eliminate comparability problems that arise out of the excess metaphysical baggage theories may seem to carry with them.

5. CONCLUSION

There are, then, no very clear connections between the doctrines of partial interpretation and meaning-variance, between those of meaning-variance and incommensurability, between those of incommensurability and anti-realism. Given the only clear account of partial interpretation available, most rival theories are meaning-compatible. But even if they are not, it is quite in order for the anti-realist of the Carnap–Putnam persuasion to be content with the evident commensurability of Ramseyfied theories, since these contain all the factual content on the underlying positivist assumptions.[17]

NOTES

[1] That this doctrine (of verification transcendence) is at the heart of the realist/anti-realist debate receives strong support in Smart (1986).
[2] For a more detailed account of the necessity for realists to give an explication of truthlikeness, see Oddie (1986), pp. 1–4, and 10–20.
[3] Devitt (1979), p. 29.
[4] Musgrave (1979), p. 336.
[5] English (1978), p. 58.
[6] Popper (1963), p. 234.
[7] As is now well known, Popper's simple idea turned out to be inadequate, in the sense that it fails to deliver intuitively obvious judgements on false theories. Rather ironically it has the consequence that *no* two false theories are comparable for truthlikeness: unmitigated incommensurability. See Tichý (1974), and for a longer discussion, Oddie (1986), Chapter 2.
[8] See Carnap (1966), Chapter 28. For an extremely clear exposition together with some interesting additional results, see Winnie (1975).
[9] That Carnap's proposal does entail meaning-variance is argued in English (1978), as well as Musgrave (1979), but the most detailed technical paper on these lines is Williams (1973). Williams proves what appears to be the exact opposite of my claim here, and there is nothing wrong with that proof. For further clarification, see note 14.
[10] For an extension of the Ramsey–Carnap method to infinite postulate systems, see Tichý (1970).
[11] See the excellent exposition of Winnie (1975).
[12] Musgrave (1979), p. 342 (quoting Hempel).
[13] Carnap (1971), p. 56: note that Carnap uses the term 'model' for what I here call a 'world'.
[14] For further arguments for the distinction between worlds and interpretations, see Oddie (1968), pp. 34–8, 60–6, 152–5.
[15] It is by virtue of this definition of meaning compatibility that results which might appear to conflict with those in Williams (1973) are derived. There is no formal contradiction however – it is just that Williams would probably consider this definition too weak. Certainly his conditions for *intertranslatability* (p. 360) are much more stringent. However, my definition of meaning *compatibility* is not meant to capture *meaning identity*. And once we take the *partiality* in Carnap's account of partial interpretation seriously, then this seems entirely natural. The question of strict intertranslatability arises only with respect to two *complete* specifications of meaning, that is, two determinate interpretations of the relevant vocabulary.
[16] See Oddie (1986), pp. 34–8, 60–5.
[17] This paper overlaps in some of its results with the first half of Oddie (1988). In that paper I also argue that realists, by virtue of core realist doctrines, are committed to quite a large degree of incommensurability, and may even be in a worse position than are Carnapian anti-realists. I am grateful to the editors of this volume, to Robert Nola (the editor of the volume in which Oddie (1988) appears) and, to Kluwer (the publisher of both volumes) for allowing

these two papers to overlap in content. I am also very grateful to the organisers of the conference for inviting me to give a paper.

REFERENCES

Carnap, R., *Meaning and Necessity*. University of Chicago Press, 1947.
Carnap, R., in *Philosophical Foundations of Physics*, ed. by M. Gardner. New York: Basic Books, 1966.
Carnap, R., in *Studies in Inductive Logic and Probability*, Vol. 1, ed. by R. Carnap, and R. Jeffrey, Berkeley: University of California Press, 1971.
Devitt, M., 'Against Incommensurability', in *The Australasian Journal of Philosophy* **57**(1) (1979) 29–50.
English, J., 'Partial Interpretation and Meaning Change', in *The Journal of Philosophy* **75** (2) (1978) 57–76.
Musgrave, A., 'How to Avoid Incommensurability', in *The Logic and Epistemology of Scientific Change* (*Acta Philosophica Fennica*, **30** (2–4) (1979) 336–46)
Oddie, G., *Likeness to Truth*. Dordrecht: Reidel, 1986.
Oddie, G., 'On a Dogma concerning Realism and Incommensurability', in *Relativism and Realism in Science*, ed. by R. Nola. Dordrecht: Reidel, 1988.
Popper, K., *Conjectures and Refutations*. London: Routledge and Kegan Paul, 1963. References are to the 1972 edition.
Smart, J. J. C., 'Realism v. Idealism', in *Philosophy* **61** (1986) 295–312.
Tichý, P. 'Synthetic Components of Infinite Classes of Postulates', in *Mathematische Logik und Grundlagenforschung* **14** (1971) 167–78.
Tichý, P., 'On Popper's Definitions of Verisimilitude', in *The British Journal for the Philosophy of Science* **25** (1974) 25–42.
Williams, P., 'On the Logical Relations between Expressions of Different Theories', in *The British Journal for the Philosophy of Science* **24** (1973) 357–408.
Winnie, J., 'Theoretical Analyticity', in *Rudolf Carnap: Logical Empiricist*, ed. by J. Hintikka (ed.) Dordrecht: Reidel (1975).

Massey University

NANCY J. NERSESSIAN

SCIENTIFIC DISCOVERY AND COMMENSURABILITY OF MEANING

INTRODUCTION

No one issue has so dominated the landscape of contemporary philosophy of science as has the "problem of incommensurability" of scientific theories, so named by Kuhn and Feyerabend in 1962. The "problem of incommensurability" in fact comprises several interrelated problems, among which are: logical compatibility and comparability, justification and validity, meaning, rationality, and progress. My concern here is with incommensurability of meaning. Succinctly put, the problem, as posed by Kuhn and Feyerabend, is this: If a scientific concept derives its meaning from its place within a theoretical network (T) and a substantial change of theory occurs (T'), then the interconnections among meanings of the concepts (including the introduction of new concepts) change. It is this "meaning variance" in cases of theory change which causes the "problem of incommensurability of meaning": How can the adherents of theory T and those of theory T' communicate with one another since, in a very real sense, they are speaking different languages?

In reviewing the papers in *Criticism and Growth of Knowledge* for this conference, I was struck by the fact that none of the contributors, apart from Kuhn and Feyerabend, address this problem. The omission in the case of Lakatos is especially puzzling since 'Falsifiability and the Methodology of Scientific Research Programs' is an attempt to apply results of his earlier analysis of mathematical theories in 'Proofs and Refutations' to scientific theories. The main thesis of that work is that a change of mathematical theory involves, centrally, a change of concepts. As a consequence, understanding a change of mathematical theory requires addressing the problem of why and how changes were made in specific concepts. Thus, it is rather surprising that Lakatos did not address the problem of conceptual change for scientific theories, and thus, the Kuhnian "problem of incommensurability of meaning". What he claimed for mathematical theories is true of scientific theories as well: understanding a change of

K. Gavroglu, Y. Goudaroulis, P. Nicolacopoulos (eds.), Imre Lakatos and Theories of Scientific Change. 323–334. © *1989 by Kluwer Academic Publishers.*

scientific theory requires an analysis of how and why concepts changed and of the interaction between the development of the concepts and the development of the formalism of a theory. Willliam Berkson has discussed this difference between "Lakatos one" (of 'Proofs and Refutations') and "Lakatos two" (of 'Falsifiability. . . .'), and speculates that it occurs because "Lakatos two" was unable to work out the problem of how to combine conceptual change with questions of validity for scientific theories.[1]

Lakatos is not the only one to fail to deal with the "problem of incommensurability of meaning". After an initial flurry of activity, this aspect of the more general problem has receded into the background of philosophical discussion. The most widely accepted response – the "causal theory of meaning" – bypasses the heart of the problem, incommensurability of sense or intension, and attempts to provide for commensurability of reference or extension after theory change. This response has developed out of an attempt to transfer the results of analysis of a problem in the philosophy of language to what is perceived to be a similar problem in the philosophy of science: how to maintain a fixed reference in the face of change of description about the entity referred to. While this approach may succeed with proper names and even natural kind terms, the differences between these and theoretical terms in science are significant enough to create substantial problems when transferring the "causal theory" to the domain of scientific languages.[2] And, even if we were successful in providing for continuity of reference, we would still be conceding the main aspect of the "problem", that of incommensurability of sense, which makes impossible true communication and understanding.

My intention in this paper is to examine how the process of formation and change of scientific concepts can create *commensurability* of meaning. What is needed is to establish what is meant by continuity of meaning between the concepts of T and T'. This requires an analysis of (1) how "the meaning" of a scientific concept is to be represented and (2) what features of the processes through which scientific concepts emerge and are subsequently altered contribute to continuity of meaning.

Before addressing these questions, though, it must be acknowledged that the characterization of the process of concept formation and change in science put forward by Kuhn and Feyerabend – and which lies at the heart of the problem – is seriously distorted. Their characterization is that scientific theories change through "revolutionary overthrow"; i.e., in a discontinuous manner such that the concepts of the new theory completely replace those of the previous. Thus, there is no *development* from previous theories and their concepts. Conceptual change is likened to a *Gestalt*-switch: first we see the duck, then the rabbit, but never both together. The change is abrupt, with no analyzable process leading from one to the other. This characterization of the process of theory change forces the problem of incommensurability upon us. It is, however, a mischaracterization. The source of this mischaracterization is the failure to examine the "fine-structure" of theory change, i.e, the period of transition between the old and new theory. For example, the development of electromagnetic theory in the 19th century played a crucial role in the passage from classical to relativistic mechanics. The transition was not made directly or abruptly from one to the other, but required passing at least through the theory of electrons. If we make the mesh fine enough in the historical grid, considerable commensurability of meaning from one period to the next in science can be shown. There is genuine novelty and discontinuity in change of theory, but there is also continuity which needs to be accounted for. The problem to be addressed, then, is not how to live with "incommensurability of meaning"; but rather, how to account for "meaning variance" *with* commensurability. In the remainder of this paper I will draw upon material from my study of the development of the electromagnetic field concept from Faraday to Einstein to address the two fundamental questions posed above.[3]

HOW IS "THE MEANING" OF A SCIENTIFIC CONCEPT TO BE REPRESENTED?

A determination of what is meant by "the meaning" of a scientific concept is a prerequisite to any characterization of how such concepts change. Yet, this question has seldom been addressed explicitly. The most widely assumed form of representation for a scientific concept is that of a "definition", i.e., a set of conditions each of which is necessary and all of which are jointly sufficient to define it. The "definitional view" of the representation of a concept is so deeply ingrained in our intellectual tradition that it has been acting as a metatheoretical prescription. For the most part, philosophers and historians of science have accepted uncritically – either tacitly or explicitly – this form of representation.[4] What few challenges there have been have come mostly from those concerned with sciences where classification by definition has proven to be an obstacle.[5]

It is widely acknowledged, however, that the problem of how to select the necessary and sufficient conditions which define a scientific concept has yet to be resolved. Central among the difficulties is how to distinguish between a necessary or "essential" feature and an "accidental" one; i.e., between one all instances must have and one which most instances have. In the case of 'electromagnetic field' this difficulty is apparent when trying to decide, for example, whether "is the state of space" is essential to the concept or not. If it is essential, then there is no electromagnetic field concept before that in the special theory of relativity, since none of the other historical candidates examined has this feature. We can see now how the definitional form of representation has contributed to the "problem of incommensurability": if features *essential* to the meaning of 'Y' in T differ from those *essential* to the meaning of 'Y'' in a later theory T', how can the concepts be related?

Even if it were possible to distinguish clearly between essential and accidental features, the problem of how to select those features which represent a particular concept would remain. Taking once again the electromagnetic field concept, in the period between Faraday and Einstein there were at least four different candidates from which to select the features. Whatever course we take, we end up in the paradoxical situation of either having to make all of the concepts the same in at least their essential features – which is a serious distortion of the historical data – or having four different concepts, with no means of specifying the relationship between them that has been uncovered by historical analysis. In short, the definitional form of representation of the meaning of a scientific concept cannot accommodate a substantial body of historical data. Rather than trying to fit scientific concepts to an *a priori* form of representation, it would be more productive to start from the historical data on various concepts and see what form of representation can accommodate these. Using the data from my case study of the electromagnetic field concept, I will sketch the requirements for a form of representation that can allow for change *with* continuity of meaning. To allow for a change of meaning, either intra- or intertheoretically, the representation must be both synchronic and diachronic. That is, it must enable us to stipulate what constitutes "the meaning" of a scientific concept at a point in its history and as it changes over time. We need, also, to make the distinction between a "concept-type", such as 'electromagnetic field', and a "concept-token", such as the electromagnetic field concept of the theory of electrons. For each concept-token, i.e., synchronically, its meaning can be represented by a multicomponent "vector", the components of which will be determined by examining the role of that kind of concept within a theory; and not by considering what is or is not essential to its meaning. These components (or "features") will be different for different kinds of concepts, e.g., substantive concepts and dispositional concepts. For a quantitative substantive concept such as "electromagnetic field', salient components would include: mathematical structure, causal properties, function, and ontological status. To exhibit the concept-type, i.e., to obtain a diachronic representation, the vector can be expanded into a "matrix" (what I have elsewhere called a "meaning schema"[6]), which exhibits how the various components changed or did not change over time (see Table I). Thus, the concept-type, rather than being represented by a fixed set of necessary and sufficient conditions, is represented by a set of salient features of concept-tokens, whose earlier and later forms bear a "familial" relationship to one another. The overlapping set of features makes the concept into a unit and entitles us to call it *the* '*Y*'. Additionally, it is possible for the representation to change over time. Thus, such a form of representation allows for the representation of development, change, and continuity in a way the definitional form cannot. The actual line of descent from one concept-token to the next, e.g., from the Faradayan electromagnetic field concept to the Maxwellian, is to be found in the reasoning leading from one to the other. If we impose these "chain-of-reasoning connections" on the meaning schema, we then have a full representation of change *with* continuity of meaning, which was our objective.

HOW ARE SCIENTIFIC CONCEPTS FORMED?

In this section I will discuss how examination of the processes through which scientific concepts emerge and are subsequently altered adds another dimension to our account of the continuity of meaning from one comprehensive scientific conceptual framework to the next. Taking again the electromagnetic field case study, we can see not only how each successive field concept was constructed out of the previous, but also where the points of connection with the conceptual framework of Newtonian mechanics were made in the transition from it to relativistic mechanics. In broad outline, mid-19th century Newtonian mechanics had two distinct mathematical representations for the transmission of forces: as actions at a distance and as continua. An analysis of electric and magnetic actions as actions at a distance was undergoing development within the Newtonian framework. In opposition to this, Faraday argued for a continuous-action representation of these actions. Aspects of

Table I. Concept type: 'electromagnetic field'.

	Ontological Status	Function	Mathematical Structure	Causal Power	
Faradayan	Substance (preferred) or state of aether	Transmits electric and magnetic actions continuously through region surrounding bodies and charges	Unknown	Certain electric and magnetic effects	Concept tokens
Maxwellian	State of mechanical aether	Same plus optical	Maxwell's equations	All electric and magnetic effects, optical effects, radiant heat, etc.	
Lorentzian	State of non-mechanical aether	Same	Same plus Lorentz force	Same	
Einsteinian	State of space (ontologically on a par with matter)	Same	Same but relativistic interpretation	Same	

Faraday's own field conception are non-Newtonian, but his analysis remained incomplete and essentially qualitative. In an attempt to provide a quantitative analysis *while* incorporating electric and magnetic phenomena into the Newtonian framework, Maxwell took Faraday's general field speculation and applied techniques and forms of representation from continuum mechanics to it. Lorentz tried to incorporate the charged particles of the action-at-distance representation into the continuous-action representation by having an immobile aether and the charged particle exist in the same place. This "forcing together" of the two frameworks culminated in Einstein's formulation of relativistic mechanics, which eliminated the notion of a substantial aether from physical description. However, the problem of how to resolve the "field-particle dualism" introduced here is still a fundamental conceptual problem of modern physics.

It is clear from the electromagnetic field study that existing conceptual structures provide the basis for the construction of new structures. Besides the continuity between the various electromagnetic field concepts, there are two main points of connection with the Newtonian framework: (1) in the application of the forms of representation of continuum mechanics to electromagnetism and (2) in the incorporation of the charged particles of action at a distance into the continous-action representation of electromagnetism. The results of this study support my contention that it is through examination of the actual methods through which scientific concepts are constructed that what continuity of meaning there may be between successive theories is uncovered. In the time remaining the role of analogical reasoning in concept formation, as it pertains to (1) in particular, will be discussed. However, I do want to take the time to point out the complexity of the problem of unpacking and analyzing the specific methods employed in concept formation and change in science. One must keep in mind that these are always embedded in a context. They are employed within frameworks of beliefs and in response to specific problems. These beliefs and problems are theoretical, experimental, methodological, and metaphysical in nature. When we examine the reasoning leading from one phase of development to the next, we see that changes in meaning come about in response to changes in beliefs and problems. For example, Maxwell altered the Faradayan field concept in response to his belief that the new conception must be incorporated into the Newtonian framework. This belief, and the problems of finding a suitable mathematical representation that it led to, provided the basis for

Maxwell's selection of suitable analogies. These changing "networks of beliefs and problems" are in part communal and in part individual. They overlap to such an extent as to provide continuity and diverge enough so as to create fundamental changes. To have a full understanding of the processes of concept formation and change in science, these networks must be included in our analysis.

Analogical reasoning is a key method in the construction of scientific concepts. A general characterization of this method is that it performs "assimilating" and "articulating" functions. That is, analogical reasoning provides temporary meaning for the new conceptualization by assimilating the phenomena under investigation (in the "target domain") to known structures (in the "base domain"). In the process of mapping the known structures onto the target domain and fitting them to the requirements of the target phenomena, a new concept is articulated. Analogical reasoning thus creates a bridge from existing conceptual frameworks to new.[7] What the new conception incorporates, and thus shares with the domain of the analogue, are certain abstractions expressed in the relations that are mapped from the base to the target domain.

Analogical reasoning figured centrally in Maxwell's construction of a quantitative electromagnetic field concept. It is not possible to go into the details of Maxwell's analogies here. To provide some idea of the method, though, a sketch of the analogy used by Maxwell in his dynamical analysis of the elctromagnetic field will be given. The analogy provided a means of expressing certain potential stresses and strains in the electromagnetic medium in terms of well-formulated relationships between known mechanical phenomena. It was designed to take the following into account: (1) a tension along the lines of force, (2) a lateral repulsion between them, (3) the occurrence of electric actions at right angles to magnetic, and (4) the rotation of the plane of polarized light by magnetic action. Maxwell maintained that a mechanical analogy consistent with these four constraints is that of a fluid medium composed of elastic vortices under stress. The problem of how to construe the relationship between electricity and magnetism required a modification of the analogy. Since contiguous parts of the vortices must be going in opposite directions, mechanical consistency requires the introduction of "idle wheels" to keep the rotation going. Maxwell conceived of these "idle wheels" as small spherical particles, surrounding the vortices and revolving in the direction opposite to the motion of the vortices, without slipping or touching each other. There is a tangential pressure between the spherical particles and the vortices. The dynamical relationships expressed here between the particles and the vortices are those between a current and magnetism. Finally, in his analysis of electrostatic induction, the analogy was once again modified to express the necessary relationships. Now the entire medium was treated as an elastic solid. This part of the analysis was to provide Maxwell with a key element of his representation – the "displacement current" – from which it followed that electromagnetic actions are transmitted at approximately the speed of light. Maxwell's pictorial representation of this analogy is given in Figure 1.

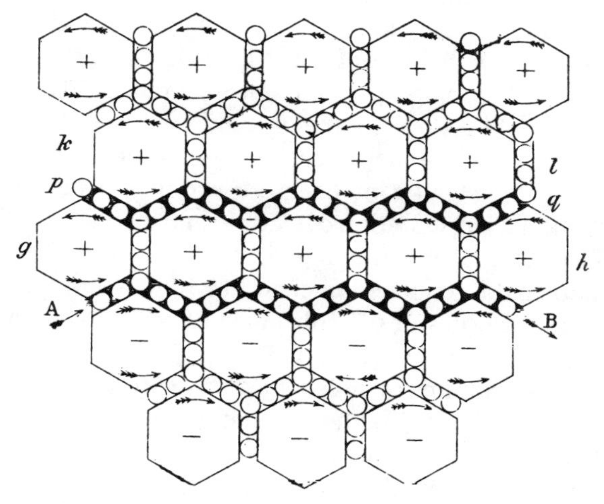

Fig. 1. *Maxwell's pictorial representation of the assumed stresses and strains in the aether.*

Thus, Maxwell's specific applications of what he called the method of "physical analogy" consisted of working out a possible continuous-action mathematical representation for electric and magnetic actions on the basis of a partial isomorphism between the laws of certain phenomena in the domain of continuum mechanics and some relationships assumed to hold for electromagnetic phenomena. As employed by Maxwell, a "physical analogy" consists of a mathematical formalism plus a pictorial representation of the relationships to be applied in the analysis. The key to success in applications of this method lies in the recognition of the *structural* similarity between the two domains. As Maxwell said:

In many cases the relations of the phenomena in two different physical questions have a certain similarity which enables us, when we have solved one of these questions, to make use of our solution in answering the other. The similarity which constitutes the analogy is not between the phenomena themselves, but between the relations of these phenomena.

... We must not conclude from the partial similarity of some of the relations of the phenomena ... that there is any real physical similarity between the causes of the phenomena. The similarity is a similarity between relations, not a similarity between things related.[8]

That is, in constructing a mapping between the base and target domains, no assumption should be made that the actual physical phenomena in the two domains are the same. For example, a current does not have the attributes of the spherical particles; rather, the dynamical relationships between the balls and the vortices are those assumed to hold between an electric current and magnetism. By employing this method in the construction of a quantitative field concept, Maxwell was able to exploit the powerful representational capabilities of generalized dynamics without making any hypotheses as to the specific nature of the underlying processes, other than the quite general assumption that they are mechanical in nature. Thus, significant progress was made in the quest for a continuous-action representation of electric and magnetic actions without full articulation of the electromagnetic field concept. The meaning of 'electromagnetic field' remained unclear in a fundamental way: the actual process in and properties of the substance of which the electromagnetic field is a state, i.e, the aether, remained unspecified. The unclarity results from conflating the notions of 'mechanical' and 'dynamical'. The relationships common to both the base and the target domains are those of what Maxwell called a "connected system", i.e., a general dynamical system. As was worked out in the period between Maxwell and Einstein, 'dynamical' is a broader notion than 'mechanical'. Newtonian mechanics is just one instance of a general dynamical system, relativistic mechanics is another. That is, the relationships expressed in the electromagnetic field equations are those of a dynamical system, but one which is non-Newtonian.

It can be seen from the preceding that the Maxwellian electromagnetic field concept was constructed by first assuming the field to be a state of a mechanical medium, and then finding a "common abstraction" between the laws of other mechanical continua and the assumed stresses and strains in the electromagnetic medium. This "common abstraction" provides for continuity of meaning between the Newtonian framework and the developing electromagnetic framework. As Oppenheimer said of a similar case of analogical reasoning, "it is not the fact that one can use the mathematics but the fact that the structure and relations are the same that is the decisive discovery."[9]

CONCLUSION

In this paper I have tried to sketch an account of conceptual change in science which allows for meaning variance with commensurability. The requirements are (1) the development of a form of representation for "the meaning" of a scientific concept which can accommodate *change* rather than just replacement and (2) the incorporation into our analysis of those aspects of the actual processes of concept formation and change that provide continuity of meaning between the conceptual frameworks of science. What has been shown, in outline, is that the meaning-schema representation of a scientific concept makes sense of the notion of overlap in meaning in cases where a concept undergoes a change of meaning, as with the different electromagnetic field concepts. Additionally, a step has been made towards accounting for continuity of meaning in the case where a new scientific concept is introduced and articulated to the point where it is scientifically viable (as it stands, Faraday's field concept is not). Where analogical reasoning is the primary means of concept formation, "shared meaning" is to be found in the abstractions, created in the mapping process, that are common to both the base and target structures.

I have argued elsewhere that the analysis of the issues raised by these requirements can be furthered by using a "cognitive–historical" method.[10] If we grant the assumption that the cognitive mechanisms at work in the meaning-making dimension of science are not fundamentally different, i.e., different *in kind*, from those we employ in "ordinary" contexts, then recent work in the cognitive sciences can be brought to bear on both the problem of how to represent the meaning of concepts and that of how to understand the process of concept formation by analogical reasoning. And, of course, more historical data need to be gathered, i.e., more case studies of conceptual change in science must be conducted.

ACKNOWLEDGEMENTS

Research for this paper was funded in part by Office of Navel Research Grant NR 667–545 and Institutional Grant for the OERI Center for the Study of Learning G-86-0005. I would like to thank Richard Grandy for his comments.

NOTES

[1] Berkson (1976) and in conversations with me in preparation of this paper.
[2] For discussion, see Shapere (1982) and Nersessian (1984), pp. 23–29.
[3] Nersessian (1984), pp. 153–9.
[4] Notable exceptions are, of course, Wittgenstein (1953), especially sections 65–88, and Putnam (1975).
[5] See, for example, Whewell (1847), I: 66–91.
[6] Nersessian (1984), pp. 153–9.
[7] This point was made by Sellars some time ago. Precisely what gets mapped in the analogical reasoning process was the subject of an unresolved dispute between Hesse (1966, 1970) and Sellars (1965). For an analysis of this issue and of Sellars' views on conceptual change, see Brown (1986).
[8] Maxwell (1888), pp. 51–2.
[9] Oppenheimer (1956), p. 131.
[10] Nersessian (1987).

BIBLIOGRAPHY

Berkson, W., 'Lakatos One and Lakatos Two: An Appreciation,' in R. S. Cohen *et al*. (eds.), *Essays in Honor of Imre Lakatos*, 1976, pp. 39–54.
Brown, H., 'Sellars, Concepts, and Conceptual Change,' *Synthese* **68** (1986) 275–307.
Hesse, M., *Models and Analogies in Science*. Notre Dame: University of Notre Dame Press, 1966.
Hesse, M. 'An Inductive Logic of Theories,' in M. Radner and S. Winokur (eds.), *Minnesota Studies in the Philosophy of Science IV*. Minneapolis: University of Minnesota Press, 1970, pp. 164–180.
Maxwell, J. C., *An Elementary Treatise on Electricity*. Oxford: Clarendon Press, 1888.
Nersessian N., *Faraday to Maxwell: Constructing Meaning in Scientific Theories*. Dordrecht: Martinus Nijhoff Publishers, 1984.
Nersessian, N., 'A Cognitive – Historical Approach to Meaning in Scientific Theories,' in N. Nersessian (ed.), *The Process of Science: Contemporary Philosophical Approaches to Understanding Scientific Practice*. Dordrecht: Martinus Nijhoff Publishers, 1987.
Oppenheimer, R., 'Analogy in Science,' *American Psychologist* **11** (1956) 127–35.
Sellars, W., 'Scientific Realism or Irenic Instrumentalism,' in R. Cohen and M. Wartofsky (eds.), *Boston Studies in the Philosophy of Science 2*. Dordrecht: D. Reidel, 1965, pp. 171–204.
Shapere, D., 'Reason, Reference, and the Quest for Knowledge,' *Philosophy of Science* **49** (1982) 1–23.
Whewell, W., *Philosophy of the Inductive Sciences Founded Upon their History*, 2 vols. (London), 1847.
Wittgenstein, L., *Philosophical Investigations*. Oxford: Oxford University Press, 1953.

Princeton University

PART V

D. A. ANAPOLITANOS

PROOFS AND REFUTATIONS: A REASSESSMENT

Since the appearance in print of Lakatos' *Proofs and Refutations*, much lively discussion has taken place between philosophers and mathematicians. It is by now generally accepted that it was a fine piece of work which, though it did not by any means settle all of the issues it raised, helped us to redirect our attention toward the dynamics of mathematical discovery. We came to realize that the study of the dynamics of mathematical discovery is equally important for both the historian and the philosopher, as well as for the mathematician. Such a study, though, is extremely difficult to carry out, and models which are meant to be rational reconstructions (or "distilled" histories[1]) of the process of mathematical discovery can only be partially successful. This is mainly due to the enormous difficulties stemming from the complexity of the issues one is faced with when one attempts both to describe such a process and to provide us with explanatory schemata about it. This is, I think, the problem Lakatos' *Proofs and Refutations* suffers from. My aim in this paper is to discuss or – at least – indicate what I consider to be some of the weak points of the work.

I

The main problem is related to the doubt expressed by mathematicians and philosophers alike: the method Lakatos describes is extrapolated in an almost *ad hoc* manner from a very narrow set of historical data. More specifically "that while the method of proof-analysis described by Lakatos may be applicable to the study of polyhedra, a subject which is 'near empirical' and where the counterexamples are easily visualizable, it may be inapplicable to 'real' mathematics".[2] The editors of the book version of Lakatos' work seem to think that "Chapter 2 of the main text and appendix I[3] should allay such a doubt".[4] But this disclaimer is insufficient. The issue is not whether the method of proof-analysis described by Lakatos can be applicable to some cases of "real" mathematics or not, but whether or not it is universally applicable or, in other words, whether or not it could play the role of the bare ultimate backbone for every plausible descriptional model of what we call mathematical discovery. My position is that the method

K. Gavroglu, Y. Goudaroulis, P. Nicolacopoulos (eds.), Imre Lakatos and Theories of Scientific Change. 337–345.

described by Lakatos though valuable in the description of some cases of mathematical discovery, cannot play this broader role. That is, it was extrapolated from a data basis so narrow that important patterns of mathematical discovery were ignored.

The method is, e.g., inadequate for dealing with problems related to structural characteristics of crisis periods, such as sudden shifts of conceptual frameworks. These periods are marked, among other things, by the transformation of loose and potentially fruitful philosophical (especially metaphysical) ideas into focused ideas which allow one to see an area no longer through a glass darkly, but with completely new eyes. The overall mathematical activity taking place during and immediately after a foundational crisis (activity which is mainly spurred by such a crisis) is usually centred around the construction of a new conceptual framework with which one hopes to remedy or avoid the problems which caused the crisis. Competing views appear as to how one could get rid of or resolve the problems which appear in such cases as inconsistencies or paradoxes. The successful attempts towards the resolution of the paradoxes have as their usual byproduct the transformation of the source of the foundational crisis into a fountain of exceptional conceptual enrichment and creative innovation at both the technical and methodological levels.

It is not surprising that the foundational problems, in one way or another, come from the contrasting characteristics of the finite and the infinite. Such problems are usually due either to a hasty extension of concepts, methods and definitions from the domain of the finite to the domain of the infinite, or to hidden, non-terminating algorithmic procedures (in a framework which was not supposed to contain them) the existence of which nobody suspected before the occurrence of the foundational crisis.

There are at least two cases, one of each sort, which are worthy of mention. The first concerns the foundational crisis which allegedly destroyed the Pythagorean school. The discovery of the first irrational number or, better, of the first incommensurable pair of line segments, as they are the hypotenuse and one of the other two sides of an isosceles right triangle, seems, if one looks at it in a superficial way, to have revealed a source of crisis not immediately connected with the infinite. This is far from true. It is enough to see the problem as stemming from the non-existence of a common measure for one of the sides and one of the diagonals of a regular pentagon, as the terminating point of the algorithm of ἀνθυφαίρεσις,[5] very well known to the Ancients.

The second case concerns the foundational crisis which followed the Cantorian introduction of the concept of *set* in mathematics, which is unproblematic in the case of the domain of the finite. The story is well known. It is enough to say that the paradoxes appeared as the result of applying the Cantorian diagonal method of comparing cardinalities as, e.g., in the case of the Cantorian set of all sets A and of the set of all the subsets of A. The usage of the notion of one-to-one correspondence for the comparison of infinite cardinalities was a generalization from the domain of the finite to the domain of the infinite which, as was proved to be the case, had to be adopted, but only *after* an appropriate modification of the domain of the infinite had been made.

The proposed ways out of the paradoxes and the ensuing competing theories which came as their result had as their basic *common* element the need to straighten out the foundational deviations and to repair the fundamental conceptual framework used. Yet, they were philosophically oriented and motivated in different ways. Why were they philosophically motivated the way they were, and not otherwise? Why did the Zermelo–Fraenkel approach finally dominate the scene? How could one predict, for instance, the intuitionistic approach as a possible way out of the paradoxes before its initial conception by Brouwer? Such questions have to be central to any research program which has as its aim the construction of a model capturing the essential features of the dynamics of mathematical discovery. Lakatos' model or the Lakatosian attempt to produce a model in *Proofs and Refutations* is too weak to answer such questions. Moreover, it is a model which tells us nothing about "why" questions in general. As a model, it has a purely descriptional character with a narrow descriptional basis and no explanatory features of the sort needed for the detailed examination of theory-formation, especially in times of foundational crisis. Additionally, it says nothing about the criteria which could play a role in the process of choosing between competing and equally consistent theories. As it turns out, competing theories are not chosen merely for reasons of adequacy or internal consistency.

The proposed various alternative ways out of the trap of set-theoretic paradoxes and the corresponding theoretical frameworks which emerged (with Hilbert's approach and program as the only exception) were not chosen or discarded in accordance to whether or not they were adequate or internally coherent. The fact, of course, that they were not proved to be internally inconsistent or inadequate was of paramount importance for their legitimacy as candidates for the final choice of the one which would replace the old, inconsistent, naive Cantorian framework and play the role of the new safe foundation for the ongoing activity of doing mathematics. As we know the final choice was that of the Zermelo–Fraenkel approach which dominated, the scene of set-theoretic research from then to now.

The method of proofs and refutations proposed by Lakatos does not give us an adequate explication of why the Zermelo–Fraenkel approach finally dominated the scene of set-theoretic research. The reason for it is that Lakatos' method, because of the narrow data basis it was extrapolated from, is insensitive to or ignores the fact that aesthetic reasons, reasons of familiarity connected to a sense of continuation, reasons of plausibility, reasons of fertility, etc., can play the role of the basic criteria for the choice between equally consistent, competing mathematical theories. This is so because Lakatos' model seems to have been tailored only to meet cases of processes of mathematical discovery which concern the posing, transforming and the final proving or refuting of concrete conjectures. Such conjectures appear as problems for solution in the framework of informal mathematics where the existing possibilities are exhausted by the description of literally visualizable, trial and error procedures. And this is not the whole story. The insensitivity of the Lakatosian approach to quite important patterns of mathematical discovery extends to further cases.

The method of proofs and refutations is inadequate, for instance, for dealing with problems concerning more or less successful attempts of reconceptualization which take place in normal periods of mathematical activity. These attempts are thought experiments characterized by the element of surprise. As we said they occur in normal periods of mathematical activity. That is, they occur during periods which are not foundationally turbulent; i.e. when there is not any obvious need traceable in the overall mentality of the members of the mathematical community for some sort of fundamental conceptual innovation. They are due rather to the philosophical worries or curiosity of exceptional individuals and not to an appropriately conceptually saturated prior period of mathematical research.

A fine example of such a case is the invention of non-standard analysis by A. Robinson.[6] Robinson's attempt came out of the blue with no obvious need leading to it. It was a quite successful and technically articulated attempt to relabilitate the viciously attacked almost discarded and nearly forgotten Leibnizian infinitesimals.

What can one say about such cases? Is the Lakatosian method of proofs and refutations broad enough to include them? The answer is unfortunately negative. What sort of already existing conjecture (i.e. already posed by the mathematical community, that is) was Robinson trying to answer, and how can Lakatos' proof-analysis method be applied in the case of Robinson's rehabilitation of the Leibnizian infinitesimals? Of course, after any new conceptual start a process is taking place which only in the last analysis and when stretched to the point of triviality can be thought of as described by the Lakatosian model. But then Lakatos' model becomes too general. Abstracted from the concrete case of the "nearly empirical" subject of the study of polyhedra it gets transformed into an empty general schema which is trivial in the sense that we could do without it, since it has neither detailed descriptional or informative content, nor any explanatory force. As we have already insisted, the Lakatosian method, or "simple pattern of mathematical descovery" as he calls it, is an answer to "how" questions concerning mathematical activity extrapolated from a very narrow data basis. It does not answer "why" questions. Additionally it is a pattern which only in the last analysis and taken with a grain of salt can be thought of as applicable to the vast majority of cases of processes involved in mathematical discovery. *After all, a process of mathematical discovery does not always start with a primitive conjecture*.

But let us return to the example of Robinson's rehabilitation of the Leibnizian infinitesimals. As we said it belongs to the category of thought experiments in mathematics which are triggered not by a primitive conjecture but by philosophical considerations of a quite revolutionary nature. What does the Lakatosian model tell us about them? Very little. And more generally, what does the Lakatosian model tell us about the birth and the development of theoretical frameworks which are technical fixations of loose philosophical (metaphysical) ideas? Moreover, what does it tell us about the factors that make a mathematical problem interesting or a certain approach to a problem worthy or promising? Consider for instance the sort of mathematical activity spurred by the need to specify a new conceptual framework which is easier to work with in a specific area. Such activity is characterized by the introduction of stronger axioms or more powerful tools for the exploration of the given area. An example is the introduction by Solovay and Scott of Boolean-valued model-theoretic techniques as a substitute for P. Cohen's method of *forcing* for the production of consistency and independence results in set theory. Another example is the relatively recent adoption of infinitary combinatorics and of philosophically controversial set-theoretic axioms, like the Continuum

Hypothesis or its negation, large cardinals axioms, Martin's axiom and the like, for fruitful research in classical areas like topology, descriptive set-theory and functional analysis. What can the Lakatosian method of proof-analysis tell us about them? How can it be applied in those cases? Where is the initial primitive conjecture that was supposed to have triggered such activity? These are the sort of questions which have to be answered first, in a convincing way, if one wants to seriously hold the position that Lakatos' method is not as narrow and inadequate for the description and explanation of the processes of mathematical discovery, as we insist it is.

II

The examination of the dynamics of mathematical discovery is a task which has to be faced and a job which has to be carried out by both the historian and the philosopher of mathematics. Analytical philosophy, influenced by the Vienna circle in this respect, was practised so that for a long time the internal examination of the structure of scientific theories was the central point of the research done in philosophy of science. The dynamics of mathematical discovery and more generally of scientific progress was considered as a subject which was not purely philosophical. It was a subject for interdisciplinary research and philosophers, especially of the analytic breed, were not ready to abandon or forget their prejudices against research which had to be done in such a way that historical and sociological considerations would play an important role. For them, as for most philosophers, philosophy should be the study of the eternal and not of the ephemeral. Lakatos' work (as well as the work of others, like Kuhn's for instance) helped the philosophical community to get rid of such prejudices and consider the study of the dynamics of scientific discovery as a philosophically worthy activity. In this respect Lakatos' work concerning the dynamics of mathematical discovery was of paramount importance. Yet it was not and could not be the final word in the area. There are aspects of the dynamics of mathematical discovery which were not touched by Lakatos' work. In what follows I will describe a tentative and, quite possibly, not exhaustive classificatory list of issues which I think have to be dealt with before any serious attempt is made toward the construction of a general model.

It seems to me that the issues one has to face in dealing with the problem of the dynamics of mathematical discovery can be separated into at least four categories:

(a) The first includes issues related to the examination of specific examples of patterns of mathematical discovery which can be gathered on a tentative basis. Such a gathering should take place, first, in a way that attention is paid even to the most minute detail, as far as this is possible, and, second, without a preconceived theoretical model in mind of how these patterns should look like. There are, of course, two problems each related correspondingly to one of the two conditions mentioned. First, it is almost impossible to have all the historical data involved in each specific example of mathematical discovery. In the process of gathering (and for each particular case) one is forced to attempt a rational reconstruction and create a kind of "distilled" history out of the available historical data. Here the task of the gatherer or the historian is formidable. The rules for "objectivizing" the given data are not always available. Second, one is never free of preconceived ideas in mind of what sort of pattern or patterns "genuine" examples of mathematical discovery should follow. It is of paramount importance for the scientist or the philosopher who makes the attempt toward such a gathering to try to minimize the side-effects his theoretical prejudices could have upon the gathered historical material. The two problems mentioned are not completely solvable. Yet we should not get discouraged. Standards of objectivity, though relative to the specifics of the historical period and the individuals involved, always exist and we should follow them. They constitute what we consider to be our scientific alibi.

(b) The second category of issues one is faced with in dealing with the problem of the dynamics of mathematical discovery is of a classificatory nature. That is, one is always faced with issues which are related to the classification of the gathered cases of mathematical progress or discovery into groups according to their common characteristics. A classification of this sort could again take place without a preconceived theoretical idea of how it should be carried out. It is more fruitful to start from the gathered material and be led by it to general classificatory schemata than the other way round, having in mind, of course, the obvious truism that such a process is never absolutely free or independent of one's already acquired conceptual background. That is, we never start from scratch and therefore our classificatory attempts are always in a sense biased. Yet it is correct to say that, starting from the gathered material and being led mainly and consciously by it to general classificatory schemata, we at least try to follow the path of the empiricist doing our best to avoid the sometimes overbiased rationalistic tendencies and the pitfalls to which these tendencies lead us whenever we attempt to move from the general to the particular.

(c) The third category includes issues relevant to questions about mathematical discovery and progress which are not of a "how" but of a "why" nature. Questions of this sort are the most difficult to be answered. To answer "why" questions concering, for instance the choice between competing, equally consistent mathematical theories in periods of foundational crisis is sometimes a formidable job. That is so because the final winner of the competition is not always chosen for reasons which we come to think as having an objective status, as, for instance for reasons, of internal consistency or of adequacy in solving the problems created by the occurrence of paradoxes or inconsistencies in the old conceptual framework. It is, of course, true that a theory has to be internally consistent and adequate in solving the problems created by the occurrence of paradoxes or inconsistencies in the old conceptual framework, in order to be a legitimate candidate for the title of the winner. But this is not enough. Sometimes the winner-theory is chosen for aesthetic reasons as, e.g., reasons of simplicity, or of elegance of the new proof-structure introduced, for reasons of familiarity as, e.g., reasons which are connected to a sense of continuation with the previous mathematical practice, for reasons of prolificity, i.e. reasons connected with the relative richness in results of one theory over another, etc. Such reasons, in the form of conditions shaping mathematical discovery and progress, could be called *conceivability conditions or constraints*. They play, an important role especially when some kind of choice is involved concerning alternative paths which could be followed in the process of experimenting in mathematics with various conceptual frameworks. Additionally "why" questions concerning e.g. the choice of a conceptual framework over alternative ones could be legitimately asked and, may be, satisfactorily answered in the light of a prior analysis of what would count as a conceivability condition or constraint and of a specification of such conditions as they refer to particular examples of patterns of mathematical discovery.

(d) The fourth category includes issues having to do with the specification of similarities and differences between patterns of discovery in different scientific fields. The comparative study of patterns of discovery in mathematics and physics, for instance, is quite central to the study of the dynamics of discovery and theory-formation in the exact sciences. The specification of their similarities and differences could be and should be the product of a research program conceived as a project for the formation of a general theory of the dynamics of discovery across the scientific board.

NOTES

[1] See I. Lakatos, *Proofs and Refutations: The Logic of Mathematical Discovery* (Ed. by J. Worral and E. Zahar). Cambridge University Press, Cambridge, U.K., 1976, p. 5.
[2] *Ibid*., Editors' Preface, p. ix.
[3] Concerning Cauchy's proof of the theorem that the limit of any convergent sequence of continuous functions is itself continuous.
[4] *Op cit*., Editors' Preface, p. ix.
[5] See Kurt Von Fritz, 'The Discovery of Incommensurability by Hippasus of Metapontum', in D. J. Furley and R. E. Allen (Eds.), *Studies in Presocratic Philosophy* Routledge and Kegan Paul, London, 1970, pp. 401–404.
[6] See A. Robinson, *Non-standard Analysis*. North-Holland, Amsterdam, 1966.

University of Athens

V. RANTALA

COUNTERFACTUAL REDUCTION

The view that explanations are arguments has been frequently challenged in the current philosophy of science. For instance, the assumption that in scientific changes earlier theories, or laws, are deductively explainable by, thus reducible to, their successors has appeared controversial and – in view of actual scientific developments – very open to criticism. Various arguments have been presented against this optimistic conception of scientific change, but in this paper I shall only consider the following type of objections. It has been argued that an old theory is often *replaced* by a new one, which would mean that the two theories are incompatible or that the old one is false if the new one is true, or that they are incommensurable. How could the new theory in such a case deductively explain its predecessor? Is there anything to be explained? If there exists an inference from the new theory to the old one and the former is true and the latter false, the inference presupposes connecting assumptions which are at variance with facts. But if the theories are incompatible, there may not even exist any explanatory argument – except in some approximate or limiting sense. If, on the other hand, they are incommensurable, but there is a connecting inference, it is only formal, that is, its conclusion is not *really* a law of the old theory – even though it is syntactically identical with one – since the scientific terms contained in the conclusion still represent concepts of the new theory and they are not identical with those of the old.

Objections of that kind are familiar; they constitute a part of the criticism which was put forward by Kuhn, Feyerabend, and others, and which has resulted in attempts to advance less empiricist and more relativistic, philosophy of science and in attempts to find new metatheoretical tools. In particular, a number of efforts have been made to redefine intertheory explanation so that the notion would correspond to what happens in actual science. Instead of surveying different suggestions made, I shall in this paper discuss a particular notion of intertheory reduction and explanation and its amendments. It is a notion which has been introduced – in its most general form – in earlier papers by D. Pearce and myself, but which can be traced back to still earlier suggestions made in logic and philosophy of science.[1]

K. Gavroglu, Y. Goudaroulis, P. Nicolacopoulos (eds.), Imre Lakatos and Theories of Scientific Change. 347–360.

According to that conception, a reduction of one theory to another is a combination of syntactical and semantical correlations satisfying appropriate conditions – thus it is not purely syntactical or purely structural thing. From these correlations a logical argument can be derived which – on relevant assumptions concerning its qualities – can be construed as a generalized covering-law explanation of the reduced theory in terms of the reducing one. Except, perhaps, for one thing: such an argument may presuppose connecting principles which are counterfactual, whence its status as an explanation is not clear. Such is the case, for instance, when one theory is in a limiting-case correspondence to another. This is a trouble I already mentioned above. I have preliminarily discussed this matter in my (1989) where I pointed out that such arguments may support (in the sense to be discussed below) relevant counterfactual conditionals – whether or not the arguments can be counted as explanations proper. In what follows, I shall more thoroughly study their supporting role; and I attempt to argue that this *supporting role* is explanatory, too – so that their status as intertheory explanations is somewhat more complex than what is assumed in the covering-law model. But first I shall summarize the main results of my earlier discussion and the technical definitions needed.

1. REDUCTION

In this section, I shall briefly describe a notion of intertheory relation which can be regarded as a generalized notion of reduction. I shall not discuss in detail what kind of entities theories are assumed to be in this exposition. For the purposes of the present paper, the most relevant component of a *theory* in a reconstructed sense is its class of *models* M. The elements of M are abstract, model-theoretic representations of the phenomena – actual or merely possible – to which the theory is applicable, i.e., in which the laws of the theory are true after they have been formulated in a suitable syntax.[2] If the models in M are of type τ, then τ is also the *type* of the theory. We assume that M is definable in some logic L. If $Sent_L(\tau)$ is the class of all sentences of L of type τ, this assumption means that there exists a sentence θ in $Sent_L(\tau)$ such that M is the class of all models in which θ is true:

(1) $\boldsymbol{m} \in M \Leftrightarrow \boldsymbol{m} \models_L \theta$.

This condition will also be expressed as follows:

(2) $M = Mod_L(\theta)$.

Such a logic L will be called *adequate for* the theory in question. Thus an adequate logic is a logic in which the axioms of the theory can be represented. In (1) and (2), the sentence θ is such a representation of axioms.[3]

Let T and T' be theories such that T is of type τ and T' of type τ', let M and M' be their classes of models, respectively, and let L and L' be logics such that L is adequate for T and L' is adequate for T'. The general reduction relation referred to above is called 'correspondence'. Such a relation reducing T to T' is defined in terms of two mappings assigning to specified models of T' models of T and to each sentence of L a sentence of L'. More precisely, a *correspondence of T to T'*, relative to $\langle L, L'\rangle$, is a pair $\langle F, I\rangle$ of mappings.

(3) $F: K' \rightarrow^{onto} K$,
$I: Sent_L(\tau) \rightarrow Sent_{L'}(\tau')$

such that K and K' are nonempty subclasses of M and M', respectively; K is definable in L and K' in L'; and F and I are *truth preserving* in the sense that the following condition holds for all $\boldsymbol{m} \in K'$, $\varphi \in Sent_L(\tau)$:

(4) $F(\boldsymbol{m}) \models_L \varphi \Leftrightarrow \boldsymbol{m} \models_{L'} I(\varphi)$.

F is a *structural correlation* and I is a *translation*.

Correspondence is a general (reconstructed) notion of reduction and it covers important special cases, the most controversial of which is the so-called limiting case-correspondence. It is often argued, contrary to Kuhn, that even in scientific revolutions theories are conceptually connected. A new theory is a generalization of a classical one in the sense of yielding it as a limiting case. Thus it is said, for instance, that classical particle mechanics is a limiting case of relativistic particle mechanics.

In the present framework, a correspondence $\langle F, I\rangle$ is called a *limiting-case correspondence* if K' consists of *nonstandard* models of T satisfying appropriate conditions, some of which are considered counterfactual, and $F(\boldsymbol{m})$ is a *standard approximation* of $\boldsymbol{m}$. Roughly, a model is nonstandard if it involves a nonstandard topology in one form or another; a standard approximation of a nonstandard model is a standard model which in a relevant structural sense is very 'close' to it.[4]

2. COUNTERFACTUALS

I shall next summarize, mainly in formal terms, some basic features of Nelson Goodman's and David Lewis' analyses of counterfactual conditionals.[5] As I indicated above, I am here concerned with the relation of such conditionals and valid arguments. Accordingly, I shall mainly consider features pertaining to that relation.

According to Goodman's theory, and other theories which Lewis calls 'metalinguistic', a counterfactual conditional is true if there exist appropriate conditions from which in conjunction with the antecedent the consequent can be inferred. Hence, a main problem of counterfactuals is to define such conditions. A key notion here is that of 'cotenability': the conditions must be cotenable with the antecedent in order to be relevant. Cotenability is compatibility, but in a strong sense of the word. This notion will also play a central role in my approach to explanation in the next section.

The problem is handled by Lewis in the following way. Cotenability is a relative notion: it is relative to how the notion of possibility and that of similarity of phenomena – or, rather, of possible worlds – have been defined. To discuss it and related notions in more precise terms, we shall consider a Kripke model of the form

$$\boldsymbol{W} = \langle W, R, \leqslant, V \rangle$$

where, as it is commonly said, W is a nonempty set of possible worlds; R is an accessibility mapping assigning to each $w \in W$ a subset of W; $\leqslant$ is a comparative (three-place) similarity relation of worlds; and V is a valuation function on an appropriate language.[6] If a triple of worlds $\langle w, v, u \rangle$ belongs to $\leqslant$, this is denoted by '$v \leqslant_w u$' and reworded by saying that v is *at least as similar* to w as u is. If a sentence φ of the language is *true at* a world w; I denote '$w \models \varphi$'. If $u \models \varphi$ for some $u \in R(w)$, then φ is *possible at* w, and if $u \models \varphi$ for all $u \in R(w)$, then φ is *necessary* at w. It is assumed here that

(i) if $u \in R(w)$ and $v \leqslant_w u$, then $v \in R(w)$;
(ii) $u \in R(u)$;
(iii) $w \leqslant_w u$ for all $u \in W$.[7]

A model of that kind will be called a *Lewis model*.[8] In what follows, I shall assume that requisite concepts are defined in relation to a Lewis model which is somehow fixed and that the sentences considered belong to a fixed language. If I want to emphasize that a notion is related to a model $\boldsymbol{w}$, I shall add the expression 'relative to $\boldsymbol{w}$'.

In this framework, cotenability is defined as follows. Assume first that a sentence φ is possible at a world w. Then a sentence χ is *cotenable with φ at w* if and only if for some $u \in R(w)$ such that $u \models \varphi$, the following holds: $v \models \chi$ for every $v \in W$ such that $v \leqslant_w u$. If φ is not possible at w, then χ is cotenable with φ at w if χ is necessary at w. Thus in the former case – which is the nontrivial alternative – the sentences φ and χ are jointly true at some world u accessible from the given world w and χ is true at every world v which is more similar to the given one than u is. Thus such a world v is also accessible from the given world w, and χ is true at w. Intuitively, then, the sentence χ is 'so strongly' possible at w that it is compatible with φ, no matter how 'implausible' the latter is. One may want to say, with Lewis, that e.g. laws of nature tend to be in this sense cotenable with counterfactual suppositions. We can also see that the above definition accords, intuitively speaking, with Goodman's (counterfactual) description of cotenability, according to which the cotenability of χ with φ means that it is not the case that the former would not be true if the latter were.

Consider now a counterfactual whose antecedent is φ and consequent ψ:

(5) $\varphi \Box\!\!\rightarrow \psi$

and a valid argument

(6) $\chi_1, \ldots, \chi_k, \varphi \mid \psi$

where the stroke '|' signifies logical inference.[9] Then the counterfactual (5) is said to be *backed*, *at* the world w, by the argument (6) if each of the premisses $\chi_1, \ldots, \chi_k$ is cotenable with the antecedent φ at w. A truth condition for counterfactuals can be defined by means of the notion of backing:

(7) A counterfactual (5) is *true at* a world w iff there exists a valid argument of the form (6) backing it at w.

This condition is a precise formulation of the idea that a counterfactual is true if there are suitable additional premisses from which together with the antecedent the consequent can be derived. There is a more straightforward but equivalent truth condition, which, however, seems to correspond to a different intuition. According to that intuition, (5) is true at w if the consequent ψ is true at every world accessible from w at which φ is true, but which otherwise resembles w 'as much as possible'. More precisely:

(8) A counterfactual (5) is true at a world w iff, if there exists an accessible world u at which the antecedent φ is true, then for some such world u, the material implication $\varphi \rightarrow \psi$ is true at every world which is at least as similar to w as u is.

If, in particular, the antecedent is not possible at w, (5) is true at w in a trivial sense.

Instead of speaking of 'similarity' of worlds, one can use the term 'closeness' – which may in some contexts express a different intuition. It is more important to notice, however, that similarity and closeness are relative notions, relative to a point of view. In this sense they resemble the notion of relative identity.[10]

I have discussed counterfactuals by means of Lewis' theory, and thus used the term 'possible world'. It is a standard term which, in my view, must be given a very general and abstract scope. It may refer to such things as situations and events of various kinds, phenomena, physical systems, model-theoretic structures; to anything that is not such a comprehensive entity as the term may mistakingly suggest. It may refer to entities which are, in some specified sense, actual or merely possible, standard or nonstandard; and which are always considered as *conceptual* things, somehow determined and structured by our theories. They are theory-relative things rather than given. In the next section, possible worlds are, formally, model-theoretic structures which, however, are thought of as theoretical representations, that is, reconstructions, of more or less vague empirical systems and phenomena.

3. COUNTERFACTUAL EXPLANATION

Let us return to the correspondence relation defined in Section 1. Assume $\langle F, I \rangle$ is a correspondence of a theory T to a theory T', relative to $\langle L, L' \rangle$. Assume, furthermore, that the classes M, $M\varphi$, and $K\varphi$ are defined as follows:

(9) $M = Mod_L(\theta)$;
$M' = Mod_{L'}(\theta')$;
$K' = Mod_{L'}(\theta', \sigma_0, \ldots, \sigma_n)$.

Thus the sentences θ and θ' represent the axioms, or laws, of the theories, and the sentences $\sigma_0, \ldots, \sigma_n$ the requisite auxiliary assumptions. From (3), (4), and (9) the following result can be inferred:

(10) $\theta', \sigma_0, \ldots, \sigma_n \models_{L'} \mathrm{I}(\theta)$.

Here '$\models_{L'}$' means logical consequence in L'[11]. If L' is axiomatizable and complete, (10) amounts to a deduction in L'. I shall simplify notation by rewriting (10) in the form

(11) $\theta', \sigma_0, \ldots, \sigma_n \mid \mathrm{I}(\theta)$.

Thus, if there is a correspondence of T to T', then the (axioms of the) latter theory in conjunction with the auxiliary, connecting assumptions logically implies the *translation* of the (axioms of the) former theory. The argument (11) does not directly connect the new theory with the old one itself but with its transform in the new language. This feature is a straightforward consequence of the above model-theoretic reconstruction of reduction, which involves both syntactical and semantical correlations. The correlations together take care of possible changes in meanings. It seems that this feature has not been very extensively discussed in connection with intertheory explanation. That theories can be conceptually connected in this way is noticed by Kuhn when he says that an out-of-date theory can be regarded as a special case of its up-to-date successor, but that it must be *transformed* for the purpose. But Kuhn does not seem to acknowledge that such transformations would be of any importance with regard to intertheory explanations.[12]

To discuss these matters conveniently, I shall use the following terminology in the rest of this section. Given a theory T, a *transformed theory* $I(T)$ is a theory whose axioms are represented by a translation of the form $I(\theta)$, as indicated above; it is, of course, of the same type as T'.[13] To explain a theory is to explain its axioms. I may also speak about explanation and reduction (correspondence) in connection with *any laws* in the same sense as in connection with theories.

Can an argument of the form (11) be of any explanatory consequence? Such an argument may qualify as a covering-law explanation of a transformed theory $I(T)$ – in the sense of some DE-model proposed in the literature – if it satisfies relevant theoretical and pragmatical conditions. It is not, strictly speaking, such an explanation of the theory T itself. But perhaps (11) can be conceived of as an explanation of T in an *amended* sense – on appropriate conditions, of course. Some of the conditions concern the translation I. For instance, it must be workable and instructive and the result $I(\theta)$ must be empirically or conceptually meaningful. Then, *if* all the requisite conditions of adequacy are satisfied, we may say that the amended explanation of T consists of

(12) (i) interpreting T in the conceptual framework of T' by means of the translation I;
(ii) explaining the result by means of the covering-law explanation (11).

When a law is explained by subsuming it under a more general law, the former is necessarily looked at from a point of view of the latter. The earlier covering-law approaches to reduction, which have been criticized by Kuhn and others, are unsatisfactory in that they do not account for this feature in its full generality. For instance, suggestions made by Nagel, Hempel, and others, concerning how the terms of a reduced theory should be connected with those of a reducing one are not general enough to *really* take care of the most intricate scientific changes – those involving idealizing, limiting, or other counterfactual assumptions.[14] On the other hand, the framework introduced in Section 1, above –yielding the schema (11) – is designed to deal with such intricate connections.[15]

How concepts are connected across a radical change of that kind is a serious problem concerning intertheory explanation. The nature of auxiliary assumptions is another, though closely related. Can there – in such a case – be any adequacy conditions for the sentences $\sigma_0, \ldots, \sigma_n$ occurring in (11) such that the process (12) could be thought of as *explaining* something? In what sense can they be explanatory? If the theory T is considered true in some sense, or highly confirmed or probable, then why it *holds*, or under what conditions it holds, can perhaps be explained by embedding it in the more comprehensive theory T', according to (12). But if it is definitely considered false, in view of the new theory, different questions must be asked – whence the auxiliary assumptions, some of which are now counterfactual, play a different explanatory role. That role is described by Clark Glymour by saying that in such a case a theory is explained by

(13) (i) showing under what conditions it *would hold*;
(ii) *contrasting* those conditions with the conditions which actually obtain.[16]

In other words, the theory would hold if certain conditions obtained; and it does not in fact hold since those conditions do not obtain. The syntactical aspect of intertheory explanation in Glymour's sense is derivational. Roughly, the (laws of the) primary theory, together with definitions connecting the two vocabularies, counterfactual 'special' assumptions, and possibly other devices, entails the (laws of the) secondary theory. On the other hand, without counterfactual assumptions the secondary theory is not generally entailed by the primary theory; or the negations of some of its laws can be entailed by means of the true special assumptions negating the given counterfactual ones. These two aspects which are involved in the derivation of the secondary theory now yield the counterfactual explanation as described in (13).[17]

Thus the syntactical aspect of Glymour's model consists of derivation which is similar to that in Nagel's model, except that clearly counterfactual auxiliary assumptions are accepted.[18] But, as suggested by Glymour, its explanatory import is in that it yields something like (13). Glymour supports his view by providing examples from science. I shall next study, by means of the theoretical framework I have introduced above, whether something like that view can be *generally* maintained and whether it can be improved. In view of what we learnt in Section 2 about the relation between arguments and counterfactual conditionals, the view may not be perfectly obvious. But investigating under what conditions a derivation from one theory to another would support a relevant counterfactual conditional – in the sense of backing it – can be of explanatory importance, too.

To discuss these questions in terms of our framework, assume that in the argument (11) the special assumption σ_0 is considered counterfactual. Let it be simply denoted by σ and let (11) be written into the form

(14) $\theta', \sigma_1, \ldots, \sigma_n, \sigma \mid \mathrm{I}(\theta)$.

Let us also consider the counterfactual

(15) $\sigma \,\Box\!\!\rightarrow \mathrm{I}(\theta)$.

Assume that there obtains a correspondence of T to T', as indicated above; whence (14) is valid. Let us assume, furthermore, that a sentence is *true* if it is true at a specified world w of a given Lewis model $\boldsymbol{W}$; thus w is considered as an 'actual' world.[19] We now ask whether (15) is true in the sense that it is *backed* by (14) at w (relative to $\boldsymbol{W}$).[20] If this is the case, it is roughly what Glymour argues for above. Thus we must ultimately ask whether the law θ' and the special assumptions $\sigma_1, \ldots, \sigma_n$ are *cotenable* with the antecedent σ at w (relative to $\boldsymbol{W}$). To study whether they are is to be considered as an explanatory task in an attempt to explain T. If we, in addition, take account of the roles of (14) and (15) in this attempt, it seems appropriate to say that all these things together form a *counterfactual explanation* of T (*relative to* $\mathbf{W}$). More precisely, the counterfactual explanation in question consists of

(16) (i) interpreting T in the conceptual framework of T' by means of the translation I;
(ii) deriving the result by means of the argument (14);
(iii) showing that the conditional (15) is backed by the argument (14) at w (relative to $\mathbf{W}$).

The third step is related to a given world. Since cotenability and backing depend on similarity and possibility, they are relative to context. Informally speaking, they are relative to adopted scientific principles – which are, determined by the explaining (primary) theory and the respective paradigmatic presuppositions. Thus it seems likely that the laws of the primary theory are considered cotenable. According to Lewis, laws tend to be cotenable since they are so important to us.[21] But there can also be other, more personal criteria of cotenability. Formally speaking, Lewis models provide contexts for cotenability and backing.

4. EXAMPLES

Let us first illustrate the third step of (16) by a simple example of limiting-case correspondence. Consider a classical law saying that mass is independent of velocity and the corresponding relativistic law:

$$\theta: \; m = m_0;$$
$$\theta': \; m = m_0(1 - v^2/c^2)^{-1/2} \qquad (v < c).$$

That θ is a limiting case of θ' – so that in the limit as $c \to \infty$ the latter becomes the former – can be paraphrased by saying that there is a limiting-case correspondence $\langle F, I \rangle$ of θ to θ'. Instead of showing in detail that there is such a correspondence in the above formal sense, I shall make some informal remarks on the respective argument of the form (14).[22]

Let us use the following informal notation.[23] If an entity a is infinitesimally close to b, denote '$a \approx b$'. Write '$a = \infty$' or '$a < \infty$' according to whether a is a positive infinite or non-negative finite number.

The members of the argument (14) which is yielded by the correspondence $\langle F, I \rangle$ are θ' and the following:

σ: $c = \infty$;
σ_1: $v < \infty$;
$\sigma_2, \ldots, \sigma_n$: required theorems of analysis, etc.;
$I(\theta)$: $m \approx m_0$.

Assume that an 'actual' world w to which truth, cotenability, and backing are related is a world at which θ' and $\sigma_1, \ldots, \sigma_n$ are true and $c < \infty$.[24] Thus w is a world which is in agreement with relativity theory; whence it might be considered as a piece of our real world – if relativity theory is considered true. We ask whether $\theta', \sigma_1, \ldots, \sigma_n$ are cotenable with σ at w, i.e., whether the argument (14) implies that the counterfactual (15) is true. If they are, we can say with good reason something like this:

(17) If the velocity of light were infinite, then mass would *almost* be velocity-independent.

Here *almost* means infinitesimal accuracy, hence identity for all practical purposes.

If σ is considered impossible, this would make the conditional trivially true. If one thinks that σ is not possible in any important *conceptual* sense, that is, in any sense that would be theoretically relevant, then one would presumably hold the schema (16) to be of no explanatory importance. Usually, however, it is thought that the assumption σ, though contrary-to-fact, provides an explanatory link between classical and relativistic theories, so that it is conceptually relevant. Let us assume, then, that it is possible at w. If we think that accepted laws tend to be cotenable, then it is justifiable to consider θ' cotenable with σ at w. If θ' is true at a world u, then no world at which it is not true can be more similar to the actual world than u is. This view gives a special importance to accepted laws, and it emphasizes the fact that θ is being explained *from* θ'. One way to argue against the view would be to claim that appropriate *structural* similarity criteria would provide more objective notions of cotenability, that is, criteria which more directly pertain to the structure of worlds. It is not quite evident, however, in what sense the result would be more objective, if the structure of the 'actual' world is dependent on our theories.

The assumptions $\sigma_1, \ldots, \sigma_n$ are also more or less like laws, physical or mathematical, whence they can be considered cotenable on similar grounds as θ'. If they are not laws, they are at least important truths. Thus it seems that we are entitled to maintain that the counterfactual conditional is true. We may also see that investigating the conditions under which its truth follows from (14) is of explanatory importance since it pertains to adequacy questions and to presuppositions involved.

To further illustrate the explanatory importance of the step (iii) of (16), let us consider another example. Write now '$a \geqslant b$' if $a - b$ is a finite, but not infinitesimal, positive real number or zero. Assume, just for the sake of illustration, that there is a correspondence $\langle F, I \rangle$ of θ to θ' such that θ, θ', σ, $\sigma_2, \ldots, \sigma_n$ are as above, but σ_1 and $I(\theta)$ are as follows:

σ_1: $v/c \geqslant 0$;
$I(\theta)$: $m \geqslant m_0$.

Then the steps (i) and (ii) of (16) can be taken, but not (iii). At least it seems obvious that σ_1 cannot be considered cotenable with σ since it would not be true at w if σ were (for v is finite in w). On the other hand, even though the argument (14) is valid, it does not seem to provide any adequate explanation for the relation of the effective mass and the rest mass, indicated by $I(\theta)$. This is simply because the special assumptions σ and σ_1 are not adequate. Thus this example appears to sustain my view that the step (iii), when successful, provides adequacy conditions for the step (ii), that is, for the respective 'deductive explanation'.[25]

NOTES

[1] See e.g. Pearce and Rantala (1984a, 1984b, 1985).

[2] For a more extensive and exact discussion of the notions of this section see the works listed in Note 1.

[3] I make the simplifying assumption that a single sentence represents the axioms of a theory. If a class of models K is defined by a finite number of sentences $\theta_1, \ldots, \theta_n$, I shall write: $K = Mod_L(\theta_1, \ldots, \theta_n)$. By a logic I mean a logical system in a generalized sense.

[4] Details cannot be presented here, but see Pearce and Rantala (1984a).

[5] Goodman (1947, 1979); Lewis (1973).

[6] As can be seen, I am simplifying things by only considering Kripke models on the level of propositional logic. But the most essential matters of the present paper are not affected by that assumption.

[7] For the purposes of the present paper, we need not consider what other conditions should be satisfied.

[8] If one wants to consider a specified 'actual' world to which truth and other notions are related, a distinguished world can be added to **W** to play the role of such a world. But it would not make much difference here.

[9] Logical inference is of course relative to the logic used; but in what follows, this does not cause any confusion.

[10] See Lewis (1973) for a discussion of that feature. Cf. also the next section. See Niiniluoto (1984) for a discussion of closeness in the context of verisimilitude.

[11] That is, K' is a subclass of $Mod_{L'}(I(\theta))$.
[12] See Kuhn (1962), pp. 102–3.
[13] We can neglect here the question whether the transformed theory can be really regarded as a theory or the transformed law as a law.
[14] For a discussion and examples of intertheory translations in such cases, see the works referred to in Note 1; Pearce (1987); Balzer (1985). The case that the *terms* of the old theory are preserved over the change but their *meanings* change cannot be dealt with by adding syntactical definitions to auxiliary assumptions. Though terms can always be renamed in a purely logical sense, such a renaming may not correspond, in intuitive sense, to what happens in science. A case in point is e.g. the relation of classical particle mechanics and relativistic particle mechanics.
[15] See the works mentioned in Note 1, and Note 14.
[16] Glymour (1970).
[17] Glymour distinguishes between adding counterfactual special assumptions and taking limits. But by using nonstandard models, in the sense of Section 1, the latter procedure can be considered as a special case of the former.
[18] The syntactical side of Glymour's approach seems to have Nagelian features, e.g. concerning translation, which restrict its generality, but this is not the most important thing here.
[19] The general structure of the worlds of $\boldsymbol{W}$ is determined by the type of the theory T.
[20] Thus we do *not* merely ask whether *there is* a valid argument backing (15).
[21] See Lewis (1973) for a more comprehensive discussion.
[22] That there is follows, in fact, from the results of Pearce and Rantala (1984a). For another example, see my (1987).
[23] For details, see Pearce and Rantala (1984a).
[24] If one wants, it can be taken to be a standard model of θ' in the sense mentioned in Section 1.
[25] It is clear that counterfactual explanation illuminates only one side of revolutionary scientific change. A very different picture is given e.g. by approximate explanation. I am indebted to Ilkka Niiniluoto for reminding me of their different nature.

REFERENCES

Balzer, W., 'Incommensurability, Reduction and Translation', *Erkenntnis* **23** (1985) 255–267.

Glymour, C., 'On Some Patterns of Reduction', *Philosophy of Science* **37** (1970) 340–353.

Goodman, N., 'The Problem of Counterfactual Conditionals', *Journal of Philosophy* **44** (1947) 113–128.

Goodman, N., *Fact, Fiction, and Forecast*. Sussex: Harvester Press, 1979.

Kuhn, T., *The Structure of Scientific Revolutions*. Chicago: University of Chicago Press, 1962.

Lewis, D., *Counterfactuals*. Oxford: Blackwell, 1973.

Niiniluoto, I., *Is Science Progressive?* Dordrecht: D. Reidel, 1984.

Pearce, D., *Roads to Commensurability*. Dordrecht: D. Reidel, 1987.

Pearce, D. and Rantala, V., 'A Logical Study of the Correspondence Relation', *Journal of Philosophical Logic* **13** (1984a) 47–84.

Pearce, D. and Rantala, V., 'Limiting-Case Correspondence between Physical Theories', in W. Balzar *et al.* (eds.), *Reduction in Science*. Dordrecht: D, Reidel, 1984b, pp. 153–185.

Pearce, D. and Rantala, V. 'Approximate Explanation is Deductive-Nomological', *Philosophy of Science* **52** (1985) 126–140.

Rantala, V., 'Scientific Change and Counterfactuals', in *Proceedings of the 5th Joint International Conference on History and Philosophy of Science, Veszprém, Hungary, 1984*, forthcoming, 1989.

University of Tampere

ARIS KOUTOUGOS

RESEARCH PROGRAMMES AND PARADIGMS AS DIALOGUE STRUCTURES

INTRODUCTION

Ever since I read the papers in the Lakatos–Musgrave edition of the Symposium 'Criticism and the Growth of Knowledge' the issue of normativeness and the relevant debates decisively influenced my further study in the philosophy of science.

Most arguments in those papers seemed to me directly or indirectly expressing a critical stand towards some normative implications of Kuhn's supposedly descriptive account. This I considered an important as well as an inevitable turn in the philosophy of science; as long as historiography was kept at a distance from the main normative issues concerning the demarcation criteria for science, questions of such a sort did not even arise. The sort of arguments which were flourishing in the philosophy of science, elaborated, often in a highly technical manner, the normative claim that science should be kept apart from metaphysics. This was the case whether discussing general principles of demarcation such as verifiability, or difficulties in Carnap's programme for a purely observational language.

But also other classical debates such as the endless realism vs empiricism one, carried behind their vague ontological implications, the normative issue. For example, clarity in ontology for realists simply represents the view that it is "things" out there that do really exist and not structures, otherwise empty, of the processes by which relevant knowledge is obtained. But the study of those structures is what motivates empiricists. As a result empiricists avoid "unnecessary metaphysical commitments" to "things" proposing honesty and/or even modesty in ontology, and demand clarity instead, in the analysis of the apprehension of (a non-metaphysical) knowledge. Further, normativeness can be traced too in other issues of direct relevance such as progress and objectivity in science, and further, to theory ladenness and incommensurability; for what is the hot issue with incommensurability if not its implications for objectivity which allow the diagnosis of *Progress* (a necessary concept for realists if they wish to enjoy

K. Gavroglu, Y. Goudaroulis, P. Nicolacopoulos (eds.), Imre Lakatos and Theories of Scientific Change. 361–374.

the possibility of "approximation" to truth)? But I will come to incommensurability later.

At this point I must force the closing of the background comments with the following: normativeness was fundamentally underlying the main issues and was clearly separated from descriptive historiographic accounts of science up to a certain point in time which obviously cannot be later than the appearance of Kuhn's views. But the beginning of the end of the purely normative philosophy of science was Popper's alternative to logical positivism. For the first time philosophers of science were given to compare not just properties of scientific theories but normative criteria for the comparison and evaluation of such properties as well as criteria of demarcating science in general. The search for criteria of comparison led to a paradoxical but inevitable reversal of roles between actual history of science and normative metascience; normative theories were meant to criticize actual science but in the end historiographic evidence was used to judge normative theories. Imre Lakatos was obviously reflecting upon this need to control normative theories when proposing his "methodological" version of sophisticated falsificationism to remedy Popper's austerity to actual history of science.

This interrelation of normative and descriptive elements was therefore a basic feature of the symposium arguments which were marked with the discomfort Kuhn's critics felt because of the possible normative implications of his descriptive account; and Kuhn himself when forced to declare whether he meant to be normative or descriptive he replied that he meant to be both.[1]

1. NORMS AS INTERNAL PARAMETERS

I believe I have said enough to justify the posing of the following problem: Is it possible to give an account of the scientific enterprise so that the criteria in use for comparison of theories as well as other norms and values which appear in various processes of decision making would be *internal* parameters of an overall descriptive model?

Let me begin this investigation by looking first at some properties of Lakatos' Research Programmes and Kuhn's Paradigms considered as possible approximations to the required model.

Kuhn describes science as the practice of scientists in the context of communities where leading theories and the need to stick with them as much as possible seem to depend mainly on the specifics of the scientists' training through exemplars, that is, through application of standard solution types to all problems, which are treated for that reason as puzzles for the leading theories.

This norm, the preference to *puzzle solving*, could be considered because of its dependence to the training specifics as an internal parameter of Kuhn's descriptive model; a parameter P will be considered in this context internal if its value at t, $V(t)P$, depends on the current state of the model $X(t)$ which in turn is at least partially determined by $V(t-1)P$.

However the duration and scope of Kuhn's puzzle solving norm as well as its resistance to extra difficulties are obviously varying since change ultimately does occur through crises and revolutions. But this variability cannot be accounted for by the explicitly described processes. Therefore much is left to the undescribed by Kuhn processes of communication, or, of Nornal Science, or, of a paradigm since the concepts are practically interchangeable; they are interchangeable because the concept of scientific community is equivalent to the "world view" shared in its context and these "world views" are in one to one correspondence with paradigms. What therefore seems to be produced in these processes of communication are fixations and demolitions of world views accompanied by the embarassing in many respects phenomenon of incommensurability, meaning in this case the alleged overall non-comparability of paradigms.

Now, the interesting aspect of Kuhn's model always in relation to the structure of the required one is the following:

Normative elements are produced internally; the norm which expresses the preference to standard solution types, is being stabilised through the process of communication of scientists in the early periods of the new normal science. Thus the internality property in this case is exhibited by the paradigm's generating this preference norm by which it further survives (as long as it does, that is, until all applications of the dominant theory have been fully utilised). This I shall compare now with an equivalent property of Lakatos' RP and then look at another property of Kuhn's model.

Lakatos attempted with one stroke both to make Popper's falsificationism historically realistic and show, as an answer to Kuhn's challenge, that rational descriptive models of science are possible. He considered for that purpose in his analysis of RP several properties which would serve as explicit indicators of the quality state of the developing series of theories constituting a Research Programme. These properties – indicators – were in fact indicating the outcome, first, of all possible results of comparing any two successive theories in the series in terms of their experimental performances, and then the resulting status of the whole series, of the Research Programme.

Insofar as there is a marked difference with Kuhn's internality property, for although the RP parameters (I am referring to the terms "progressive", "degenerating", "theoretically progressive" etc.) obtain their specific state-values in the process of the development of a RP, their nature (the set of possibilities of those values) is predetermined. Also predetermined and rationally so is the norm taken as a function of those state-values, the norm according to which the replacement or not of the RP is to be decided.

However this predetermination of the rejection condition of a RP before the actual values of the parameters (whose function define the rejection norm) are obtained, is bound to produce serious deviations from the actual history of science. It seems that rationality advances over Kuhn are secured at the expense of internality, a situation which Lakatos proposes to remedy by admitting processes of decision making which will generate their own norms, the well known "appeal procedures", which he considers as they may represent the new era of "non-instant rationality".[2]

The question now for both Kuhn and Lakatos is whether there are any processes fulfilling such requirements; can they be processes of rational communication of scientists while norms or standards or appraisal are *internal variables*? From Kuhn we can draw some further evidence. Although we know nothing about the structure of communication processes during crises and revolutions out of which "Gestalt switches" may result, at least this we are told, that these processes do lead to non-comparable phases of equilibrium (the different periods of normal science). Therefore non-comparability must be compatible with some modes of communication if the above processes are to extend over the complete Kuhnian cycle (crisis, revolution, normal period etc.). Now, given that norms are internal products of this cycle, we can conclude that internality may also be compatible with communication over the non-comparable; as a matter of fact the latter is necessary for processes of change with internality, because internality means that the possibility of change extends over the complete set of parameters and therefore no fixed references are in principle available as required by standard conceptions of communication. And this is not an entirely new idea, it has its short history in Feyerabend's theses concerning the possibility to compare incommensurable theories in dialogue processes where new meanings may emerge.[3] This compatibility of communication with non-comparability together with the internality property represent the structure of the required model which I will further examine as a "dialogue" structure.

Research programmes and Paradigms would represent such "dialogues" if they could consistently incorporate the appropriate processes of communication which would make explicit the Gestalt switches in Kuhn and the appraisal procedures in Lakatos.

3. PRESUPPOSITIONS FOR THE POSSIBILITY OF COMMUNICATION VS CLASSICAL PRECONCEPTIONS ON MEANING AND REFERENCE

The question that remains is whether these communication processes already determined by means of the internality and comparability properties are possible at all; for communication over the non-comparable (the non-available to be objectively referenced) is at first sight paradoxical.

The sort of evidence logically required to establish the possibility of such a communication would be any real phenomenon which could be shown to have this "paradoxical" structure; and we don't have to search very far, for this phenomenon is part of the usual everyday discourse between any two individuals. Let me explain: When I disagree with a speaker who utters a sentence S it seems at first sight that I do a rather strange thing. I understand first (I have to) the meaning of S and then I correlate this meaning with something in the world which in my eyes does not fit with it. But is this really strange? Wouldn't it be more strange to believe instead that disagreements are illusions or a false concept? For there is no other way for disagreement to occur unless such a comparison is adopted.

The reasons for this seeming strangeness have to do with a serious preconception about the concept of correspondence and the ways it relates meaning with reference. More deeply than this we may find some preconceptions about objectivity.

What looks strange when I correspond S, that is, its meaning conceived perhaps as a possible state of affairs, with an actual state of affairs which misfits S, is that it seems as if I use information indicating to me what I should correspond S with, information which I cannot possibly have access to; I cannot normally have access because the only information I am allowed to use is what only the purely linguistic event of the sentence gives me, and this is the possible state of affairs I have conceived or understood by S, a sort of ideal conditions for it to be true. This reminds in a way the old socratic type of puzzle, the puzzle of the false sentence; if a and b differ, then it is false to say that $a = b$, yet it is impossible for one to produce this falsity whether one knows both a and b (it then makes no sense to mistake them), or not (which makes a complete reference to them, if any, impossible) (Theaetetus, 187). What really prevents falsity here is not the hypothesis that we may have either perfect knowledge of something or none at all (this hypothesis is above all a simplifying reduction of partial knowledge to perfect knowledge of some only parts of an entity), but the objectivistic preconception that the criteria which decide the actual (or practical) reference of a sentence are represented simply by its meaning which indicates what the sentence should ideally (or theoretically) be referenced by.

This resulting analytic relation between meaning and reference is a standard requirement for objectivity in communication; it is believed that people refer to the same things by the same sentences and they also may as a result think of the same things by the same sentences. However important as it may this be for objectivity and/or comparability in communication, it nourishes socratic type puzzles and leaves out of account falsity and disagreement. Yet, these are facts of our everyday discourse and we have to accept them together with any possibly associated "non-comparability".

Further, it is worth noticing that the way by which the above analytic preconception perplexes falsity and disagreement destroys the possibility of a healthy correspondence account of truth because it reduces the concept of correspondence to its sense as congruity and ignores its function also as correlation. By this distinction[4] we can effectively separate what I called the ideal conditions (the conditions which stand to an ideal congruity relation to S), or, the *ideal* reference of a sentence, SI, and its *actual* reference, SA, (which represents what has been simply correlated to SI without assuming any congruity relation), the latter conceived as the set of circumstances actually chosen out of the possible matching candidates (or domains of application of SI). On the basis of this distinction it becomes possible to define truth (or falsity) as a congruant (or a non-congruant) correspondence between SI and SA; all we need is to realize that the criteria by which we arrive at the simple correlation of SI with SA do not have to be, and as a matter of fact they are not, identical in nature with the criteria which decide their congruity.

Frege discarded any possible correspondence account of truth simply by considering the two senses of correspondence antagonistic instead of complementary:

> It is said to be possible to establish the authenticity of a bank-note by comparing it stereoscopically with an authentic one. But it would be ridiculous to try to compare a gold piece with a twenty-mark note stereoscopically. It would only be possible to compare an idea with a thing if the thing were an idea too. And then if the first did correspond perfectly with the second they would coincide. But this is not at all what is wanted when truth is defined as the correspondence of an idea with something real. For it is absolutely essential that the reality be distinct from the idea.[5]

As a matter of fact Frege in this passage makes an explicit use of correspondence only as congruity relation (the stereoscopic comparison is meaningful only in case of expected congruity between things or even in case of possible resemblance allowed among elements constituted of the same quality such as ideas) which he obviously finds impossible to hold in cases where by definition the relative elements constitute distinct qualities (ideas and reality) which only simple correlations may connect them. Frege's argument is perfect, yet only because he assumes that the one of the two senses of correspondence has to accomplish what only both can; simple correlation has to connect as a first step the two things to be compared and then let congruity be decided on the basis of appropriately defined criteria. This is also the logical order to define falsity: falsity requires, first, correlation of S with its actual reference and then negative results of similarity checking to follow. Also, in the context of communication it must be possible that the hearer corresponds (that is, simply correlates) what he first understands as the meaning of the sentence with a reference which he then judges to be a non-congruent (a mis-matching) one.

Wittgenstein's approach to the problem with the Picture theory of meaning seems to allow for this necessary use of the two senses of correspondence. He considers that we must be able to understand a proposition without knowing if it is true or false (NB., p. 93). This clearly is equivalent to a separation of what is understood as sense, or, as "truth possibilities", from the "truth conditions" which are fixed through expressing agreement or disagreement with the "truth possibilities" (*Tractatus*, 4. 431).

Therefore Wittgenstein's "truth conditions" seem to contain both a positive and a negative element (agreement, disagreement, or, truth, falsity). These two elements – components of sense, the conditions of truth (in the literal sense) and the conditions of falsity, leave indeed enough space for a correspondence (in both its senses) account of truth and possibly for a solution to the problem of the false sentence; the independence of the meaning's truth possibility allows its correlation with either its truth (congruent) or falsity (non-congruent) component thus resulting to either its truth or to its falsity.[6]

However, Wittgenstein's solution remains in the context of an analytic correspondence between sense and reference and this is crucial for it allows only a very narrow solution, if at all, to the falsity problem; the information contained in the sense of the sentence, the "picture", which for Wittgenstein is sufficient to make a correlation even with a falsifying reference, relates to a structure and it is a structural similarity and not a complete fact that we try to find, that we try to locate as reference (this is not made at all clear when it is stated that the picture informs us about what fact to look for BB. p. 31)). But when a structural similarity is recognised only a possible reference is determined, especially when more than one such similarities are found in the referential context. What then will force the correlation of a particular falsifying (dissimilar) fact upon the meaning of S? Would that be the understanding of its negation? If we take a sentence S in Wittgenstein's terms as expressing the claim that a given structure s is satisfied by a set of values (a model) $\text{Mod}\,w$ (in symbols, $S = s \text{ sat Mod}\,w$), then we see that its negation which would certainly be considered as falsifying S, is not, as Wittgenstein would have it, determinable: let this negation be expressed as $-(s \text{ sat Mod}\,w)$ which in natural language is equivalent to $(s \text{ sat} -\text{Mod}\,w)$. However, $-\text{Mod}\,w$ cannot as a real entity, whatever it is, be precisely determined by its negative logical relation to $\text{Mod}\,w$ for there are no negative entities.

All this leaves us with very narrow, almost useless, accounts of falsity and disagreement; a negation is only practically determinable if there are no more than one candidates for the reference role in a given context. On the other hand, anything which could be considered as directly determined (practically or theoretically) as reference from the sentence itself simply provokes the socratic puzzle: If I know the structure (the Picture) and the model which satisfies it how can I mis-arrange them with others (and if I don't how can I make the statement at all)?

Does Austin's definition of true statement escape analyticity? Austin defines the truth of a statement by distinguishing two kinds of conventions; (1) the demonstrative, which decide the reference, (2) the descriptive, which decide meaning.[7]

Looking at the surface of this definition it seems that nothing of the sentence's meaning – determined by the descriptive conventions – has anything to do with the determination of reference; the latter is performed by the demonstrative conventions which appear as Austin defines them – as correlating the statement to the world (="historic situations") – to be part of the statement. This may explain why the statement (and not the sentence) being thus responsible for the link with the world (the determination of reference) is what only can be true or false. However, looking closer at the structure of this definition it reminds very much Wittgenstein's picture theory; the clue for this analogy is to "read" the pair statement – sentence, as sentence – picture. Austin's sentence provides what is to be examined for congruity with what the demonstrative conventions locate as reference. The same does Wittgenstein's picture; it provides a structure s which discloses the entities (values) Mod$w$ to be checked for congruity with some Modx located by the sentence as what should satisfy s in reality, (that is, as what completes together with s the reference of S). The only difference is that Wittgenstein gives no explicit choice mechanism for Mod$x$ (this is what the reference choice mechanism reduces to in this case) as Austin does by giving this explicit role of reference choice to the demonstrative conventions. However, what is important is that in both cases the reference pointers seem to belong to the interior of the larger units, the sentence for Wittgenstein and the statement for Austin. And this is also the only point where the analogy may not hold; it would not hold only if the demonstrative conventions were not part of the truth value bearer, the statement. This would indeed prevent analyticity; It would also force upon us the task to investigate the nature and origin of these "conventions" in the processes of communication.

In what sense then disagreement is meaningful given that falsity is either impossible (as it is within the context of classical analyticity preconceptions), or, indeterminable (as it is the case where reference is freed from being exclusively determined from the symbolic form of the intended communicated event)? Of course it is not difficult for one to see that people very often come to disagree. What is difficult is to account for disagreements which are not necessarily results, either of brain misfunctions, or of careless mistakes, or of irrational behavior of the one or both communicating parties.

To allow for this we must assume the existence of flexible contexts of reference choice which provide room for alternative arrangements of their content and thus for the possibility of alternative views.

When I choose to correlate an uttered sentence with the world I have to remember the simple fact that the world is first of all my accepted arrangement of things, my "view" of it. In other words, what I understand from a speaker's statement is that I take him as making the claim that the "meaning" of *S* matches by comparison to some element, part of the actual referential context.

All these elements or "facts" represent various levels of correlations of more "immediate" facts. The degrees of immediacy of a fact can be defined in terms of the degree of agreement produced whenever this fact is intended or taken as reference of a sentence. This variability of levels of correlations in terms of immediacy (a parameter which modifies its value within a recursive process) shows that facts (derived from correlations) of less than a certain degree of immediacy, represent not usual facts but possibilities of facts, or, possibilities of recognition of facts. Therefore the speaker's claim that *S* matches some fact of the referential context, means that he "sees" (the meaning of) *S* as a possibility *SX* in this context and disagreement simply means that the hearer does not "see" any possibility, which more precisely means that: (a) the hearer understands by *S* the possible fact *SI* (an ideal reference of *S*) which he believes that the speaker "sees" as a possibility in the given context, (b) the hearer attempts a first ordering of the referential context (of the set of available references), *SA*(1), in terms of similarity of its elements with *SI*, and (c) he finds the closest to *SI* element of *SA*(1) not "similar enough" with *SI* (this criterion of more or less similarity is not necessary to define disagreement; theoretically it would suffice to use an identity or non-identity one although it is more natural to assume the former for it makes agreement more realistic (less coincidental. Notice also that knowledge of the relation between *SX* and *SI* has no significance whatsoever).

The above formulated disagreement can be interpreted as a difference in view since it results from the difference in the recognition of factual possibilities in the given context.

This flexibility of the dialogue context is the direct result of the non analytic, indirect, connection of meaning (as ideal reference *SI*) and reference (as actual reference *SA*) (it is also recursively redefined within communication processes since the recognition of factual possibilities mainly depends on the overall agreement generating perspectives).

4. COMPARABILITY AND NORMATIVENESS IN DIALOGUES

Let me now review the comparability problem in the light of the previous analysis.

What may be considered as non comparable is the epistemic status of sentences which use common terms with different empirical meanings. For example let $S1$, $S2$ be two such sentences, $ts1$, $ts2$ their truth values, and two individuals A, B who assign to the common terms of $S1$, $S2$ correspondingly empirical meanings a, b. The evaluations of $S1$, $S2$ in terms of $ts1$, ts 2 which are supposed to be non-comparable if a, b differ, are results of comparison between ideal references $SI1$, $SI2$ and actual references $SA1$, $SA2$ performed correspondingly by A and B. In other words, these results of comparison represent agreements or disagreements between A and B with respect to the sentences $S1$ and $S2$. For example when A says $S1$, B's evaluation of $ts1$ will be the result of his comparing $SI1$ with $SA1$. Let us call this evaluation $(BeS1)A$. Also when B claims $S2$, A's evaluation of $ts2$ will be the result of comparing $SI2$ with $SA2$, or as above $(AeS2)B$.

Let us suppose now that we could obtain measure reports for the degrees of agreement $(BeS1)A$ and $(AeS2)B$, correspondingly, $mB(s1)$ and $mA(s2)$. These quantities which measure how much B agrees with $S1$ and how much A agrees with $S2$ are obviously comparable giving a very small but conceptually clear indication of the relative acceptances of $S1$ and $S2$. More reports from communication processes would improve estimates of these relative acceptances.

This is clearly a communication over the non-comparable by means of agreement measures $mX(s)$ with no signs of paradoxical features. The different empirical meanings a, b, acted as hidden variables bearing no significance in the actual processes of comparison. This is because the corresponding "understandings" or ideal references $SI1$, $SI2$ which A and B assign to S participate in the process to be compared only with the actual references to which they have been correlated; only if they were directly related in terms of a purely metaphysical psychologistic presupposition the question of their objective evaluation would bring to the surface the lack of any possibility of their common empirical interpretation and thus raise a problem of their comparability.

Therefore the fact that ideal references are not directly comparable in any obvious sense of the term (A has no access to $SI1$ and B has no access to $SI2$) has no real significance; the Fregean type of objections to interpretations of sense as possibly including non-communicable mental entities seem to have been overrated in the history of the issue.[8]

The alternative view proposed here was the disengagement of reference from a strict analytic dependence on sense and their "pragmatic" reconnection in the above dialogue structure. This move demands among other things to transfer any objective significance from meanings in themselves to communication, that is, to $(XeS)\,Y$ estimates of agreement.

These estimates can be objectively interpreted as "resistances" in the "three-corner fights" of communication between

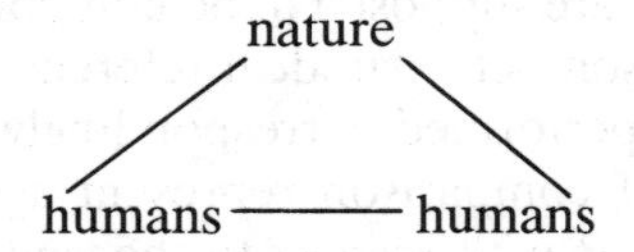

representing knowledge inasmuch as the Lakatosian form of it, the three-corner fight model between

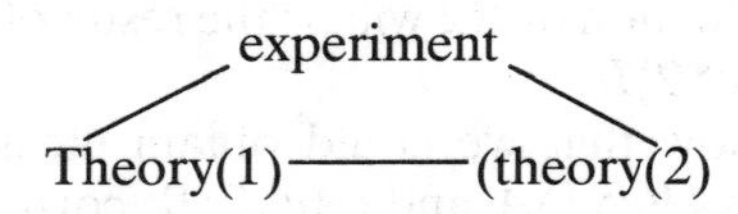

produces scientific knowledge. Within such a perspective not only major comparability issues are replaced with technical problems of a lesser philosophical complexity (as for example comparing, more precisely, producing theoretical reconstructions of actual comparisons of, *SI* with *SA*) but the interpretation of normativeness as an internal systemic concept becomes possible too.[9]

NOTES

[1] Cf. I. Lakatos, and A. Musgrave, (1970), p. 237.

[2] *Ibid*., pp. 154–7, where also "new emphasis to the hindsight element . . . lead to a further liberalisation of our standards . . . (and) on the whole, stress the importance of methodological tolerance . . .".

[3] Feyerabend (1962), p. 39.

[4] Cf. G. Pitcher, (1964), pp. 9–10.

[5] G. Frege, 'The Thought: A Logical Inquiry', in P. F. Strawson (1967), p. 19.

[6] Eric Stenius, one of the main advocates of the Picture Theory, holds that Wittgenstein does bring back correspondence and a solution to the "false sentence' problem (cf. entries 8.7 and 8.10, pp. 117–118, in Stenius (1981).

[7] J. L. Austin, 'Truth', in Pitcher (1964), p. 22.

[8] As M. Dummett remarks referring to Frege's attempt to restore communicability over sense, "... sense is not psychological; it is not because a speaker's utterance of a word occurs in response to some inner mental process and that utterance triggers off a similar process in the mind of the hearer, and they share a common psychological make up, that the one understands the other. Their agreement over the sense of the utterance consists in an agreement on the conditions under which it is true ... (Dummett, 1981, p. 38). This type of objection to sense as such a psychological entity affirms indirectly what it attempts to deny: the solution (agreement over the sense is agreement over the truth conditions), given the criticisms to the disappearance theory of truth which (as Dummett, *ibid*., p. 40) interprets the claim made by Wittgenstein in the 'Remarks on the Foundations of Mathematics', appendix on Gödel's theorem and in PI 304, 317, 363), rules out an account of meaning in terms of truth conditions, amounts to "a disappearance theory of agreement"; agreement becomes a mystery unless the psychologistic structure denied above is possible.

[9] What is interesting for further research is the analysis of factors and/or of processes which may determine the actual reference SA, since its correlation to the sentence rests only on a vague set of pragmatic considerations.

Some results from my earlier attempts in relation to this issue (1983) have indicated that the main factor out of all these pragmatic considerations is a *Prestige* parameter which is the basic link among several feedback relations of processes which reduce disagreement. Therefore the distinction between *SI* and *SA* becomes the source of internal systemic normativeness expressed by concepts such as the *tendency* to reduce disagreement and the prestige parameter P, a function of which together with the size of the initial disagreement D, determine the size of final disagreements through process of minimization, $D\text{min} = f(D, P)$. P may be defined recursively in such a way that it will tend to express at any given time the past history of communication successes in terms of comparing aggregates of Dmin. Thus an internal connection between normative and descriptive elements is plausible since disagreements express cognitive differences through which the norm motivating their reduction processes is generated; these processes contain interpersonal evaluations in terms of aggregates of these cognitive differences which determine the P factors.

It is precisely the role of the prestige P in this cycle which I would like to see further investigated because its internal determination seems to allow for more accurate models of rational consensus. This model may for example compare favorably with rational consensus as weighted average proposed by Lehrer and Wagner where the weights are purely external prestige parameters (Lehrer and Wagner, 1981).

A substantial part of this work is already completed in manuscripts with the title "Truth and Consensus'.

REFERENCES

Austin, J. L., 'Truth', in Pitcher, G. (ed.), *Truth* (Perspectives in the Philosophy Series), Prentice Hall, 1964.

Dummett, M., 'Truth', in P. F. Strawson (ed.), *Philosophical Logic*, Oxford University Press, 1967.

Dummett, M., 'Frege and Wittgenstein', in I. Block (ed.), *Perspectives on the Philosophy of Wittgenstein*, Basil Blackwell Publ., Oxford, 1981.

Feyerabend, P. K., 'Explanation Reduction and Empiricism', in Feigl and Maxwell (eds.) *Minessota Studies in the Philosophy of Science*, 3, 1962.
Feyerabend, P. K., 'Against Method', in Radner and Winokur (eds.), *Minessota Studies*, 4, 1970.
Frege, G., 'The Thought: A Logical Inquiry', in *Philosophical Logic* (reprinted from *Mind* **65**, 1956) P. F. Strawson ed., Oxford, 1967.
Matthews, Gwynneth, *Plato's Epistemology and Related Logical Problems*, Faber and Faber, London, 1972.
Hacker, P. M. S., 'The rise and fall of the Picture Theory', in I. Block (ed.). *Perspectives on the Philosophy of Wittgenstein*, Blackwell, 1981.
Kuhn, T. S., *The Structure of Scientific Revolutions*, Chicago University Press, 1962.
Kuhn, T. S., 'Second Thoughts on Paradigms', in F. Suppe (ed.), *The Structure of Scientific Theories*, University of Illinois Press, 1979.
Koutougos, A., 'Meaning Relativism and the Possibility of communication', 7th intern. congress of LMPS, Salzburg 1983 (Abstracts of Section 11), 1983.
Lakatos, I., 'History of Science and its Rational Reconstruction', in *Boston Studies in the Philosophy of Science*, 8, Reidel, Dordrecht, 1971.
Lakatos, I. and Musgrave, A. (eds.), *Criticism and the Growth of Knowledge*, Cambridge University Press, 1970.
Lehrer, K. and Wagner, C., *Rational Consensus in Science and Society*, Reidel, Dordrecht, 1981.
Schmitt, F. F., 'Consensus, Respect, and Weighted Averaging', *Synthese*, **62**(1) January, 1985.
Stenius, E., 'The Picture Theory and Wittgenstein's Later Attitude to it', in I. Block (ed.), *Perspectives on the Philosophy of Wittgenstein*, Blackwell, Oxford, 1981.
Wittgenstein, L., *Tractatus Logico – Philosophicus*, Routledge and Kegan Paul, London, 1922.
Wittgenstein, L., *Philosophical Investigations*, Blackwell, Oxford, 1953.
Wittgenstein, L. (BB) *The Blue and Brown Books*, Blackwell, Oxford, 1958.
Wittgenstein, L. (NB) *Notebooks* 1914–1916, Blackwell, Oxford, 1961.

National Technical University of Athens

PETER KROES

PHILOSOPHY OF SCIENCE AND THE TECHNOLOGICAL DIMENSION OF SCIENCE

Modern science forms an inseparable whole with modern technology. A good deal, if not the greater part of present-day scientific research takes place in industrial research laboratories where science is practised in a technological setting and is exploited for technological ends. In The Netherlands, for instance, 60 to 70% of all the research in physics takes place in, or is financed by industry. For other highly industrialised countries the situation is not very much different. The foregoing means that scientists working on unified field theories, cosmological models, the foundations of quantum mechanics, evolution theory etc. can hardly be said to represent the whole of modern science. Clearly they form a minority among the community of scientists. Nevertheless, philosophers of science tend to direct their attention to the work of this minority and to base their models of science on it. As a consequence, the technological dimension of science, and the methodological problems connected with it, have stayed in the dark. However, in order to arrive at an adequate picture of modern science and of the way it develops, its technological dimension should be taken into account. For instance, the fact that solid state physics has become, in the course of this century, one of the major fields of research within the physical sciences, can only be understood on the basis of the technological relevance of the kind of knowledge produced by this type of research.

With a few exceptions, philosophers of science have paid almost no attention to this technological dimension of science, particularly to the methodological problems concerning the so-called technological or engineering sciences. At present, the methodology of the technological sciences is still a virtually non-existent field. The dominant picture of technological knowledge is that it is essentially of the same nature as scientific knowledge and that therefore the same methodological rules and principles apply to both types of knowledge. Technology is considered to be *applied* science and technological knowledge to be nothing more than *applied* scientific knowledge, that is, scientific knowledge *adapted* to the

K. Gavroglu, Y. Goudaroulis, P. Nicolacopoulos (eds.), Imre Lakatos and Theories of Scientific Change. 375–382.

realisation of an artefact. According to this view, this adaptation of scientific knowledge to practical ends takes place through inserting special boundary conditions into scientific theories. These boundary conditions are derived from the practical requirements which the desired artefact has to fulfil. The idea is that artefacts can be *deduced* from scientific theories together with special boundary conditions. The following quotation from Rapp (1974b, p. 98) illustrates this line of thought:

> Thus, for instance, the *findings* of physical investigations of the high vacuum and thermal emission of electrons led to the construction of the amplifier tube. Only by this means did present-day radio, radar and television technology become possible, though the electron tubes have recently been replaced by transistors, which are derived from solid state physics.

Such remarks strongly suggest that there is a direct, linear path from scientific knowledge to the application of this knowledge, that is to the concrete technological artefact. From this point of view, the methodology of the technological sciences hardly deserves any special attention.

This picture of technology as applied science, widespread as it is, is nevertheless a myth (Agassi 1974). Clearly, science plays a dominant role in modern technology and it is one of the important factors in technological development. An engineer cannot do his job properly without an appropriate background in the natural sciences; he has to be familiar with basic scientific theories and with standard scientific procedures for solving problems. But this crucial role of science in modern technology does not imply that the task of an engineer can be reduced to simply inserting specific, technologically relevant boundary conditions into extant scientific theories for the design, the development and construction of useful artefacts. This is not the way the transistor was actually discovered and developed; the first operating transistor, the point-contact-transistor, was not *derived* from solid state physics, as Rapp suggests in the above quotation (see e.g. Hoddeson 1981, Sarlemijn 1987). The following considerations show that much more is involved in passing from scientific knowledge to a technological product, and that the technological relevance of scientific knowledge manifests itself in much more indirect and subtle ways.

First of all, there is in most cases a wide gap between the results of 'pure' (or 'fundamental', or 'academic') research and the kind of knowledge needed for technological purposes. Casimir, former director of the Philips Research Laboratories, remarks that the "results of academic research that are potentially applicable are often not sufficiently precise or do not relate to the most usable materials or structures" (1983, p. 299). The main reasons for the existence of this gap are the *wide-ranging nature* of scientific theories (the more phenomena a theory explains, the better it is), and the widespread use of *idealisations* in science. As regards the first point, the comprehensive nature of scientific theories, there is a striking difference between scientific and technological theories. In contrast with scientists, technologists tend to develop theories which have only a very limited domain of application. Often, technological theories are only valid for a specific kind of artefact (e.g. technological theories about steam engines, internal combustion engines, radar, antennae etc.) or some type of materials (e.g. semiconductor theories). For technologists the comprehensive nature of theories is not necessarily a virtue as it is for scientists.

The frequent use of idealisations in science also severely hampers the direct technological application of scientific knowledge. Scientific theories often are only valid under ideal conditions (no friction, ideal gas, perfect crystal, boundary conditions at infinity etc.). An engineer, of course, cannot afford to make such idealisations; he has to deal with the 'real' world in which there is always friction, in which every material contains impurities, no crystal is perfect etc. In other words, the conditions under which scientific knowledge is valid are commonly different from the conditions which the engineer faces in practice. So he is forced to do 'additional' research in order to bridge the gap between the theoretically possible and the practically feasible. According to Casimir, again, "the primary task of industrial research laboratories [is] to complement the results of academic research that are applicable in principle so that they become applicable in practice" (*ibid*.). This kind of research is strongly 'mission-oriented' and produces knowledge which is primarily technologically relevant.

In the second place, engineers have to deal with certain kinds of technological problems for which science does not offer any solution at all. Let me just mention here problems with regard to scaling-up, product-design, and product- or process-efficiency. Such problems are typical for the technological sciences. In general, key notions with regard to technological problems are 'feasibility', 'reliability', 'efficiency', 'cost–benefit analysis', 'optimalisation' etc. In scientific problems these notions play no significant role at all. Consequently, the engineer usually cannot turn to science for ready-made solutions. Science is not just tailor-made for these kinds of problems. It is therefore no surprise to see that engineers have developed and make use of methods of inquiry which are particularly suited to solve technological problems and which have no real counterpart in the pure sciences. Let me just mention here the theory of similitude with the closely related field of dimension analysis, on which there exists an extensive technological literature (e.g. Szücs 1980, Langhaar 1965); these theories give a theoretical foundation to the use of scale models, which is of immense importance in technology. In two excellent case studies, Vincenti has drawn attention to some of the differences between scientific and technological methods of inquiry (1979, 1982).

In view of the foregoing, the conception of technology as applied science is much too simplistic. Seldom if ever the design of a technological product can be deduced in a straightforward way from scientific theories by feeding these theories with special boundary conditions. This idea gives a distorted picture of the nature of technological research and it is misleading because it suggests that technological development is totally dependent upon scientific progress. In spite of the overall importance of science for technology, this is not the case (see e.g. Vincenti 1984).

From the foregoing we may also conclude that technological problems are generally of a different nature than scientific ones, and that technological knowledge cannot simply be equated with applied scientific knowledge. In view of the different aims associated with science and technology, these conclusions are not really surprising; science is primarily oriented towards the understanding and explanation of nature, whereas technology is oriented towards the control and domination of nature.

It is, however, not difficult to understand why science and technology, in spite of their difference in orientation, are nevertheless mutually dependent. As far as the dependence of technology on science is concerned, it is clear that in most cases, better understanding leads to better control. Conversely, science heavily depends on technology because in most branches of science the study of natural phenomena takes place within an experimental set-up which requires advanced equipment for controlling these phenomena. What would, e.g., modern elementary particle physics be without the, from a technological point of view, highly sophisticated particle accelerators?

Indeed, it is the *experimental* nature of modern science that is crucial for the fruitful cooperation between science and technology. The common denominator, which brings together science and technology, is the experiment in which the manipulation of nature takes place. Since modern natural science is experimental, it is ipso facto technologically relevant. For Aristotelian physics this is not the case.

Now, given that the conception of technology as applied science is a myth, the problem of the technological relevance of scientific knowledge remains open. In what way does science contribute to the solution of technological problems? In my opinion, philosophers of science cannot afford to ignore this problem any longer, because, as I remarked at the beginning, science and technology form an inseparable whole. This problem becomes even more pressing as soon as one realises that it is virtually impossible to distinguish coherently between scientific and technological knowledge. According to Mayr any boundary between scientific and technological knowledge will be arbitrary (1982, pp. 157–158):

> If we can make out boundaries at all between what we call science and technology, they are usually arbitrary. [. . .] Traditionally we regard physics as a science and the manufacture of diesel engines as a technology. But what is thermodynamics, when textbooks are available in all shades of emphasis, ranging from purely practical concerns to the most esoteric theory? How should we classify subjects like aerodynamics, semiconductor physics, or even medicine?

Instead of a sharp boundary, there is a smooth, gradual transition from scientific to technological knowledge. Of course, it is worthwhile to characterise the two extreme cases which surely are different. But there is no reason for the philosopher of science to restrict his attention to the so-called 'pure science' which is just one extreme end of a whole spectrum of different research activities generating different kinds of knowledge. Such an attitude runs the risk of enforcing mythical conceptions concerning the nature of technological knowledge, but it may also lead to a distorted picture of science itself, in which the technological dimension of science completely stays in the dark. In most current models of (the development) of science, the technological relevance of scientific knowledge plays no significant role at all!

Let me indicate here briefly some topics which within this perspective deserve closer attention from the side of the philosophers of science. In the first place a systematic comparison is needed of the role of experiments in scientifically and technologically oriented research. Even in case a scientist and an engineer are interested in the same kind of phenomenon, the actual experiments performed by both may be completely different. This means that the nature of an experiment is determined to a large extent by the context in which it takes place; of course, the same applies, mutatis mutandis, to the kind of knowledge produced by experiments. Since experimental data are an important starting-point for theory formation, the context will also have a significant influence on the kind of theories produced.

This brings me to a second point. Not only in scientific but also in technological research, the construction of theories plays an important part, especially in the kind of research that is intended to bridge the gap between fundamental theories and practical applications. Often the situation is such that the fundamental theories are known, but that it is impossible to use these theories under conditions of interest to the engineer (because under those conditions the fundamental equations become too complex). In such cases, engineers tend to develop some kind of 'intermediary theories', which are partly based upon the fundamental theories but describe the behaviour of systems which are technologically relevant in terms of 'technological variables' (e.g. theories about turbulence in hydro- and aerodynamics). Little is known about the general properties of these intermediary theories, about how they are related to the fundamental theories and to the intended applications.

Last but not least, the question poses itself whether or not engineers make use of the same methods of inquiry as scientists. Until now, opinions on this matter have been divided. Some claim that there may be differences with regard to motivation and choice of subjects, but that there are no real methodological differences between scientific and technological research (Casimir 1983, Rumpf 1973). According to others, there are significant methodological differences (Skolimowski 1974, Rapp 1974b). Apart from the excellent studies by Vincenti mentioned above, I do not know of other concrete case studies into this matter. His case studies clearly indicate that engineers employ certain methods of inquiry which are characteristic of technological research. In this respect a detailed study of the use of similitudes (analogies) in technology would be highly desirable; as I remarked above, the theory of similitude is characteristic for technology. Philosophers of science have paid much attention to the use of analogies in scientifically oriented research (cfr. for instance Hesse (1966) and Nersessian (1987)), but not to the use of analogies in technological-scientific research. A first attempt to analyse the role of analogies in technological thinking from a methodological point of view is contained in Sarlemijn and Kroes (1985).

Since these topics fall outside the scope of traditional methodology of science, they have been largely neglected by philosophers of science. A better insight into these questions, however, will contribute not only to a better understanding of science, but also of technology and the way they develop.

To conclude, the fact that much of present-day scientific research takes place in industrial research laboratories, is in my opinion not just an institutional matter or just a contingent fact 'external' to science. These laboratories are not a facade behind which science develops autonomously according to its own rules. This embedding of science in an industrial context, which started at the end of the 19th century, has affected the scientific enterprise in deeper ways. It has brought science into contact with another problem-solving tradition, namely technology, by which both science and technology were transformed. A clear insight in this transformation of science and of technology is still lacking.

ACKNOWLEDGEMENT

Parts of this essay have previously been published in *Methodology and Science* **20**(3), 1987. I am grateful to the Editor, Piet H. Esser, for permission to use them in the writing of this work.

REFERENCES

Agassi, J., 'The confusion between science and technology in the standard philosophies of science', in Rapp 1974a.

Casimir, H. B. G., *Haphazard Reality*, Harper and Row, New York, 1983.

Hesse, M., *Models and Analogies in Science*, University of Notre Dame Press, 1966.

Hoddeson, L., 'The discovery of the point-contact transistor', in *HSPS* **12**(1) 1981.

Langhaar, H. L., *Dimensional Analysis and Theory of Models*, John Wiley & Sons, New York, 1965.

Mayr, O., 'The science–technology relationship', in B. Barnes and D. Edge (eds.), *Science in Context*, MIT Press, Cambridge, Mass., 1982, p. 155–163.

Nersessian, N., 'Scientific discovery and commensurability of meaning', this volume, 1989.

Rapp, F., (ed.) *Contributions to a Philosophy of Technology*, Reidel Publ. Co., Dordrecht, (1974a, 1974b).

Rapp, F., 'Technology and natural science – A methodological investigation', in Rapp 1974a.

Rumpf, H., 'Gedanken zur Wissenschaftstheorie der Technikwissenschaften', in H. Lenk and S. Moser (eds.), *Techne, Technik, Technologie*, Verlag Dokumentation, Pullach bei München, 1973, pp. 82–107.

Sarlemijn, A., 'Analogy analysis and transistor research', *Methodology and Science*, **20** (3), 1987.

Sarlemijn, A. and Kroes, P., 'Technological analogies and their logical nature', in P. T. Durbin (ed.) *Technology and Contemporary Life*, D. Reidel, Dordrecht, 1988.

Skolimowski, H., 'The structure of thinking in technology', in Rapp 1974a.

Szücs, E., *Similitude and Modelling*, Fundamental Studies in Engineering, Vol. 2, Elsevier Scientific Publishing Co., Amsterdam, 1980.

Vincenti, W. G., 'The air-propellor tests of W. F. Durand and E. P. Lesley: a case study in technological methodology', *Technology and Culture* **20** (1979), 712–751.

Vincenti, W. G., 'Control-volume analysis: a difference in thinking between engineering and physics', *Technology and Culture* **23** (1982) 145–174.

Vincenti, W. G., 'Technological knowledge without science: the innovation of flush riveting in American airplanes, ca. 1930–ca. 1950', *Technology and Culture* **25** (1984), 540–576.

Eindhoven University of Technology

GERARD RADNITZKY

FALSIFICATIONISM LOOKED AT FROM AN "ECONOMIC" POINT OF VIEW*

What may be gained by applying concepts generalized from economics such as, e.g., cost–benefit analysis ('CBA' for short), to methodological problems? It is claimed that the "economic" perspective may help the researcher to see what sorts of questions he should take into account when dealing with particular methodological decision problems. This claim is supported by applying the "economic" perspective to two standard problems of the philosophy of science.

1. THE CONCEPTS OF 'ECONOMIC' APPROACH AND OF METHODOLOGY

1.1. *The Economic Approach or the Expanding Domain of Economics*

Economics has been viewed as a general analytic machine for maximizing behavior. Yet, economics interpenetrates all of the other social sciencees, and it is reciprocally penetrated by them. What gives economics its imperialist power is the fact that *its key concepts are universal in applicability: scarcity, preference, opportunities, cost, choice, competition, etc., apply to all problem solving and, hence, to all life, not only to human phenomena* (Hirshleifer 1985, Ghiselin 1986, Posner 1987, Radnitzky and Bernholz 1987). Only during the last three decades economics began its imperialist age spreading the economists' theory of behaviour to the entire domain of the social sciences. Concepts generalized from economics and the pattern of economic theory have been successfully applied to fields of study that have traditionally been thought to lie beyond the scope of economic thinking, where 'success' is defined as achieving higher explanatory and predictive power. To forestall possible misunderstandings let me mention that the economic approach has nothing to do with explanations in terms of 'economic interest' or financial matters. For every act there is a cost. Money is just *one* means of measuring costs; there is also the time invested and the benefits forgone when using resources in the act, and there may be psychological costs, political costs, etc.

Popper proposed the method of 'situational analysis' – or, as I would

K. Gavroglu, Y. Goudaroulis, P. Nicolacopoulos (eds.), Imre Lakatos and Theories of Scientific Change. 383–395.

prefer to call it, 'the rational-problem-solving approach' – to the MS of *The Poverty of Historicism* in 1938 and he conceived it as "an attempt *to generalize the method of economic theory (marginal utility theory) so as to become applicable to the other theoretical social sciences*." (Popper 1976, pp. 117f., italics in the original). All rational choices involve the weighing up of benefits and costs. Hence, CBA is the core of the economic approach/rational-problem-solving approach. Strictly speaking, what is compared are benefits and forgone benefits, i.e., opportunity costs. If we generalize the concepts of CBA, we recognize that not only in economic life but also in daily life we operate within the CBA-frame, whenever we act rationally.

When CBA is applied to the methodology of research, we have to distinguish between the situation of an individual researcher and the methodology of research as a special discipline (a discipline which, perhaps, is even more imperialistic than the economic approach, since it deals with methods, criteria, argumentations, etc., in *any* scientific discipline). In the case of an *individual researcher*, CBA faces the same sorts of problems as it does in any individual decision-making process. It may very well be the case that his utility function includes such aims as earning money, getting promoted, etc., but does not include producing new knowledge. However, it may turn out that he can realize his primary aims only by producing new knowledge and, hence, this becomes his interim aim. In this case, we get an *indirect* utility function, which includes scientific progress as a means.

The researcher may be viewed as *rational 'discovery-maximizing producer'*, who selects and processes problems and problem solutions and, in particular, selects projects on the basis of their expected returns in new knowledge, in discovery. His time preference or "impatience" exactly corresponds to the choice of the discount rate. One of the problems of the use of CBA in economics is the problem of which benefits and costs are to be included in the analysis. In the application of CBA-thinking in methodology this problem has a counterpart which leads to the problem of rational theory preference. The alternative benefits forgone at the moment of choice are not actually experienced. The decision maker has to imagine them. He compares descriptions or representations of possible situations with expected payoffs. Hence, as already mentioned, the expected effects of the sacrificed alternatives are *subjectively* evaluated when the comparative evaluations are made "vicariously', when imagined or expected benefits are being compared in thought experiments.

It is illustrative to distinguish two types of success: success in the sociological sense, i.e, an increase in one's reputation, and success in the abstract sense, i.e., achieving scientific or intellectual progress. Hence, from the viewpoint of an individual researcher who is striving for success in the sociological sense an action may be rational, even if it does not realize the scientific progress that he would have been capable of achieving. Outlays for having learnt a technique are certainly past outlays (and in this sense bygones are forever bygones). However, they are not necessarily sunk costs relative to the current decision making of the individual researcher, but may influence his entrepreneurial choice, in particular, if past outlays are high. Thus, it may be rational for him to stick to the old "paradigm', even if he recognizes that the new "paradigm' holds more promise in the long run. Retooling to the new techniques may be too time-consuming for him, so that his reputation is best served by continuing to work with the old "paradigm". All this is so, because, for the individual, the effects of the sacrificed alternatives are *subjectively* evaluated.

When CBA is *applied to the methodology of science as a discipline*, the situtions is different. If we imagine an actor, it is the researcher as *ideal type*. Personal preference scale and psychological motivation of any particular, real individual are irrelevant to the methodological issues. Methodology does not aim at describing or explaining the actions of particular scientists or of the scientific community, but deals with methodological problems. *The basic aim of research is scientific or intellectual progress, and this aim is constitutive of the meaning of research*, and, hence, it cannot be put into question in that context. Therefore, methodological rules, criteria, decision, etc., are appraised in terms of their *instrumental* value for an approximation to the basic aim, to scientific progress. The methodological appraisals associated with the ideal type of researcher or, speaking less metaphorically, with methodology as discipline, as a sort of technology, are objective evaluations. The changing body of scientific knowledge evolves like a spontaneous order (in Hayek's sense), and through the continuous whittling away of problem solutions that are less good than their competitors, it can achieve progress. It is claimed that CNA applied to the methodology of research may be based on evaluations of opportunity costs that are *objective*, in spite of the fact that CBA in personal decision making is bound to the fact that cost as it influences choice is subjective.

1.2. *The Concept of a Methodology of Research*

Methodology is here conceived as a sort of technology for improving the researcher's chances to achieve his aim *qua* researcher: scientific progress. Hence, it constitutes an epistemic resource. It is embodied in a set of *hypothetical recommendations*. A typical rule of general methodology may have the following form: 'Presupposing that your aim is scientific progress, then, if you find yourself in a problem situation of type S, follow rule R, because by doing so your *chances* of success will be better than what they would be if you followed any of the rules which compete with R.' Thus, the rules of methodology are basically forward-looking, technological (i.e., hypothetical) rules of action. They include, as a subset, rules for the *ex post* appraisal of results, of competing problem solutions. Of course, methodology has also to provide rules for the appraisal of competing methodological rules.

Methodology deals with abstract entities (with Popper's W-3 entities); hence, it proceeds argumentatively. It does *not* attempt to describe or to explain the behavior of scientists. It is neither an empirical discipline nor a normative enterprise; it is very similar to technology. (Cf. Albert 1982, pp. 26–31. This idea is elaborated, e.g., in Radnitzky 1981, esp. pp. 45–50, in Radnitzky 1982 and Radnitzky 1985.)

Methodological rules are embedded in a set of global assumptions and may, in turn, influence these assumptions: general assumptions about the human cognitive apparatus (e.g., fallibilism), as well as global assumptions about the aspects of the world that one wishes to investigate (the assumption of order, the assumption of epistemological realism, etc.). Methodology also interacts with relevant results of empirical inquiry. This interplay of methodology on the one hand, and assumptions of philosophical cosmology and anthropology combined with results of empirical science on the other, does not lead to a vicious circle or to an infinite regress, if the quest for knowledge with a truth guarantee has been abandoned. Methodology attempts to articulate the rules that may increase the researcher's chances of achieving scientific progress. Only if the rules have been articulated, is it possible to examine them critically, and, sometimes, to improve them. The aim of this essay is to argue that methodology can profit from explicitly applying the CBA-frame to its problems.

2. CBA APPLIED TO THE PROBLEM OF THE EMPIRICAL 'BASIS'

2.1. For a Popperian to tentatively accept a statement means that he regards it to be worthy of being subjected to further criticism, to severe empirical testing (Popper 1959, New App. 9, p. 419). However, with the so-called basic statements this is different, because they are but raw material in research. Logical positivism and the philosophical tradition in general have been unable to solve the problem of the empirical 'base', because they have placed it in the context of justification philosophy. Popper has solved the problem, at least in principle (Popper 1979 (MS 1930–1933) *Grundprobleme*, and Popper 1934/1959.) According to Popper a basic statement is a report of observation asserting that an observable event is occurring in a certain region of space and time. (The concept is clarified, e.g., in Popper 1979/1930–1933, pp. 122, 127, 132; Popper 1959, Ch. 5; Andersson 1984a, pp. 54, 57, 62.) A basic statement must be able to contradict a general statement, or else it could not function as a potential falsifier. From the viewpoint of *epistemology* there is no difference between a basic statement and a general statement: both are fallible, criticizable and revisable. From the viewpoint of *methodology* there is an important difference: basic statements are easier to test than theories, and in this sense closer to experience. Perceptual experience constitutes the 'cause' and the 'evidencing reason' for our believing and asserting a basic statement.

The tradition that has dominated the philosophy of science has projected a dilemma: Either basic statements provide a secure source of knowledge, or they are nothing but the outcome of arbitrary decision or convention. Popper restructured the problem by placing it into a non-justificationist context. He stresses that observation is used to test, and that the process of testing is potentially infinite. Since no attempts are made to prove a theory or a basic statement to be false (let alone to prove it to be true), no infinite regress need arise. From the viewpoint of logic a singular statement can falsify a theory. From the viewpoint of methodology a theory is falsified only by a statement that describes a reproducible effect, by a 'falsifying hypothesis'. If a falsifying hypothesis B is regarded as less problematic than a theory T (plus the statement of the initial conditions and, of course, the relevant auxiliary hypothesis [often low-level hypotheses]), then T is falsified by B.

2.2. in the *practice* of research, as well as in daily life, the handling of basic statements does not constitute any problem (cf. Popper 1959, p. 93,

Andersson 1984a, p. 58). From a certain moment we regard a particular statement as unproblematic simply because we are convinced. We cannot decide to be convinced. Experienced certainty is, of course, epistemologically irrelevant, because everybody may be wrong. If a basic statement is not unproblematic at the moment, then a CBA is called for. However, in the practice of research, it will remain implicit. How much effort is invested in re-checking a particular statement depends upon the agent's subjective valuation of how much is at stake, i.e., it depends upon 'the logic of the situation'. However, to describe the practice of research is not the task of methodology. Therefore, let us turn to the methodological problems.

2.3. In the *methodological reconstruction* it is recommendable to reconstruct the handling of basic statements as a *decision* process. Being unproblematic at the moment means that, at the moment, there appears to be no reason to test it. However, a statement can be unproblematic only in the light of certain other statements that are presupposed; and these are presupposed, because on previous occasions it was decided that they are unproblematic, and so far no concrete reasons have emerged for revising these decisions. Since we do not search for an Archimedian point, this need not lead to an infinite regress.

The decision process involved may suitably be reconstructed as a CBA in which one takes into account the estimated marginal utility of re-checking the statement in question and of the costs involved, in particular, the opportunity costs of the re-checking. From the methodological point of view the important questions are: How can a particular basic statement be criticized? and: When is it rational to put it into question?

Basic statements are indispensable in testing a theory; they are not incorporated into the evolving body of scientific knowledge, but serve as a point of departure for producing a falsifying hypothesis. In order to maximize criticism and thereby the chances of error elimination, basic statements must be intersubjectively testable. Such testability is a matter of degree. Consider, for example, the report of a pointer reading such as the statement 'At t on this dial the pointer is opposite the figure 2.' ('b_0' for short). A certain perceptual experience is a motive for our asserting b_0, and if the statement is put into question, a report of the perceptual experience – preferably by several observers – will be used to defend it. A statement of the type b_0 is unproblematic *only* in the light of certain assumptions, e.g., that the observer's vision is normal, that parallax is under control, etc. Statements of the level b_0 are of interest only in a problem situation in which from b_0, in combination with statements about the initial conditions of the experimental set-up and auxiliary hypotheses about the functioning of the measuring apparatus used, etc., further test statements are derived such as, e.g., the statement 'At *t* the pressure in atmospheres in this container is 2.' ('b_1' for short). The *predictive implications* of a b_1-statement depend upon the state of the art. Rejecting b_1 while retaining the report about pointer readings b_0 would force us to reject one or more of the other premises of the argument, which in turn will involve certain "costs".

What does it mean to re-check a basic statement like b_1? It will mean measuring again, getting others to check the functioning of the various measuring apparatus, and so forth. Whether or not re-checking a particular basic statement is a good investment decision depends in large measure on the *objective problem situation*. The "costs" of re-checking the statement are opportunity costs: what results *could* have been achieved if, instead of re-checking the statement, time and effort were invested in another enterprise. In general, re-testing a b_1 statement is a profitable enterprise only when we wish to produce a falsifying hypothesis, i.e., a statement of the form: 'Whenever certain initial conditions are fulfilled, then b_1.' ('*B*' for short). A statement of the *B*-type is of interest if it contradicts the theory that we wish to test. Only a *B*-type statement possesses sufficient intersubjective testability to function as a falsifying hypothesis.[1] Therefore, the methodological problem of the empirical 'base' turns out to be essentially the problem of whether or not to *invest* time and effort into *processing a particular basic statement into a falsifying hypothesis for the theory we wish to test*. Like any other investment decision, this decision can rationally be made only on the basis of a CBA. The objective problem situation is decisive. Re-checking absorbs scarce resources of time and effort; hence it is rational to embark on such an enterprise *only if* there are *concrete* reasons for challenging that particular global assumption of the background knowledge.

The problem of the empirical 'base' does not constitute any major problem of methodology. It is an epistemological problem that cannot be avoided, because we test theories by means of test statements and test test statements themselves by deducing from them other basic statements. Knowledge in the objective sense is possible only because there are basic statements, test statements, that are regarded as unproblematic, at the moment (Popper 1979/1930–1933, p. 132).

3. CBA APPLIED TO THE PROBLEM OF RATIONAL THEORY PREFERENCE

3.1. In technological *applications* of scientific theories and in everyday life the pertinent question is, indeed, whether or not to accept a theory. We accept a theory in the sense that we trust it will help us to realize a particular practical goal. This expectancy is rational only if the theory is sufficiently well-tested for the practical purpose at hand. For economic reasons it is often rational not to use the best (in the methodological sense) of the available theories.

In *research* the situation is completely different. The question of when it is rational to stop problematizing a particular position is a reasonable question if asked with respect to a basic statement, because basic statements are but raw material of science. If the question is asked with respect to theories, it is misleading because it is based on the false assumption that research has a 'natural' end. Basic research is open-ended, because solved problems create new problems (Popper 1959, p. 104; Popper 1934, pp. 69 f.). In the context of basic research to speak of 'accepting' a theory is innocuous only if by 'accepting' we mean that we regard the theory to be worthy of further empirical testing (Popper 1959, New App. 9, p. 419). In basic science predictions are but a means of testing a theory. Exempted from criticism, *pro tempore*, are only those theories that we use as auxiliary hypotheses when testing a theory or when producing an explanation. So long as a theory is taken seriously as a possible solution to a scientifically interesting problem, it has to be tested.

When a theoretical system (consisting of the theory to be tested, auxiliary hypotheses, and statements of initial conditions) has been falsified, the researcher has the choice between blaming the theory for the falsification or blaming one or all of the other premisses of the falsification argument for the falsification. He may, for instance, opt for modifying one of the auxiliary hypotheses and try again. In that case, he has, however, to show that there are good reasons for modifying the auxiliary hypothesis in that way, e.g., by demonstrating that he is now able to derive testable consequences from the set of new premisses (the theory and the old initial conditions plus the modified auxiliary hypothesis) and that these testable consequences are corroborated, i.e., have withstood falsification attempts. Or the researcher may decide that the theory is to blame. In that case he has the option of investing time and effort in modifying the 'old' theory or in replacing it by a radically new one. In each case there is competition of theories: either between the 'old' theory and the improved successor, or between the 'old' theory and the radically new theory. Hence, the issue is never one of acceptance or rejection of a single theory, but of rational theory *preference*. We always start with some background knowledge, some first step towards a theory, and in this sense there is some competition of theories in every problem situation. Of course, the question of whether a particular theory is preferable to another arises only if the two theories are competitors. Theories are competitors (in a narrow sense of the word) if they offer incompatible solutions to the same, or highly overlapping, sets of explanatory problems; they are competitors in the wide sense if their realms of applicability overlap partially, but one theory has greater explanatory power than the other.

The style of thought that has dominated the philosophy establishment, logical empiricism, has projected an ideal of science according to which progress consists in probabilifying a theory more and more. Those who recognized that the logical positivists' program is a will-o'-the-wisp but could not free themselves from the justificationist ideal of knowledge, either adopted the instrumentalist view of theories or joined the sociological turn in the "philosophy of science"' (cf., e.g. Munz in Radnitzky and Bartley 1987). The latter is at the same time a relativistic turn: what matters is success in the sociological sense; there is no other criterion, because theories belonging to one 'paradigm' are incommensurable with those developed under another 'paradigm'. Thus, theory instrumentalists and relativists have lost or abandoned the problem of rational theory preference. Popper replaced the question 'How can we prove a theory?' by the question 'How can we improve a theory?' and thereby placed the problem in a non-justificationist context.

3.2. *From the Problem of Rational Theory Preference to the Problem of a Research Policy that is Internal to Science*

Is CBA relevant in the response to a falsification? A successful falsification attempt of a theory T results in the insight that from certain premises, which are *unproblematic* at the moment, it follows that non-T. This is all *logic* can tell. Given this insight and the basic aim of research, the rational response to the problem situation is governed by the rule: Propose a theory that is not hit by the falsifying hypothesis. This is all *methodology* can tell.

The researcher is now faced with a problem of a research policy that is internal to science: to allocate scarce resources of time and effort *within* a given field of study. For this sort of *risky investment decision* it is in principle impossible to give any methodological rule, if only because it is impossible to know in advance whether a modified version of the falsified theory will do, whether we will be able to create a radically new problem solution, etc. The CBA-frame can help the researcher to see which sort of questions he should take into account when comparing competing investment projects and what good reasons can be given for a particular problem preference.

What *good reasons* can be given for the methodological decision that a theory T' is to be *preferred to a theory* T? Time and again the researcher has to answer this question. He has to function as a methodologist of his own. (Sometimes, he may not be aware of this fact (cf. e.g., Radnitzky 1981, pp. 45–50). Incidentally, those who deny that they operate within some methodological framework and claim that they do not need any methodology are as a rule adherents of naive positivism.) To clarify what should be meant by 'good reasons' in this context one must, firstly explicate the idea that a theory T' is better than T in certain respects, i.e., make explicit the relevant dimensions of criticism;[2] second, produce arguments for the claim that it is these criteria that matter, i.e., that they follow from the basic aim of science; third, state what empirical support can be given for the conjecture that T' fulfills these requirements in a higher degree than T does.

Comparison of past performance in explaining and predicting enables us to make a conjecture about the problem-solving capabilities of the competitors. That means that we have to *compare the two balance-sheets of past performance*. Thereby all available information should be used, including appraisals of the 'scientific importance' of the problems the theory can solve or has failed to solve. It is rational to conjecture that the theory that has so far survived all the tests or more of the tests than the rival theory possesses a greater problem-solving capability than the rival that performed less well. On a realist view of theories we attribute this to the fact that it is more truthlike than its rival. Since the logical problem of induction cannot get a positive solution, conjectures about future performance of theories on the basis of their differential past performance must remain risky.

Since in the methodological appraisal of the *past* performance of theories not only the logical point of view is relevant, it would not be rational to consider equivalent a failure to explain a fact that is scientifically fairly unimportant and a failure to explain an observed regularity of great scientific importance. Hence, successes and failures should be weighted by the scientific importance of the problems concerned. The concept of 'scientific importance' can be explicated objectively in terms of the contribution which a correct solution to a problem makes for progress in the discipline concerned. (Cf., e.g., Radnitzky 1980, pp. 221–227.) This explicatum has nothing to do with the subjective interests or predilections of individual scientists. The appraisal *ex ante* of the relative scientific importance of competing problems is likewise a problem that confronts every scientist who works on his own responsibility. In making these risky investment decisions, both the researcher and the peer group may be helped by the CBA-frame. It draws their attention to such problems as the costs of defending a theory whose past performance is less good than that of its rival(s) and the costs of acquiring certain epistemic resources such as methods, skills, etc. In those fields where theorizing is closely monitored by empirical testing, theory change is an objective process, and eventually the better theory will drive out the less good one and do so for "economic" reasons.

3.3. *CBA can Show that Theory Change is an Objective Process, in which the Evaluation of Benefits and Costs is Objective*

The best way of arguing for this thesis is to show CBA in operation. I adapt the simple example that George Schlesinger uses in his review of Thomas Kuhn: the round-earth theory superseding the flat-earth theory (Schlesinger 1981, pp. 458 f.). Since ancient times there were observations (basic statements) that provided the raw material for a falsifying hypothesis of the flat-earth theory, above all that the sails of an approaching ship can be seen before the ship's hull can be seen. This observation may not be independent of theory, but it is sufficiently neutral with respect to the two competing theories to serve as a crucial experiment, and it is unproblematic in an obvious sense. The round-earth theory explains the observation. The flat-earth theorist can immunize his theory against falsification, but only at a cost, e.g., by modifying the auxiliary hypotheses about light rays. The round-earth theory explains why lunar eclipses are round. The flat-earth theory can be defended only at the cost of denying that lunar eclipses are due to the shadow that the earth casts on the moon. This immunizing is costly, because it forces one to reject also the theory that explains why lunar eclipses occur in the middle of the lunar month, etc. The first sailing around the earth can readily be explained by the round-earth theory. The costs of defending the flat-earth theory become increasingly high: the fact has to be explained away. Eventually, the costs of defending the falsified theory become unbearably high, e.g., when confronted with pictures of the earth taken from satellites. In all these cases the *valuation of the costs is objective*.

It is easy to cull from the history of science examples that illustrates this process: the Ptolemaic theory versus the Copernican, the hollow-earth theory versus the Copernican theory (cf., e.g., Sexl 1983), the theory of the ether (at least in the form given it in a Newtonian framework) versus the theory of relativity, and so forth.

The fact that during an initial period it may not be possible to muster good reasons for preferring one of the competing theories is (*pace* Feyerabend) not a valid objection against the thesis that the valuation of costs of defending the less good theory are objective and, eventually, will become prohibitive.

In summary, the rational response to falsification and rational theory preference are guided by CBA-consideration. One cost of defending a theory that is less good than its competitor is the resource cost of the activity itself, the time and effort needed to produce the various immunizing devices. However, far more important are the *opportunity costs*: the results which *could* have been achieved if the time and effort were invested in other undertakings. Opting for the less good theory, forces one to reject not only the explanations that can be constructed with the help of the better theory, but eventually also that theory itself. This time the "cost" is not an opportunity cost, but a "cost" in the sense that epistemic resources which have been available suddenly become unavailable to the defender of the theory with less explanatory power than its (rejected) competitor. We certainly make use of CBA whenever we act rationally, but making CBA-considerations explicit helps us to conceptualize the problem situation, to articulate the various steps in the decision-making process and thereby to make them more criticizable. Hence the CBA-frame constitutes itself an epistemic resource for the researcher whenever he has to act as a methodologist of his own.

NOTES

* The ideas of this paper originated in the preparation for a Colloquium on the applications of the Economic Approach, the results of which are presented in Radnitzky and Bernholz, 1987.

[1] Popper 1979 (1930–33), p. 132: '... what matters in it (in scientific method) ... is primarily not particular, *singular statements*, but *general regularities*; i.e., statements that ... can and must be tested again and again ...' (italicized in the original, Eng. trans. GR). See also Andersson 1984a, p. 63.

[2] The various dimensions of criticism have to be stated and ranked in a way that fits the particular problem situation. Popper recommends bold conjectures, an *entrepreneurial spirit*, because in science, like in business, risk willingness and possible gains are correlated.

REFERENCES

Albert, H., *Die Wissenschaft und die Fehlbarkeit der Vernunft*. Tübingen: J. C. B. Mohr, 1982.

Andersson, G., 'How to accept fallible test statements? Popper's criticist solution', in Andersson (1984), pp. 47–68.

Andersson, G. (ed.), *Rationality in Science and Politics.* (*Boston Studies in Philosophy of Science*, Vol. 79). Dordrecht: Reidel, 1984.

Andersson, G., *Kritik und Wissenschaftsgeschichte. Kuhns, Lakatos' und Feyerabends Kritik des Kritischen Rationalismus*. Tübingen: J. C. B. Mohr, in press, 1987.

Ghiselin, M., 'The economy of the body', *The American Economic Review* **68** (1978) 233–237.

Grmek, J., Cohen, R. and Cimino, G. (eds.), *On Scientific Discovery. The Erice Lectures 1977.* (*Boston Studies in the Philosophy of Science*, Vol. 34). Dordrecht: Reidel, 1981.

Heinrich, B., *Bumblebee Economics*. Cambridge, MA: Harvard University Press, 1979.

Hirshleifer, J., 'The expanding domain of economics', *The American Economic Review* **75** (1985) 53–68.

Popper, K., *The Poverty of Historicism*. London: Routledge and Kegan Paul, 1957. Orig. publ. in *Economica* 1944/45.

Popper, K., *The Logic of Scientific Discovery*. London: Hutchinson, 1959. German original *Logik der Forschung*, 1934.

Popper, K., *Die beiden Grundprobleme der Erkenntnistheorie*. Tübingen: J. C. B. Mohr, 1979

Posner, R., 'The law and economics movement', *The American Economic Review* **77** (1987) 1–13.

Radnitzky, G., 'Progress and rationality in research', in Grmek *et al*., (1981), pp. 43–102.

Radnitzky, G., 'Teoria della scienza', *Enciclopedia del Novecento*. Roma: Istituto della Enciclopedia Italiana, Vol. VI, 1982, pp. 370–386.

Radnitzky, G., 'Réflexions sur Popper. Le savoir, conjectural mais objectif, et indépendent de toute question: Qui y croit? Qui est à son origine?' *Archives de Philosophie* (Paris) **48** (1985) 79–108. Repr. in Radnitzky, G., *Entre Wittgenstein et Popper. Détours vers la découverte: le vrai, le faux, l' hypothèse*. Paris: Vrin, 1987.

Radnitsky, G. and Bartley, W. W., III. (eds.) *Evolutionary Epistemology, Rationality, and the Sociology of Knowledge*. La Salle, Ill.: Open Court, 1987.

Radnitzky, G. and Bernholz, P. (eds.), *Economic Imperialism: The Economic Method Applied Outside the Field of Economics*. New York, NY: Paragon House Publishers, 1987.

Schlesinger, G., 'The Essential Tension: Selected Studies in Scientific Tradition and Change', by Thomas Kuhn. The University of Chicago Press, 1977', *Philosophia* (Israel) **9** (1981) 455–467.

Sexl, R., 'Die Hohlwelttheorie', *Der Mathematische und Naturwissenschaftliche Unterricht* **36** (1983) 453–460.

University of Trier

PART VI

PETER URBACH

THE BAYESIAN ALTERNATIVE TO THE METHODOLOGY OF SCIENTIFIC RESEARCH PROGRAMMES

1. INTRODUCTION

Lakatos's methodology of science developed out of Kuhn's theory of the paradigm, and this in turn was proposed largely in reaction to Popper's philosophy of science. Popper held that the most significant scientific theories were falsifiable by "possible, or conceivable, observations." (1963, p. 39). Kuhn, however, pointed out – as Duhem and Poincaré had pointed out before him (and before Popper) – that this is untrue. Most of the really important theories of science are not falsifiable by observational statements. When such theories make empirical predictions, they do so in combination with certain auxiliary hypotheses, and if a prediction fails, one is not compelled by logic to infer that the main theory is false, for the fault may lie with one or more of those auxiliary assumptions. The history of science has many examples of an important theory leading to false predictions and that theory, nevertheless, not being blamed for the failure. We find in such examples that one or more of the auxiliary assumptions used to derive the prediction was taken to be the culprit. In this way, certain theories may come to exert a commanding influence over a whole area of scientific research.

Both Lakatos and Kuhn examined this process in detail, giving rise to their respective methodologies. The ideas of a paradigm and of a research programme are rather similar. However, Lakatos attempted to go beyond Kuhn, first, by specifying the fine-structure of a research programme (Kuhn held that the nature of a paradigm was inarticulable). Secondly, Lakatos tried to supply criteria of success for a research programme. I shall argue in this paper that Lakatos's criteria are unwarranted, but that the descriptive element of his claim is substantially correct. I shall also argue, following Dorling, that research programmes are best understood through the application of Bayes's theorem to scientific methodology.

K. Gavroglu, Y. Goudaroulis, P. Nicolacopoulos (eds.), Imre Lakatos and Theories of Scientific Change. 399–412.

2. LAKATOS'S CRITERIA OF APPRAISAL

A research programme in Lakatos's sense takes the form of a central or "hard core" theory, together with an associated "protective belt" of auxiliary assumptions. The function of the latter is to combine with the hard core in order to draw out specific predictions, which can be checked by experiment. The auxiliary assumptions are described as protective because during a research programme's life-time they, not the central theories, are discarded if a prediction is shown to be false. According to Lakatos, it is perfectly legitimate systematically to treat the hard core as the innocent party in a refutation, provided the research programme occasionally leads to successful novel predictions or to the successful explanation of existing data. Lakatos called such programmes "progressive". If, on the other hand, the programme persistently produces false predictions, or if its explanations of phenomena are habitually ad hoc, Lakatos called it degenerating. These terms are, of course, highly-coloured, and indeed, they were intended to intimate that the one kind of research programme was of greater epistemic merit than the other kind.

This claim has been much criticised. It is clear that progressive research programmes are not necessarily based on true theories, or on objectively probable, or reliable theories, nor do such theories necessarily have greater verisimilitude than those of a degenerating programme. Moreover, there is no assurance that progressive programmes will continue to be progressive. Hence, the claim that the research programme criteria are of any objective epistemic significance seems hard to interpret, let alone defend. Lakatos eventually recognised that his criteria had no epistemic force, and in the end he seems to have given up the claim that they do, declaring in 1973 in a seminar at which I was present that he wished all occurrences of the word "rational" to be expunged from his writings.

As a descriptive thesis, however, there is much to be said for the methodology of scientific research programmes, for it seems that, as a matter of historical fact, progressive programmes have usually been well regarded by scientists, while degenerating ones were distrusted and eventually dropped. (Evidence for this is to be found, for example, in the case-studies reported in Howson, 1976.) But although Lakatos and Kuhn identified and described an important aspect of scientific work, they provided no rationale or explanation for it. From their writings, one could think that the scientist's mere whim or pure chance determines that the hard core theory will be vindicated, and an auxiliary hypothesis blamed, after a refutation. Unfortunately, this suggests that it is a perfectly canonical scientific practice to set up any theory whatever as the hard core of a research programme (or as the central pattern of a paradigm) and to blame all empirical difficulties on auxiliary theories. This is far from being the case.

3. RESEARCH PROGRAMMES AND THE BAYESIAN APPROACH

What considerations determine which theories, out of a group which have been jointly refuted, should be held responsible for that refutation? In a paper that has not received the attention that is due to it, John Dorling (1979) has shown how this question should be answered within the Bayesian scheme.

Suppose a theory, T, and an auxiliary hypothesis, A, together imply an empirical consequence, which is shown to be false by the observation of the outcome e. Let us assume that, while the combination $T \& A$ is refuted by e, the two components taken separately are not falsified. We wish to consider the separate effects wrought on the probabilities of T and A by the adverse evidence e. The comparisons of interest here are between $p(T/e)$ and $p(T)$ and between $p(A/e)$ and $p(A)$. The conditional probabilities can be expressed using Bayes's theorem, in the following terms:

$$p(T/e) = \frac{p(e/T)p(T)}{p(e)} \qquad p(A/e) = \frac{p(e/A)p(A)}{p(e)}$$

In order to evaluate the posterior probabilities of T and of A, one must determine the values of the various terms on the right-hand sides of these equations. Before doing this, it is worth noting that these expressions convey no expectation that the refutation of $T \& A$, considered jointly, will have a symmetrical effect on the separate probabilities of T and of A, nor any reason why the degree of asymmetry may not be very large in some cases. Also, the expressions allow one to discern the factors that determine which hypothesis suffers most in the refutation. In particular, the probability of T changes very little if $p(e/T) \approx p(e)$, while that of A is reduced substantially just in case $p(e/A)$ is substantially less than $p(e)$.

4. AN EXAMPLE ILLUSTRATING THE BAYESIAN APPROACH TO RESEARCH PROGRAMMES

A particular historical example might best illustrate this. In 1815, William Prout, a medical practitioner and chemist, advanced the hypothesis that the atomic weights of all the elements are whole-number multiples of the atomic weight of hydrogen, the underlying assumption being that all matter is built out of different combinations of some basic element. Prout believed hydrogen to be this fundamental building-block, though the idea was entertained by others that a more basic element might exist, out of which hydrogen itself was composed. Now the atomic weights recorded at the time, though close to being integers, when expressed as multiples of the atomic weight of hydrogen, did not match Prout's hypothesis exactly. However, these deviations from a perfect fit failed to persuade Prout that his hypothesis was wrong; he took the alternative view, namely that there were faults in the methods that had been used to measure the relative weights of atoms. Thomas Thomson drew a similar conclusion. Indeed, both he and Prout went so far as to adjust several reported atomic weights, in order to bring them into line with Prout's hypothesis. For instance, instead of accepting 0.829 as the atomic weight (expressed as a proportion of the weight of an atom of oxygen) of the element boron, which was the experimentally reported value, Thomson preferred 0.875 "because it is a multiple of 0.125, which all the atoms seem to be." (1818, p. 340) (Thomson erroneously took 0.125 as the atomic weight of hydrogen, relative to that of oxygen.) Similarly, Prout adjusted the measured value of 35.83 for the atomic weight of chlorine (relative to hydrogen) to 36.

Thomson's and Prout's reasoning can be explained as follows: Prout's hypothesis (T), together with an appropriate assumption (A), asserting the accuracy of the measuring technique, the purity of the chemicals employed, etc., implies that the measured atomic weight of chlorine is a whole number, when expressed as a proportion of the atomic weight of hydrogen. Suppose, as was the case in 1815, that chlorine's measured atomic weight was 35.83, and call this the evidence e. Let us assume that the prior probabilities of T and of A were 0.9 and 0.6, respectively. In other words, I am assuming that chemists of the early nineteenth century, such as Prout and Thomson, were fairly certain about the truth of T, but less so of A, though more sure that A is true than that it is false. Contemporary near-certainty about the truth of Prout's hypothesis is witnessed by the chemist J. S. Stas. He reported (1860, p. 42) that "In

England the hypothesis of Dr. Prout was almost universally accepted as absolute truth' and he confessed that he himself "had an almost absolute confidence in the exactness of Prout's principle" when he started researching into the matter (p. 44). (Stas's confidence eventually faded after many years experimental study, and by 1860 he had "reached the complete conviction, the entire certainty, as far as certainty can be attained on such a subject that Prout's law . . . is nothing but an illusion" (p. 45)). It is less easy to ascertain how confident Prout and his contemporaries were in the methods by which atomic weights were measured, but it is unlikely that this confidence was very great, in view of the many clear sources of error and the fact that independent measurements generally did not produce identical results. On the other hand, chemists of the time must have felt that their methods for determining atomic weights were more likely to be accurate than not, otherwise they would not have used them. For these reasons, I conjecture that $p(A)$ was of the order of 0.6 and that $p(T)$ was around 0.9, and these are the figures I shall work with. It should be stressed that these numbers and the others I shall assign to other probabilities are intended to illustrate a principle; nevertheless, I believe them to be sufficiently accurate to throw light on the progress of Prout's hypothesis. As will become clear, the results we obtain are not very sensitive to variations in the assumed prior probabilities.

In order to evaluate the posterior probabilities of T and A, one must fix the values of the terms: $p(e/T)$, $p(e/A)$ and $p(e)$. These can be expressed as follows:

$$
\begin{aligned}
p(e) &= p(e/T)p(T) + p(e/{\sim}T)p({\sim}T) \\
p(e/T) &= p(e.A/T) + p(e.{\sim}A/T) \\
&= p(e/T.A)p(A/T) + p(e/T.{\sim}A)p({\sim}A/T) \\
&= p(e/T.A)p(A) + p(e/T.{\sim}A)p({\sim}A)
\end{aligned}
$$

Since T & A, in combination, is refuted by e, the term $p(e/T.A)$ is zero. Hence:

$$P(e/T) = p(e/T.{\sim}A)p({\sim}A)$$

It should be noted that in deriving the last equation but one, I have followed Dorling in assuming T and A to be independent, i.e., that $p(A/T) = p(A)$ and, hence, $p({\sim}A/T) = p({\sim}A)$. This seems to accord with many historical cases and is clearly right in the present case. By parallel reasoning to that employed above, we may derive the results:

$$p(e/A) = p(e/{\sim}T.A)p({\sim}T)$$
$$p(e/{\sim}T) = p(e/{\sim}T.A) + p(e/{\sim}T.{\sim}A)p({\sim}A)$$

Provided the following terms are fixed, which I have done in a tentative way, and shall justify presently, the posterior probabilities of T and of A can be determined:

$$p(e/{\sim}T.A) = 0.01$$
$$p(e/{\sim}T.{\sim}A) = 0.01$$
$$p(e/T.{\sim}A) = 0.02$$

The first of these gives the probability of the evidence if Prout's hypothesis is not true and if the method of atomic weight measurement is accurate. Such possibilities were explicitly considered by some nineteenth century chemists, and they typically took a theory of random assignment of atomic weight as the alternative to Prout's hypothesis; I shall follow this. Suppose it were established for certain that the atomic weight of chlorine lies between 35 and 36. (The final results we obtain respecting the posterior probabilities of T and A are, incidentally, not affected by how wide this interval is made.) The random allocation theory would assign equal probabilities to the atomic weight of any element lying in a 0.01-wide interval. Hence, on the assumption that A is true, but T is false, the probability that the atomic weight of chlorine lies in the interval 35.825 and 35.835 is 0.01. I have assigned the same value to $p(e/{\sim}T.{\sim}A)$, on the grounds that if A were false, and some of the chemicals were impure or the measuring techniques faulty, then, still assuming T to be false, one would not expect atomic weights to be biased towards any one particular part of the interval between two adjacent integers.

I have set the probability $p(e/T.{\sim}A)$ rather higher, at 0.02. The reason for this is that some impurities in the chemicals and some inaccuracy in the experimental method were moderately likely, given the acknowledged state of the art. If Prout's hypothesis were true, such imperfections are likely to have caused atomic weights to deviate somewhat from integral values. But the greater the deviation, the less likely, on these assumptions. Thus the probability that the atomic weight of an element lies in any part of the 35–36 interval is not distributed uniformly over the interval, but is more concentrated around the whole numbers. Let us proceed with the figures I have assumed for the crucial probabilities. This gives:

$$p(e/{\sim}T) = 0.01 \times 0.6 + 0.01 \times 0.4 = 0.01$$
$$p(e/T) = 0.02 \times 0.4 = 0.008$$

$$p(e/A) = \quad 0.01 \times 0.1 \quad = 0.001$$
$$p(e) \quad = 0.008 \times 0.9 + 0.01 \times 0.1 = 0.0082$$

Finally, Bayes's theorem enables one to derive the posterior probabilities in which we were interested, viz.,

$p(T/e) = 0.878$ (Recall that $p(T) = 0.9$)
$p(A/e) = 0.073$ (Recall that $p(A) = 0.6$)

These striking results show that evidence of the kind I have described may have a sharply asymmetric effect on the probabilities of T and of A. The initial probabilities that I assumed seem appropriate for chemists such as Prout and Thomson, and if correct, the results deduced from Bayes's theorem explain why those chemists regarded Prout's hypothesis as being more or less undisturbed when certain atomic weight measurements diverged from integral values, and why they felt entitled to adjust those measurements to the nearest whole number. Fortunately, these results are relatively insensitive to changes in the initial probability assignments, so their accuracy is not a vital matter as far as the present explanation is concerned. For example, if one took the initial probability of Prout's hypothesis (T) to be 0.7, instead of 0.9, keeping the other assignments, we find that $p(T/e) = 0.65$, while $p(A/e) = 0.21$. Thus, as before, after the refutation, Prout's hypothesis is still more likely to be true than false, and the auxiliary assumptions are still more likely to be false than true. Other substantial variations in the initial probabilities produce similar results though, with so many factors at work, it is difficult to state concisely the conditions upon which they depend without just pointing to the equations above. The example of Prout's hypothesis thus shows that Bayes's theorem can explain the key feature of Lakatos's research programmes in particular cases (Dorling has illustrated the point with a number of different examples) and, moreover, that the explanatory scheme it provides is an extremely promising one.

The marginal influence which we have seen an anomalous observation may exert on the probability of a theory is to be contrasted with the dramatic effect that a confirmation can have. For instance, if the measured weight of chlorine had been a whole number, in line with Prout's hypothesis, so that $p(e/T.A) = 1$, and if the probabilities we assigned are kept, the probability of the hypothesis would have shot up from a prior of 0.9 to 0.998. And, even more dramatically, if the prior probability of T had been 0.7, its posterior probability would have risen to 0.99. The existence of this asymmetry between anomalous and confirming instances was highlighted with particular vigour by Lakatos, who regarded it as being of the greatest significance in science and as one of the characteristic features of a research programme; he maintained that a scientist involved in such a programme typically "forges ahead with almost complete disregard of 'refutations'", provided he is occasionally rewarded with successful predictions (1968, p. 137): "he is encouraged by Nature's YES, but not discouraged by its NO". (1968, p. 135, footnote) As already indicated, I believe there to be much truth in Lakatos's observations; however, they are incorporated into his methodology, without explanation.

5. THE OBJECTIVITY OF SCIENCE

A number of objections might be raised against the type of analysis sketched above. First, it might be said that the probability assignments are arbitrary. I concede that in many cases there may be little evidence regarding the subjective probabilities invested in the various hypotheses by the scientists concerned, and in such cases one's conclusions about their patterns of inference must remain correspondingly uncertain. But in many cases, there is clear evidence of the extent of scientists' confidence in various theories.

Another common objection to the Bayesian approach is that it allows essentially subjective elements, especially the initial probabilities, to play a role in theory-choice. Kuhn might not have objected to this; he seems to have taken the view that the criteria of scientific appraisal are inevitably personal to scientists and to the scientific community to which they belong, these criteria depending upon aesthetic judgments, amongst other things. Lakatos, on the other hand, held that purely objective criteria of success could be described. These criteria were supposed to be objective in the sense that they would enable programmes to be evaluated by referring only to theories and to evidence, and not to the scientists. Lakatos expressed this ideal unequivocally thus:

> The *cognitive* value of a theory has nothing to do with its *psychological* influence on people's minds. Belief, commitment, understanding are states of the human mind. But the objective, scientific value of a theory is independent of the human mind which creates or understands it. Its scientific value depends only on what *objective* support these conjectures have in *facts*. (Lakatos, 1979, p. 1. The emphases appear in the mimeographed version of the paper distributed by Lakatos, but they have been removed by his editors.)

It seems to me, however, that this declaration reflects a thoroughly utopian view of science and that no such pure objectivity could possibly govern scientific reasoning.

Even the most basic aspect of Lakatos's view of scientific reasoning, namely the refutation of a hypothesis, seems to depend on beliefs whose correctness cannot be defended on objective grounds alone. For if an observation statement is to perform the task of falsifying a theory (i.e., of conclusively demonstrating its falsity), then it must itself be conclusively certain. But observation statements cannot be conclusively certain. For instance, the statement "The hand on this dial is pointing to the numeral 6" is clearly fallible – it is unlikely, but possible, that the person reporting it imagined the whole episode. The same is true of introspective perceptual reports, such as "In my visual field there is now a silvery crescent against a dark blue background". It has recently been maintained (Watkins, p. 79 and p. 248), that such statements "may rightly be regarded by their authors when they make them as infallibly true", but this is not so, for it is possible, though not probable, that the introspector has misremembered the shape he usually describes as a crescent or the sensation he usually receives on reporting a blue image. These and other sources of error ensure that introspective reports are not exempt from the rule that non-analytic statements are fallible.

Despite their fallibility, the kinds of observation statement I have mentioned, if asserted under appropriate circumstances, would never be seriously doubted. That is, although they could be false, they have a force and immediacy that prevents them from being doubted; they are "morally certain", to use the traditional phrase. But if observation statements are merely indubitable, then whether a theory is regarded as refuted by observational data or not rests ultimately on a *subjective feeling of conviction*. The fact that such convictions are so strong and so widely shared disguises the fact that they are not infallible, but it does not vitiate that fact. The idea that the "cognitive value of a theory", even as regards whether it is refuted, could be "independent of the human mind", as philosophers such as Lakatos and Popper believed, is mistaken.

6. AD HOC HYPOTHESES

6.i. *Lakatos's Notion of Ad Hoc-ness*

Another point at which subjectivity intrudes into Lakatos's principles of theory-appraisal is where they employ the notion of ad hoc-ness. It will be recalled that a programme is progressive if it explains phenomena in a non-ad hoc way. The idea of ad hoc-ness in Lakatos's terminology, can be explained as follows. Suppose a theory, T, together with the auxiliary hypothesis A were refuted by evidence e. Let A' be another auxiliary hypothesis, such that T & A' logically implies e. Lakatos called A' ad hoc if it failed also to be confirmed by some new prediction, say, e'.

One qualification and one amendment need, however, to be made to this characterisation. First, e' should be sufficiently different from e; Popper expressed this idea by saying that e and e' should be independent pieces of evidence. It would not be sufficient if, for example, e described Monday's fall of a stone and e' related to the same event on Tuesday. (Lakatos did not analyse the notion of one piece of evidence being independent of, or different to, another. There is a discussion of the question in Urbach, 1987, Chapter 6.) Secondly, Lakatos, under criticism from Zahar, came to revise the requirement that e' be absolutely new to the view that it should be independent of the evidence that led to the theory in question being advanced. This modification was intended to take account of Zahar's observation that a theory may be confirmed by existing evidence, and that many historical examples attest to scientists agreeing with this.

6.ii. *Are Ad Hoc Theories Necessarily Bad?*

The epithet ad hoc as it applies to hypotheses was intended by Lakatos and others to be pejorative. That is to say, a hypothesis that is ad hoc is, in some epistemically significant sense, inferior; the more satisfactory descriptive version of this thesis is that scientists never regard such hypotheses as valuable.

I have already criticised the normative thesis; the descriptive one also seems to me wrong. For an examination of scientific practice suggests, contrary to the methodology we are considering, that scientific theories can be massively confirmed, and established as highly acceptable, merely on the evidence which stimulated their formulation and without the necessity of further, independent, evidence. Suppose, for instance, that one is entertaining the hypothesis that an urn contains only red counters. An experiment is conducted in which counters are removed at random and then replaced, and this trial is repeated, say, 10,000 times. Let the result of this trial be that 4950 of the selected counters were red, and the rest white. The initial hypothesis, together with the various necessary auxiliary assumptions, are refuted and a natural revision would be to say that the urn contains red and white counters in approximately equal numbers. This seems a perfectly acceptable procedure, and the revised theory well justified by the evidence, yet there is no independent evidence for it: it gains its support solely from the evidence which discredited its predecessor (Howson, 1984, p. 242).

A more realistic example: the assumption is made that two characteristics of a plant are inherited in accordance with Mendel's principles and that each is determined by a specific gene, the other two genes acting independently and being located on different chromosomes. The results of plant-breeding experiments show that a surprising number of plants carry both characteristics and the original assumption that the genes act independently is revised in favour of a theory that they are linked on the same chromosome. (An example of this sort is worked out by Fisher in his *Statistical Methods for Research Workers*, ch. IX.) In each of these cases, the revised theory required no independent evidence before it is scientifically acceptable.

6.iii. *Is Lakatos's Criterion of Ad Hoc-ness Objective?*

A second objection to Lakatos's thesis is that his criterion of ad hoc-ness covertly violates his requirement that the standards of science be objective and that "[the] scientific value of a theory [be] ... independent of the human mind which creates or understands it." Indeed, whether a theory is ad hoc or not depends very much on the mind that creates it.

Suppose a scientist performs an experiment and observes e', which because it implies the falsity of the prediction e, made by T & A, refutes that combination of theories. A new theory, T & A', is advanced, but since there is no fresh evidence for A', it is ad hoc and unacceptable, according to the view we are considering. But consider the fact that only part of the observational evidence, namely not-e, is required for the refutation. Now suppose that the scientist had not recorded all the information in e' but that he had merely noticed that his prediction of e, had not been fulfilled.

In other words, his experimental datum is simply not-e. Assume that he revises his theory to $T \& A'$ on receipt of this datum, and that he then performs essentially the same experiment again, but now recording e'. According to the view we are discussing, this new theory would be perfectly acceptable and not ad hoc, since it is supported by evidence independent of that which refuted its predecessor. But the experimenter would be in precisely the same position regarding his theory and the evidence for it, whichever of the two paths he had taken. In other words, according to Lakatos's criterion, whether a theory is ad hoc, and hence unacceptable or inferior, depends very much on the human mind which created it.

7. WHY SUBJECTIVE ELEMENTS ARE UNAVOIDABLE IN SCIENTIFIC APPRAISALS

Lakatos, like Popper was attempting to direct his theory of ad hoc-ness to a particular difficulty. The difficulty arises from two facts: first, the fact that any data may be explained by an infinite number of mutually exclusive hypotheses; secondly, the fact that many such hypotheses intuitively seem very unsatisfactory. A methodology that attempts to restrict scientific reasoning to the objective relations of logical entailment (and, perhaps of probabilistic entailment) between theory and evidence seems to have no resources for discriminating between the infinity of rival explanations.

So far as I can see, the only ways to deal with this situation must employ some suitable, yet-undiscovered, objective relation connecting particular theories and observations, or else to permit an element of subjectivity into theory-appraisal. The Bayesian approach adopts the second alternative, and I think it gives a plausible account of scientific reasoning (only a fragment of that account could be given in this paper – it is presented much more fully in a forthcoming book by Howson and Urbach). I would, however, like to conclude by outlining the way in which a Bayesian sees the problem of ad hoc hypotheses.

8. A BAYESIAN ACCOUNT OF AD HOC-NESS

For the Bayesian, the characteristic of ad hoc hypotheses determining whether they are well regarded or not by scientists is their probability. In particular, if a composite theory $T \& A'$ is put forward as an alternative to the refuted combination $T \& A$, the Bayesian would assess this scientific development in terms of the posterior probability of the revised theory. This posterior probability should be relatively high – at any rate, it ought to exceed 0.5, so that it is more likely to be true than false. In the present context, we shall be interested in the theory $T \,\&\, A'$. It is easy to show, and intuitively obvious, that $p(T \,\&\, A'/e') \leqslant p(A'/e')$, so that if A' is not a credible hypothesis relative to e', then its conjunction with T will be at least as incredible. That is, the probability of the new hypothesis, A', provides a lower limit for that of the conjunction, $T \,\&\, A'$. In this account, there is no need for A' to be supported by new or independent evidence; all that is wanted is that it be credible (Horwich 1982, pp. 105–108).

9. CONCLUSION

I have argued that the central feature of Lakatos's methodology, namely, its method of evaluating the hard core hypotheses of a research programme, can be best explained and rationalised, as Dorling first showed, in the Bayesian scheme of science. The criticism of that approach, namely that it is unacceptably subjective, has been met in two ways: first, by a *tu quoque* argument addressed to Lakatos, suggesting that a subjective element is ineliminable from science and, secondly, that the Bayesian approach incorporates such a subjective element in a more plausible fashion.

REFERENCES

Dorling, J., 'Bayesian Personalism, the methodology of research programmes, and Duhem's problem', *Studies in History and Philosophy of Science*, **10** (1979) 117–187.

Horwich, P., *Probability and Evidence*. Cambridge University Press. Cambridge, 1982.

Howson, C. (ed.), *Method and Appraisal in the Physical Sciences. The Critical Background to Modern Science, 1800–1905*. Cambridge University Press. Cambridge, 1976.

Howson, C., 'Bayesianism and support by novel facts', *British Journal for the Philosophy of Science* **35** (1984) 245–51.

Howson, C. and Urbach, P., *The Nature of Scientific Reasoning: The Bayesian Approach*. Open Court Publishing Company. La Salle. Ill., forthcoming.

Lakatos, I., 'The methodology of scientific research programmes', in I. Lakatos and A. Musgrave (eds.), *Criticism and the Growth of Knowledge*. Cambridge University Press. Cambridge, 1968.

Lakatos, I., 'Science and Pseudoscience'. An Open University lecture, reprinted in J. Worrall and G. Currie (eds.), *The Methodology of Scientific Research Programmes*, vol. 1, pp. 1–7. Cambridge University Press. Cambridge, 1973.

Popper, K., *Conjectures and Refutations*. Routledge and Kegan Paul. London, 1963.

Prout, W., 'On the relation between the specific gravities of bodies in their gaseous state and the weights of their atoms', *Annals of Philosophy* **6** (1815) 321–30. Reprinted in *Alembic Club Reprints*, No. 20, pp. 25–37. Oliver Boyd, Edinburgh, 1932.

Stas, J. S., 'Researches on the Mutual Relations of Atomic Weights', *Bulletin de l'Académie Royal de Belgique*, 1860, pp. 208–336. Reprinted in part in *Alembic Club Reprints*, No. 20, pp. 41–47. Oliver Boyd. Edinburgh, 1932.

Thomson, T., 'Some Additional Observations on the Weights of the Atoms of Chemical Bodies', *Annals of Philosophy* **12** (1818) 338–50.

Urbach, P., *Francis Bacon's Philosophy of Science: an Account and a Reappraisal*. Open Court Publishing Company, La Salle, Ill., 1987.

Watkins, J., *Science and Scepticism*. Princeton University Press. Princeton, 1984.

The London School of Economics

G. CURRIE

FREGE AND POPPER: TWO CRITICS OF PSYCHOLOGISM*

1

Psychologism is widely regarded as a philosophical vice. It is not entirely clear, on the other hand, what psychologism is. Part of the reason for this obscurity is, I think, that the philosophers we label anti-psychologistic have not always had a common enemy. From Frege, through Husserl, Wittgenstein and Carnap, to Popper and beyond, the target has shifted and grown. Theories of logic, knowledge, mathematics, meaning and mind have been condemned as psychologistic; so has intentional criticism in the arts.[1] No one is safe: in the style of radical politics, the denouncers can find themselves denounced. Dummett, who tells us so often that Frege eliminated psychologism from the theory of meaning, is said by McDowell to be falling into psychologism when he identifies the theory of meaning with the theory of speaker's understanding.[2] And Frege himself is said by Popper to have founded that notoriously psychologistic activity – epistemic logic.[3]

We might wonder whether some of the gains of anti-psychologistic criticism represent inflation rather than real growth. Looking at Popper's 'objective epistemology' against a Fregean background suggests that this is, indeed, the case.

2

There is a good deal that Frege and Popper have in common; in their characterization of psychologism, in their response to it, and in their general philosophical attitudes. I want to begin with these similarities, for I think they have not been widely noticed. Indeed, it is commonly supposed that the philosophies of these two men are deeply at odds. On one influential view, Frege rearranged our philosophical priorities by turning the theory of meaning into first philosophy, a position previously occupied by the theory of knowledge. From this perspective, Frege's anti-psychologism appears as an attack on psychologistic theories of meaning.[4] Popper, on

K. Gavroglu, Y. Goudaroulis, P. Nicolacopoulos (eds.), Imre Lakatos and Theories of Scientific Change. 413–430. © 1989 by Kluwer Academic Publishers.

the other hand, tells us that he is not interested in the meanings of words, that he regards questions about meaning as philosophically sterile, and that it is the phenomenon of human knowledge that occupies him as a philosopher.[5]

There is the potential for much confusion here. When we look carefully at Popper's reasons for objecting to the theory of meaning, we see that they fall into two categories. There are objections first of all to a concern with the meanings of particular words; if there is a philosophical target here it is presumably the kind of 'linguistic philosophy' that eschews general theories in favour of detailed analyses of particular usage.[6] Then there are objections to the verificationist theory of meaning developed by the logical positivists.[7] Much work on semantics and philosophical logic falls into neither of these categories. Certainly, Frege's own theory of sense and references does not. It is a quite general theory, and it is not in the least verificationist. Furthermore, Popper himself is greatly interested in language and how it works, and much of what he says about these things would normally be regarded as a contribution to the theory of meaning, in a general sense that encompasses the function and use of language.[8]

A scrutiny of Frege's concerns holds similar promise of reconciliation. While it is undoubtedly the case that Frege's theory of sense and reference is a contribution to the theory of meaning in this general sense, I do not believe that Frege's philosophical programme is centrally concerned with meaning. Rather, his abiding concern is the status of our knowledge. Like Popper, he is concerned to defend the objectivity of knowledge.[9] As I shall show further, the anti-psychologism of both has its roots in this common project. I am hopeful therefore that we can establish the following proposition: despite what some of Frege's readers say about him, and despite what Popper says about himself, there is a large area of agreement between these two philosophers. Identifying this area will set the scene for some criticisms of Popper that I shall present later on.

3

The psychologism that Frege opposed is but one aspect of the naturalistic, scientific spirit of mid nineteenth-century thought, though it derives ultimately from the empiricism of Locke and Berkeley.[10] Frege rejected the whole programme of naturalism, with its empirical account of logic and mathematics, its genetic approach to concepts, and its construal of thinking as the having of ideas or mental pictures. Part of the clamour of the times was for evolutionary and historicist methods, and Frege complained that mathematics was disappearing under a pile of irrelevant speculations about the origins and mechanism of our mathematical ideas: 'the theory of evolution is making its victorious course through the sciences and . . . the historical conception of everything is threatening to transgress its own limits (1879–91, p. 4; see also his (1906), proposition 11). The polemical part of *The Foundations of Arithmetic* is an attack on naturalism in mathematics; not merely on the psychologistic identification of numbers with ideas, but on Mill's empiricism and the early formalist's identification of numbers with concrete inscriptions.[11]

Out of his early polemic Frege developed a more general attack on psychologism, which he deployed in the preface to *The Basic Laws of Arithmetic*, pursuing what he saw as the disastrous consequences of identifying thought with the having of ideas or mental images. He also began to develop a positive alternative to the empiricist theory of the mind, which he sketched in the late essay 'The Thought'. Frege was not sceptical about the existence of mental images, but he insisted that they are an accompaniment to thought, not constitutive of it.[12] If they were, our thoughts would not be communicable, for there can be no comparing the mental images of different subjects.[13] The picture theory – as we may call the empiricist model – makes it impossible for one to dispute or criticize the thought of another, and that is relativism.[14]

Frege also saw in the model a tendency to take ideas as the referential objects of thought and experience, particularly in the case of problematic entities like numbers and concepts. Again, we are forced towards relativism: 'If the number two were an idea, it would have straight away to be private to me only' (1884, section 27). My numbers may have quite different properties from yours, yet we should not be able to call this a disagreement, for there is nothing common to us to disagree about. And the process of absorbing outer things into the mind does not stop at abstract entities; the model has a general tendency to encourage us to suppose that what we think about and perceive is our own ideas. Thus it leads to solipsism:

> In the end everything is drawn into the sphere of psychology; the boundary between objective and subjective fades away more and more, and even real [*wirchliche*] objects themselves are treated psychologically, as ideas. . . . Thus everything drifts into idealism, and with perfect consistency into solipsism (1983, p. 17).

Further, there can be nothing true that is not an idea, nothing valid that is not a transition from one idea to another. But ideas are subject only to the associational laws of psychology. So truth and validity must be nothing but psychological notions. And that is subjectivism.[15]

Frege's anti-psychologism was an attempt to sweep away relativism and subjectivism by establishing a right view of the mind and its relation to thought and the world. We must recognize that thought involves something beyond the mind itself; to think is to grasp an eternally existent Thought (*Gedanke*), something that can be the common content of many cognitive acts. (We may call this the relational theory.) Pure solipsism – the thesis that there exists nothing but my mind and its ideas – is not a possibility for Frege; even in thought we reach beyond ourselves to something objective. Frege was quick to point out that there is no easy step from this to a proof of the existence of the external world, for Fregean propositions are not constituted out of things in the realm of reference.[16] But he suggested that once the existence of propositions is recognized there should be no objection in principle to supposing that the physical world also exists. Recently philosophers like Evans and McDowell have tried to close the gap, by insisting that a grasp of certain propositions presupposes the existence of their referential objects; that there are genuinely *de re* propositions.[17]

The opposition, as Frege understood it, between psychologism and anti-psychologism is the opposition between the picture and the relational theories of cognition. The picture theory leaves us trapped inside our own heads. Substituting the relational theory gives us an explanation for the commonality of thought and speech, a bridgehead into the external world, and a regulative, mind-transcendent conception of truth and validity.

4

This bare outline of Frege's position already suggests a number of points of contact with Popper's thought. Popper's first target was the rather more sophisticated positivism of Carnap and the Vienna Circle, and he saw in it the same naturalistic motivation that Frege identified in its ancestor. In *The Logic of Scientific Discovery* he described what he saw as the positivist's tendency to 'interpret the problem of demarcation [between science and metaphysics] in a naturalistic way; they interpret it as if it were a problem of natural science' (1934, p. 35). He has also been critical of attempts to lean on the theory of evolution for an account of norms and values.[18] Like Frege he has criticized the empiricist tradition, based as it is on the notion of ideas, and associated it with relativism, subjective idealism and solipsism.[19] He is insistent on the objective, mind-transcendent nature of propositions, and he affirms the relational theory of cognition.[20] Like Frege (and with the help of Tarski – help that Frege would have rejected)[21] he asserts the objectivity and absoluteness of truth and validity.[22] But it must be said that Popper does not analyse the notion of a proposition, whereas Frege grappled, not entirely successfully, with the question whether propositions are structured entities and with their identity conditions.[23] Since everything that we shall subsequently investigate in Popper's thought depends upon the notion of a proposition, this seems unfortunate. But I shall not pursue this point in what follows. We shall find much that is deficient in Popper's account that cannot be improved by clarification of this notion.

Both Frege and Popper have responded to psychologism by announcing an ontological thesis: that there exists a realm of the objective contents of thought; a realm of things which are neither mental nor physical in character. Frege's 'third realm' ('*drittes Reich*') and Popper's 'World 3' are alike in so far as they contain thought contents, mathematical objects and other *abstracta*, and in so far as things of World 3 are supposed to be capable of interacting with mental entities of World 2, and via the mental, with the physical things in World 1.[24] Both Frege and Popper insist that thinking must be understood as a process that involves interaction between the mind and an abstract, World 3 object, a proposition or thought content. They seem, in fact, to attribute causal powers to propositions. A proposition acts upon the one who grasps it, and a change of behaviour may result. Popper suggests that the very test of the reality of a thing is the possibility of its interaction with *prima facie* real things.[25] And Frege's German makes the connection very clear: Thoughts are real (*wirklich*) because they can act (*wirken*) upon us.[26] But the idea that Thoughts are literally active seems implausible and redundant. The causal powers that we might be tempted to attribute to propositions are surely better located in the concrete objects that represent them; perhaps in patterns of neural activity that constitute mental representations. And the representing relation need not itself be thought of as causal. To postulate mysterious influences emanating from propositions themselves would involve us in the (equally problematic) converse of a 'magical' theory of reference.[27]

In elaborating a tripartite ontology, in speaking of a 'Third Ream' of the objective contents of thought, and in ascribing activity to Thoughts, Frege seems remarkably close to the ideas that Popper presented fifty years later.

Yet while Popper speaks with general approval of Frege as a precursor of his own thought, he does not acknowledge the detailed parallels that seem so striking.[28]

There is another interesting similarity between Frege and Popper, though it is of a different kind from those I have mentioned so far. It seems to me that Frege and Popper stand in the same kind of relation to their philosophical communities. While both have spent their working lives as university teachers, publishing books and articles in scholarly journals, their relations to the established world of professional philosophy have been somewhat uncomfortable. Of course, Frege's reputation has risen greatly since his death; it may be that in the world of anglo-saxon philosophy there is no one who is currently more written about. But in 1936 Edmund Husserl could look back at the period of his correspondence with Frege around the beginning of the century and say 'At the time he was generally regarded as an outsider who had a sharp mind but produced little or nothing, whether in mathematics or in philosophy.'[29] Frege's polemical writings (which form a considerable proportion of his work) reflect his outrage at the situation. He constantly proclaims the errors of those who fail to notice his arguments or who fail to appreciate them.

Popper also thinks that he has been misunderstood and ignored, and has said so in print on a number of occasions. He has gone on record as saying that he solved the problem of induction in 1927, that as a result he has been able to solve a number of other philosophical problems, that few have noticed the solution, and that books continue to be written about it which do not mention his work.[30] And while it is true that Popper's contribution to the philosophy of science is appreciated, few philosophers agree with Popper's own assessment of that contribution. In the wider world of learning Popper is a well known and respected figure, but the direction of contemporary, broadly analytical philosophy is relatively uninfluenced by his thought. If anything, the degree of Popper's intellectual estrangement from professional philosophy has increased since he began to develop the theory of World 3. And what little reaction there has been to it has generally been hostile.[31]

It may be that Popper's professional reputation will follow that of Frege; that in 30 years time his achievement will be recognized as revolutionary. But there are, I think, reasons to doubt it. One reason – and this brings us back to our proper topic – is that Popper's attempt to extend his thought in a systematic way is closely connected with his theory of World 3. It seems to me that this theory is wrong; not merely in detail, but in its basic conception. Moreover, the flaw in the system is traceable to an idea that plays an important role in his early work on scientific method; an idea that constitutes the heart of his anti-psychologism. If I am right about this we have one less reason to expect a radical reassessment of Popper's achievement by future generations of philosophers.

5

The similarities between Frege and Popper that I have outlined are of some historical interest. But the really important question is whether Popper's treatment of psychologism goes beyond Frege's in significant ways. Popper's attack on psychologism is indeed more radical than Frege's, to the point where important contradictions between their thought emerge. But, as I shall try to show, it is not so much a matter of Popper's attack extending deeper into the enemy camp. Rather, he has turned the guns on a crowd of innocent bystanders.

The point at issue concerns Popper's elaboration of the idea of a Third Realm. First of all, Frege's Third Realm is conventionally platonic. Thoughts are timeless entities, unaffected in their essences by the activities of human subjects.[32] Popper's idea, on the other hand, is that the Third Realm (World 3) is *created*; it is the by-product of language. The development of language allows for the objectification of thought, and hence for the emergence of World 3.[33] This is certainly an important difference between Frege and Popper, but I do not want to concentrate on it here, having been critical of it on another occasion.[34]

The aspect of World 3 that I do want to concentrate on here is its relation to epistemology. For Popper claims that the recognition of World 3 reveals traditional epistemology to be based on a mistake: the mistake of concentrating on subjective knowledge, on World 2.[35] Now one might suppose that Popper has in mind the point made by Frege: that thought must be seen as having an objective, propositional content; that knowing, believing and other epistemic attitudes are relations between the believer and a proposition. Sometimes, indeed, Popper does seem to be making just this point. But his central claim about epistemology is more radical than this. It is that World 3, rather than the relation between the knowing subject and World 3 objects, is the proper concern of epistemology.[36] We are to have an 'epistemology without a knowing subject'. So Popper's diagnosis of psychologistic error is more extreme than Frege's. It is not merely the error of failing to recognize the relational character of thought;

it is the error of allowing the mind half of the relation any epistemological significance. For knowledge in the objective sense 'is independent of anybody's belief, or disposition to assent; or to assert, or to act' (1972, p. 109). Now this seems an odd position for Popper to adopt. It would be natural for Frege to say that the objects of World 3 are independent of human cognition. As we have already noted, the Third Realm it is the world of eternal propositions, thought and unthought. But Frege exhibits no tendency to call World 3 objects 'knowledge' simply in virtue of their being World 3 objects. Something isn't knowledge for Frege unless it is known to be true. And something being known to be true is (partly) a matter of what is going on in Worlds 1 and 2.

For Popper the thesis seems doubly difficult to maintain. The first difficulty is that, on his own admission, World 3 is a product of human thought and behaviour. In what sense, then, can it be independent of these things? At other times Popper claims merely that World 3 is 'relatively autonomous'. What are we to make of this claim? To say that a thing is autonomous is to imply that its states may vary independently of the states of other things. If World 3 objects are causally inert this seems unlikely. And I think the independent variability of World 3 is ruled out once we recognize that World 3 *supervenes* on the other two worlds. That is, assume that the behavioural and intentional states of all agents are fixed: then the content of World 3 is also fixed. Any possible world like our own in respect of the behavioural and intentional states of agents would surely be just like ours in respect of the contents of World 3. That is implicit in the claim that World 3 is produced by thought and speech. And that is why it is inappropriate for Popper to speak of World 3 as an emergent entity which, when produced, takes on a life of its own.

To say this is not, of course, to say that our theories can be reduced to the mental states of subjects; a point that Popper makes much of.[37] The logical content of a theory normally goes way beyond what is represented in the minds of individuals. But there can be dependence without reduction. And it does seem that on Popper's account World 3 is broadly supervenient upon the totality of thought and behaviour. In that case, claims about the 'autonomy' of World 3 will have to be very carefully formulated; more carefully than Popper has formulated them.

The other problem has already been alluded to, and it is more serious. To say that a proposition constitutes knowledge is surely to say something about its relation to the knowing subject. The epistemological tradition acknowledges this in defining knowledge as true, justified belief. But

Popper challenges the epistemological relevance of both belief and justification. Like E. M. Forster, he does not 'believe in belief'.[38] But so often when Popper discusses belief, it seems that he has in mind some pathologically tenacious attitude. He has, in fact, a deplorable tendency to argue in the following way. Belief is a state of unshakable and irrational conviction. Scientists do not (in this sense) believe their theories. Concern with the subject's relation to propositions (the traditional concern of epistemology) is a concern with belief (in this sense). Hence traditional empistemology is irrelevant to the philosophical analysis of science. Thus Popper says that the view according to which scientists believe their theories 'cannot explain ... the decisive phenomenon that ... *scientists try to eliminate their false theories. ... The believer ... perishes with his false beliefs*' (1972, p. 122, italics in the original). But one can of course have a belief that is rationally sensitive to the weight of evidence, and change one's belief as new evidence comes in. And when Popper speaks of acceptance he is describing a psychological state which must involve belief of some kind. Can a scientist rationally accept a theory without believing that the evidence supports it; without believing, if there is as yet no evidence, that the theory is sufficiently plausible to be worth testing? Without belief there can be no rationality.

As to justification, Popper holds that there is no such thing as the positive justification of any contingent proposition: for our conjectures 'can neither be established as certainly true nor even as "probable" (in the sense of the probability calculus)' (1963, p. vii). We shall return to the concept of justification in a moment.

If belief and justification are irrelevant to epistemology, what is it that makes some propositions knowledge and some not? On some occasions Popper suggests that there is no such distinction – that all propositions in World 3 count as knowledge. Thus he says that objective knowledge is 'the logical content of our theories, conjectures, guesses ...' (1972, p. 73). But if we identify knowledge in the objective sense with the contents of World 3 then we must say that myth, astrology, long refuted science and everything else that anybody has ever given voice to, counts as knowledge. So the news is good. The sceptics have said that there is no knowledge; Popper tells us that there is plenty. But this is news from nowhere. Astrology isn't knowledge; not on any plausible interpretation of 'knowledge'. If Popper's theory is not to be written off simply as a bizarre redefinition of terms, or as the products of concern that are simply incommensurable with those of traditional epistemology, it must be shown that his theory is rich enough to draw distinctions within the very large class of things that he calls 'knowledge'; to distinguish the good from the less good and the less good from the bad. And of course Popper wants to do exactly this. The whole thrust of his theory of science is that a testable theory is better than an untestable one, and that a tested but unrefuted theory is better than a refuted one. The task of the epistemologist is not, on Popper's account, simply to produce a list of the contents of World 3, but to say what kinds of things in World 3 are worth having. And Popper thinks that the rules of his methodology may enable us to justify a preference for one among competing theories: 'Can a *preference*, with respect to truth or falsity, for some competing univeral theories over others ever be justified by . . . empirical reasons? . . . Yes; . . . it may happen that our test statements may refute some – but not all – of the competing theories; and since we are searching for a true theory, we shall prefer those whose falsity has not yet been established' (1972, p. 8, italics in the original).

As one more or less innocent of probability theory, I do not want to challenge Popper's claim that we may be justified in accepting a statement without having made that statement probable to any degree. But I do want to ask, in line with the central concern of this paper, whether this justification can be unrelated to the experiences of subjects. Popper's claim to be in possession of an 'epistemology without a knowing subject' suggests that it can, and the suggestion is reinforced by remarks like this: '. . . knowledge in the subjective sense, is irrelevant to the study of scientific knowledge' (1972, p. 111). If this and like statements in Popper's work are not just rhetorical exaggerations of his thesis, we must take him to be saying something like this: the criteria we are to use in appraising the contents of World 3 should be applicable on the basis of an examination of the contents of, and the relations between, World 3 objects, and should not make reference to the subjective states of agents.

I think that this is indeed the motivating idea behind Popper's objective epistemology, and in a moment I shall say something about the development of this thought in his writings. But let me begin with a rather obvious objection to it.

If the epistemologist is to confine himself to investigating World 3, and not the relations of things in World 3 to the other worlds, he will be confined to investigating the logical relations between propositions – relations of entailment, consistency and inconsistency (together with their probabilistic generalizations like partial entailment, if one believes that there are such things, as Popper does not). And that, surely, is the task of the logician, not the epistemologist. The epistemologist is concerned with the reasons we may have, if there are any, for choosing between propositions in World 3. The propositions that belong to World 3 do so simply in virtue of having been uttered by somebody at some time. So one cannot say that its mere presence in World 3 is any reason for accepting a proposition. (From now on I shall just say 'proposition', when I mean 'proposition in World 3'.) It is not any kind of reason for accepting a proposition simply that it is entailed by or consistent with some other proposition. And it is not any kind of reason for rejecting a proposition simply that it is inconsistent with some other proposition (unless it is inconsistent with a purely logical truth and hence self contradictory).

What is a reason for rejecting or accepting a proposition is that it is either refuted or corroborated by experiment. But that is a question about the relations between World 3 and the other worlds; the outcome of an experiment is not something that exists in World 3. Of course a proposition describing an experimental outcome may exist in World 3, but we cannot rely on that when we decide about the acceptability of an hypothesis. For such a description may be in World 3 without it actually describing any experiment that has ever taken place, or ever will take place.

Popper says at one point: '... the assumption of the truth of test statements sometimes allows us to justify the claim that an explanatory universal theory is false' (1972, p. 7, italics throughout in the original), and this is said in the context of a defence of his falsificationist methodology. But of course the mere assumption of the truth of a statement can never justify anything, for we can assume things we have no reason to believe are true. A test statement will justify us in regarding as false an hypothesis that clashes with it only if there is some justification for assuming that the test statement is true. What can such a justification be? Surely nothing other than the evidence of our senses. Ultimately it is experience that justifies us in saying that a theory has passed or failed an experimental test. In that case the notion of epistemic preferability is essentially connected with considerations about the experiences of perceivers, and these experiences are, of course, entities of World 2.

This may seem like an obvious point, and Popper himself emphasises the role of experiment in science. He even says that 'Only "experience" can help us to make up our minds about the truth or falsity of factual statements' (1972, p. 12; italics throughout in the original). But the question is this: how is the notion of experience integrated into Popper's methodology of theory appraisal? And the answer is that Popper strongly resists the idea that experience plays a justificatory role in deciding the merits of competing theories. Given my interpretation of his 'epistemology without a knowing subject' thesis, that is exactly what we would expect him to say.

Popper's fullest discussion of the relation of experience to theory is actually contained in his early work, *The Logic of Scientific Discovery*, first published in 1934. It is not too much to say, I think, that in this discussion Popper takes the step which leads naturally to his later theory of World 3. Let us examine Popper's argument.

6

In a section entitled 'Perceptual Experience as Empirical Basis: Psychologism', Popper describes psychologism as 'the doctrine that statements can be justified not only by statements but also by perceptual experience' (1934, p. 94). Popper rejects this view, holding instead that the only thing that can justify a statement is another statement. So it does seem that Popper is claiming (to put it in terms of his later theory) that when we assess the contents of World 3, we need not move outside World 3 itself. Notice that here Popper interprets 'justify' in a very strong sense – to establish as true. Thus he says, in criticism of psychologism so defined: 'we can utter no scientific statement that does not go far beyond what can be known with certainty "on the basis of immediate experience"' (*ibid*.). But this possibly correct observation is utterly irrelevant to the question before us. For it is Popper's contention that, in this strong sense, nothing justifies anything. In that case he ought to conclude not only that psychologism as he defines it is wrong, but that no statement can justify any statement. But Popper is telling us that it is statements *rather than* experiences that justify statements. 'Statements', he says, 'can be logically justified only by statements' (1934, p. 43, italics throughout in the original).

What, then, is the role of experience in Popper's methodology? Addressing psychologism once again, Popper says '... the decision to accept a basic statement ... is causally connected with our experiences – especially our *perceptual experiences*. But we do not attempt to *justify* basic statements by these experiences. Experiences can *motivate a decision*, and hence an acceptance or a rejection of a statement, but a basic statement cannot be *justified* by them – no more than by thumping the table' (1934, p. 105, italics in the original).

But if Popper's view of the relation between experience and the acceptance or rejection of basic statements is that the relation is wholly causal (and that is what he says) the following is a consequence of his view. Suppose that *e* is a basic statement inconsistent with some theory *T*. The question of whether it is rational to accept *e* and hence to reject *T* must, on Popper's view, be independent of any questions about what experiences anybody has had. Now the only way in which we can become aware of the outcome of an experiment is by having certain experiences. Experience may play little or no part in the experiment itself, but the result of the experiment can be known of only by looking at something, or in some other sensory way. But then we have the conclusion that whether it is reasonable to accept *e* and reject *T* is independent of any experiments that may be conducted. And this of course flies in the face of common sense and of Popper's own methodology, which insists on the importance of experiments for the testing of theories.[39]

Any reasonable methodology must respect the requirement that basic statements are to be accepted only if the evidence of our senses suggests that they are true; only, that is, if experience justifies our acceptance of the basic statement. And it is not relevant to insist (as Popper does) that basic statements are accepted only tentatively, that they themselves have the character of hypotheses, that we may at another time have cause to revise our opinion of the truth value of a basic statement. No doubt this is so; nothing is certain in this uncertain world. But the fact that experience cannot establish the truth of a basic statement should not blind us to the fact that the only justification there can be for accepting a basic statement is, in the end, the evidence of our senses. And if there can be no justification for a basic statement then what reason could there ever be for accepting or rejecting any scientific theory?

It may be objected that I mislocate the source of evidential support which can accrue to theories. Popper speaks sometimes as if the demand that experience play a justifying role in our decisions about the acceptance or rejection of basic statements is tantamount to the claim that the basic statements against which we test our theories are themselves reports of observation during test situations – of which something like 'It seems to me that the needle reads 0.5' would be a paradigm example. But in fact, so the objection goes, scientists do not test their theories against statements such as these, but rather against statements about events in the domain of entities under investigation; statements like 'the current in the wire was 0.5 amps'. These statements are highly theoretical, in the sense that they are arrived at by a process which relies heavily on assumptions about the truth of various theories which are highly conjectural but which are not currently under test.[40]

But I have not claimed that the basic statements relevant to a scientific theory must be statements which report the subjective experience of scientists. No doubt actual scientific practice exhibits a tendency to rely heavily upon highly theoretical statements in order to express the evidence for or against a theory. The statements that may be justified by experience are not merely those that report the experience itself. The statement that the current in the wire was 0.5 amps may well be justified by the experience of meter reading, given a background of beliefs, themselves justified, about the nature of the experimental set up. What I do claim is that there must be, or must be assumed to be, some connection (of a non-logical kind) between the accepted basic statement (of whatever degree of theoreticity) and some experience on the part of the experimenter which gives grounds for accepting the basic statement. If there were no such requirement we would be free to accept whatever statements about spatio-temporally singular events that we liked. We accept basic statements because we believe – provisionally – that they are true, and we believe that they are true because of some kind of relevant experience.

Now the kind of experience that is relevant must be equally available to all under the right circumstances. That is what the demand for the repeatability of experiments amounts to. In believing (or 'accepting') a basic statement, we do so because we believe that an experimenter had a certain kind of experience (produced by looking into the eye-piece of a telescope or examining a computer print out, *etc*) and because we believe that any other person in the same situation would have had an experience which was similar in the sense that it would have conferred equal epistemic weight on the basic statement. Thus in addition to the condition that there must be a connection between the experience and the basic statement, we must suppose that there is a causal chain leading from events in the external world to the experimentalist's having that experience, such that the causal chain is independent of idiosyncratic features of the experimenter's sensory apparatus. Experience is the crucial mediator between the world and the basic statement that describes it. The connection may go wrong in both directions. The experience may be non-veridical, in that it suggests to us a state of affairs holding in the world that does not in fact hold. And the experience may not, after all, justify us in saying that the hypothesis under test has false consequences. That is why our dependence upon a basic statement must always be provisional. But without the assumption that we have experiences that connect properly with events in the world and with basic statements, we can never have reason to accept any basic statement, even provisionally.

7

If Popper is wrong in the way I think I have shown him to be, then it is not a psychologistic mistake to think that the epistemological status of our theories depends upon the experiences of subjects, for it is not a mistake of any kind. It is a requirement that any reasonable epistemology must satisfy. But if we choose to excize this element from Popper's theory of World 3 then we may suspect that his theory represents little progress over Frege's. Popper tells us that there are objective, mind independent propositions, that cognitive states have an objective content, that questions of entailment and truth are independent of questions about the subjective world of experience and belief. These are all ideas that Frege insisted upon. Where Popper goes beyond Frege is (a) in arguing that World 3 displays an evolutionary development, and (b) in arguing that the subjective world of experience and belief is irrelevant to methodological acceptability. Both these ideas seem to me untenable. In this paper I have argued only against the latter.

I began this paper by remarking on the varieties of anti-psychologism, and the extent to which psychologism has turned out to be a shifting target. By now we have acquired ample evidence for this claim. The lesson is, I think, that anti-psychologism is a philosophical weapon that can easily get out of hand. Frege's target was, I have tried to show, a quite limited set of doctrines. His criticism of them were, on the whole, cogent and effective. The difficulties that Popper encounters in trying to extend the anti-psychologistic programme illustrate the danger of too facile an identification of the mental with the epistemologically irrelevant.

NOTES

* A draft of this paper was written in 1980. The first draft was read and criticized by David Stove; later ones by Alan Musgrave. I am grateful to them both. I have also benefited greatly from reading Max Deutscher's (1968) and Susan Haack's (1979–80). Popper's account of

basic statements is criticized by John Watkins in his (1984), Section 7.2, on lines very similar to those of my Section 6. I recommend Watkins' lucid account to the interested reader.

I use 'Thought' as a translation of Frege's '*Gedanke*', distinguishing it from 'thought' in the psychological sense. For our purposes we may take 'Thought' to be roughly synonymous with 'proposition'.

[1] For a defence of intentionalism in the arts see my (1988).

[2] See McDowell (1979), Section X. For clarification of this somewhat confused debate see Karen Green's interesting (1986).

[3] See Popper (1972), p. 127.

[4] See Dummett (1973), Chapter 19. See also Dummett (1981), Chapter 4.

[5] See e.g. Popper (1972), pp. 123–4.

[6] See especially Popper (1963), p. 70.

[7] See especially Popper *ibid*., Chapter 11.

[8] See e.g. *ibid*., Chapter 12.

[9] See my (1982) and (1983).

[10] See Culotta (1974), Gregory (1977) and Sluga (1980).

[11] See Frege (1884), Parts I and II.

[12] See e.g. Frege (1897), p. 142.

[13] See e.g. Frege (1892), p. 160 and (1918), p. 362.

[14] Frege (1893), p. 17.

[15] *Ibid*., p. 15.

[16] See e.g. Frege (1918), p. 367.

[17] See e.g. McDowell *ibid*., and Evans (1981).

[18] See e.g. Popper (1957), Section 27.

[19] See e.g. Popper (1972), Chapter 2.

[20] See e.g. Popper (1972), p. 156.

[21] See e.g. Frege (1918), pp. 352–3.

[22] See Popper (1972), p. 308.

[23] See my (1985).

[24] See e.g. Frege (1918) and Popper (1972), Chapters 3 and 4. See also my (1981).

[25] See e.g. Popper and Eccles (1977), p. 10.

[26] See Frege (1918). See also my (1981).

[27] On magical theories of reference see Putnam (1980).

[28] See e.g. Popper (1972), p. 127.

[29] Letter to Scholz, 1936. Quoted in *Gottlob Frege, Philosophical and Mathematical Correspondence* (Oxford: Basil Blackwell, 1980), p. 61.

[30] See Popper (1972), p. 1.

[31] See Dennett's review of Popper and Eccles (1977) (Dennett 1979). Mackie's commento n Popper sums up, I think, the reaction of many philosophers: 'I am repelled by his gratuitous sneers at most other philosophers, his constant employment of straw men and Aunt Sallys, his persistent name-dropping, his over-use of superlative terms – so many things are "incredible", "bold", "daring', or "beautiful" – his endless repetition of the same points, even the same phrases; and I often deplore his reluctance to do detailed work. "'For heaven's sake, cut the cackle", one is impelled to say when reading Popper's writings, and there is a great deal of cackle to be cut' (Mackie 1985, pp. 117–8). While Mackie also says that Popper's World 3 'deserves attention', he concludes that 'On examination, Popper's metaphysical fanfare dies away' (*ibid*., p. 127).

[32] See e.g. Frege (1918), p. 371.
[33] See e.g. Popper (1972), p. 120.
[34] See my (1979).
[35] See e.g. Popper (1972), p. 108.
[36] See *ibid*., pp. 25, 108–9, 111.
[37] See *ibid*., p. 298.
[38] See *ibid*., p. 25.
[39] On this see especially Watkins (1984), Section 7.2. Watkins points out that in a much later reply to criticism Popper admits that our experiences are not merely causally related to the acceptance of basic statements, but also provide 'inconclusive reasons' (Popper 1974, p. 1114). But Popper does not seem to realize the consequences of his admission for his epistemology.
[40] See Popper (1934), pp. 104–5.

REFERENCES

Culotta, C., 'German Biophysics, Objective Knowledge and Romanticism', *Historical Studies in the Physical Sciences* **4** (1974) 3–38.

Currie, G., 'Popper's Evolutionary Epistemology', *Synthese* **37** (1979) 413–31.

Currie, G., 'Frege on Thoughts', *Mind* **89** (1981) 234–48.

Currie, G., *Frege. An Introduction to his Philosophy*. Brighton: Harvester, 1982.

Currie, G., 'Interpreting Frege. A Reply to Michael Dummett', *Inquiry* **26** (1983) 345–61.

Currie, G., 'The Analysis of Thoughts', *Australasian Journal of Philosophy* **63** (1985) 283–98.

Currie, G., *An Ontology of Art*. London: Macmillan. Forthcoming, 1988.

Dennett, D., 'Review of Popper and Eccles' (1977), *Journal of Philosophy* **76** (1979), 91–97.

Deutscher, M., 'Popper's Problem of the Empirical Basis', *Australasian Journal of Philosophy* **46** (1968), 277–88.

Dummett, M. A. E., *Frege. The Philosophy of Language*. London: Duckworth, 1973.

Dummett, M. A. E., *The Interpretation of Frege's Philosophy*. London: Duckworth, 1981.

Evans, G., *The Varieties of Reference*. Oxford University Press, 1981.

Frege, G., 'Logic', 1879–91, in P. Long and R. White (trs.), *Posthumous Writings*. Oxford: Basil Blackwell, 1979.

Frege, G., *The Foundations of Arithmetic*, 1884. Translated by J. L. Austin. Oxford: Basil Blackwell, 1959. Revised edition 1980.

Frege, G., 'On Sense and Reference', 1892, in B. F. McGuiness (ed.), *Collected Papers on Mathematics, Logic and Philosophy*. Oxford: Basil Blackwell, 1984.

Frege, G., *The Basic Laws of Arithmetic*, 2 volumes, 1893. Partial translation by M. Furth. Berkeley and Los Angeles: University of California Press, 1967.

Frege, G., 'Logic', 1897, in *Posthumous Writings*.

Frege, G., '17 Key Sentences on Logic', 1906, in *Posthumous Writings*.

Frege, G., 'The Thought', 1918, in *Collected Papers*.

Green, K., 'Psychologism and Anti-Realism', *Australian Journal of Philosophy*, **64** (1986) 488–500.

Gregory, F., *Scientific Materialism in Nineteenth Century Germany*. Dordrecht: Reidel, 1977.

Haack, S., 'Epistemology with a Knowing Subject', *Review of Metaphysics* **33** (1979–90) 309–335.
McDowell, J., 'On the Sense and Reference of a Proper Name', *Mind* **86** (1977) 159–85.
Mackie, J., 'Popper's Third World – Metaphysical Pluralism and Evolution', in *Logic and Knowledge, Selected Papers Volume 1*. Oxford: Clarendon, 1985.
Popper, K. R., *The Logic of Scientific Discovery*, 1934. Revised edition, London: Hutchinson, 1968.
Popper, K. R., *The Poverty of Historicism*. London: Routledge and Kegan Paul, 1957.
Popper, K. R., *Conjectures and Refutations*. London: Routledge and Kegan Paul, 1963.
Popper, K. R., *Objective Knowledge*. Oxford University Press, 1972.
Popper, K. R., 'Replies to Critics', in P. A. Schillp (ed.), *The Philosophy of Karl Popper*. 2 volumes. La Salle: Open Court, 1974.
Popper, K. R. and Eccles, J. C., *The Self and its Brain*. Berlin and London: Springer, 1977.
Putnam, H., 'Models and Reality', *Journal of Symbolic Logic* **45** (1980) 464–82.
Sluga, H., *Gottlob Frege*. London: Routledge and Kegan Paul, 1980.
Watkins, J., *Science and Scepticism*. London: Hutchinson, 1984.

University of Otago

DAVID PAPINEAU

HAS POPPER BEEN A GOOD THING?

1. POPPER'S ANTI-REALIST TENDENCIES

According to *1066 and All That* (Sellars and Yeatman, 1930), Charles II was a Bad Man but a Good Thing. Gladstone, by contrast, was a Good Man but a Bad Thing. In this paper I want to argue that Sir Karl Popper is like Gladstone. He is a Good Philosopher of Science, but he has been a Bad Thing for the subsequent development of the subject.

I regard it as uncontentious that Popper has been a good philosopher of science. He has done as much to advance our understanding of such topics as probability, empirical content, explanation and observation as anybody else this century. And even those critics, like myself, who think that his reaction to the problem of induction is fundamentally misguided, will concede that his resulting views about scientific method have given rise to many important insights about theory-change in science.

But, even so, I want to show that aspects of Popper's views about scientific method have exerted an unfortunate influence on recent thinking about theory-change. In particular, I want to show that certain anti-realist strands in Popper's thinking have captured the minds of most post-war philosophers of science, and have prevented them from developing adequate responses to the relativistic writings of Kuhn and Feyerabend.

Perhaps it seems odd to accuse Popper of anti-realist tendencies. I shall explain in a moment. But first let me clarify the terminology. When I oppose 'anti-realism' to 'realism', I'm not referring to the 'realist–instrumentalist' debate about 'unobservables', but to the far more fundamental debate about the nature of success in judgement in general. In this debate, realism is the view that success in judgement consists in *correspondence* to an independent reality, and anti-realism the view that rejects such talk of external correspondence, and holds instead (with Hilary Putnam 1981, say, or Michael Dummett 1978) that success in judgement is simply an internal matter of ideally satisfying rational epistemological standards.

These last remarks might make it seem even odder that I accuse Popper of anti-realist tendencies. Doesn't Popper explicitly say that truth is a matter of correspondence, and that the aim of science is to develop true

K. Gavroglu, Y. Goudaroulis, P. Nicolacopoulos (eds.), Imre Lakatos and Theories of Scientific Change. 431–440.

theories? (See especially his 1963, pp. 225–8.) But this scarcely settles the issue. You can't qualify as a realist just by *saying* that you are 'aiming at truth in the sense of correspondence'. You need also to manifest this aim in the way you treat your theories. Your epistemological practice ought somehow to answer to your professed desire for truth. But, notoriously, there is no attempt in Popper's thinking to tie up the principles for choosing between scientific theories with the professed aim of having true theories.[1] Indeed we seem to end up with just the opposite link: Popper is quite explicit that good theories are bold and therefore likely to be false (see, for example, his 1963, p. 237).

I don't want to get bogged down in a terminological debate as to whether Popper is entitled to call himself a 'realist'. Perhaps his ontological conception of truth as correspondence is enough to warrant the term overall. On the other hand, when it comes specifically to epistemology, he is certainly a *de facto* anti-realist, for the professed aim of truth is quite irrelevant to his epistemological doctrines about which theories to accept. In any case, the terminology does not matter. What I want to show is that, whether or not we deem Popper to be an anti-realist, certain assumptions integral to his anti-realist epistemology have had a bad influence on subsequent research. I shall focus in particular on two issues: firstly, the question of what attitude is appropriate to scientific theories, and secondly, the question of how methodologies can be evaluated.

2. ATTITUDES TO SCIENTIFIC THEORIES

Belief is the intellectual attitude that answers to truth. To believe a theory is to hold it to be true. To hold a theory to be true is to believe it.

Popper says he is not interested in belief. He says belief is a subjective, psychological notion, and therefore irrelevant to the logic of science. (See, for instance, Popper 1972, p. 25.) But this is a bad argument. Belief is no more psychological or subjective than Popper's preferred attitude of acceptance (on which more shortly). Nor is there anything in the notion of belief as such to rule out the possibility of objective normative standards about what people ought and ought not to believe.

The real reason Popper refuses to concern himself with belief is surely to do with the connection between belief and truth. If scientific methodology were thought of as laying down rules for *believing* theories, then it would be obvious that there was something wrong with a methodology that selected theories that were likely to be false. For we can't sensibly set out to *believe* theories that we think are likely to be false.

Popper wants to uphold a methodology which favours improbable theories. So he has no option but to replace the obvious realist question – which theories are true and therefore to be believed? – with a new question – which theories are worthy of 'acceptance'?

Many philosophers, and in particular most contemporary philosophers of science, seem to find the Popperian notion of 'acceptance' quite unproblematic. But in fact it is a deeply obscure notion. What exactly is one supposed to do when one 'accepts' a theory? Clearly it's something less than believing the theory, for one is supposed to be able to accept an improbable theory. But if it's not belief, then what is it? Is it just a matter of exploring the theory, elaborating it, testing it, etc? But that doesn't seem right either, for one can do that to a number of contradictory theories simultaneously, but presumably one can't simultaneously 'accept' a number of contradictory theories.

Of course we are told *when* to 'accept' a theory: we should 'accept' a theory if it is the boldest among those yet proposed, and hasn't yet been falsified. But that's just to restate Popper's methodological prescription. It doesn't do anything to answer the prior question of what intellectual attitude that prescription is supposed to govern.

Maybe there is more to be said about the Popperian notion of acceptance. I have my doubts. But it doesn't matter. For whether or not the notion can be made coherent, the fact that Popper has persuaded his followers to think of theory-choice in terms of 'acceptance', rather than belief, has certainly stopped them making the right responses to the relativist challenges of Kuhn and Feyerabend.

Kuhn and Feyerabend in effect undermine Popper's methodology from within. They show that observation is impotent even to decide negative questions of theory acceptance directly. The central assumptions of scientific theories are always insulated from the experimental data by interpretative hypotheses. And there are questions about the authority of the experimental data themselves.

Lakatos's response was to point out that there are better and worse ways of saving central assumptions. Such assumptions can be saved 'progressively' or 'degenerately', depending on whether or not their preservation leads to successful new predictions. Lakatos suggested that we should continue to accept a theory only as long as it remains progressive in this sense (Lakatos 1970a).

But then there is the problem of the 'time lag'. Scientific theories, especially new theories like Copernicus's theory in the sixteenth century, often suffer extended periods of degeneration, only recovering into progress after decades, or even centuries. Lakatos himself was aware of this, and suggested that the rejection of degenerate theories should not be too precipitate (*op. cit*., p. 178). But as Feyerabend quickly observed, this gives it all back again. If there is no definite point at which degeneration forces rejection, then what is wrong with a persistent scientist continuing to accept an unsucessful theory *ad infinitum*? (Feyerabend 1970, p. 215).

I think the way out of this mess is to stop thinking in terms of Popperian acceptance. If we have in mind that we will go on 'accepting' a theory as long as we do not reject it, then it is scarcely surprising that the impossibility of conclusive falsification seems to threaten us with an untrammelled freedom for anybody to 'accept' what they like. But suppose we focus instead on the question of which theories ought to be *believed*. Then things come out differently. Failure to falsify a theory isn't yet any reason to *believe* it, especially when that failure involves a degenerating development of that theory. Belief in a theory requires positive evidence that it is more likely to be true (or approximately true) than its competitors (evidence such as would be yielded by some actual empirical progress).

I am suggesting here that early Copernicans weren't at that stage entitled to believe their theory. Given the lack of positive evidence then available, any such belief would have been irrational. It would have been perfectly rational, on the other hand, to *wonder* whether the theory might be true, and consequently to devote time and energy to finding out if it was. And, of course, when such pursuit of the new theory did turn up positive evidence, *then* it became rational to believe it.[2]

Acceptance doesn't answer to truth, but simply to the absence of established falsity. So the possibility that evidence which vindicates a theory may be uncovered in the future seems to justify its continued acceptance. But belief does answer to truth. And thus the mere possibility of future evidence doesn't justify present belief (though it may well justify trying to discover such evidence).

At first sight Popper's truth-independent epistemology can seem attractive. It enables you to by-pass the problem of induction without much need of honest toil. But it demands the notion of acceptance rather than the notion of belief. And adherence to this notion has stopped philosophers of science making the obvious responses to Kuhn and Feyerabend. I think we'd be much better placed to deal with relativism if we hadn't got on the Popperian bandwagon in the first place.

3. THE EVALUATION OF METHODOLOGIES

In *The Logic of Scientific Discovery* Popper says that the adoption of a methodology is a matter of *convention*. The preference for falsifiable but not yet falsified theories is recommended on the grounds that it yields a 'fruitful' *definition* of the term 'science'.

This approach obviously comes under pressure as soon as we are faced with active competition amongst competing methodologies. The writings of Kuhn and Feyerabend don't just threaten relativism at the first order level of scientific theories. They also threaten a second order relativism of methodologies: how are we to decide between the competing methodological recommendations put forward by Popper, and Kuhn, and Feyerabend, and Lakatos, etc?

In Sections 5–7 of his 'Replies to my Critics' (1974) Popper gestures towards a rather different idea. He picks out Galileo, Kepler, Newton, Einstein, and Bohr as examples of great scientists, and Freud, Adler and modern marxists as paradigm anti-scientists. And then he suggests that the worth of falsificationist methodology lies in the fact that it distinguishes the intellectual practice of the former figures from that of the latter.[3]

This proposal in effect views a methodology for science as a kind of empirical meta-theory, which can then be tested against the data base of what scientists actually do. This might seem an improvement on mere appeal to conventional decisions. But there are obvious difficulties facing this kind of 'naturalistic' approach to the evaluation of methodologies. Scientific methodology is prescriptive: it specifies how science *ought* to be conducted. It is hard to see how such prescriptive claims can be derived from *descriptions* of what actual figures in the history of science have done. Perhaps, it might be said, the data base is itself already prescriptive. After all, it consists of the activities of a list of 'great scientists'. But then there is the obvious danger of circularity: how can we tell who are the 'great scientists' unless we have already fixed on some idea of the prescriptive standards governing science? Compare Popper himself, back in *The Logic of Scientific Discovery*, p. 52, on the possibility of naturalism in methodology:

> And my doubts increase when I remember that what is to be called a 'science' and who is to be called a 'scientist' must always remain a matter of convention or decision.

There is a way of evading this circularity. It is implicit in some of Lakatos's remarks on the evaluation of methodologies, and seems to me to be the only coherent strategy for developing methodological naturalism. As I shall explain in a moment, however, the price of this strategy is abandoning all vestiges of scientific realism.

But first let me interject some positive remarks about the evaluation of methodology. It is easy enough to throw accusations of incoherence and anti-realism about. But what alternatives are there? In particular, does the kind of epistemological realism I favour leave us any better off on this issue? I want to show that it does, and that the reason Popper and his followers flounder so much on this issue is precisely because their epistemology makes no use of the concept of truth.

For an epistemological realist the evaluation of methodologies is in the first instance a quite straightforward matter. We want our methodological prescriptions to lead us to true beliefs. So we should select methodological prescriptions that do this. (We also want interesting truths, reasonably quickly. But truth is the *sine qua non*.)

How can we find out which prescriptions will lead us to truths? I don't see why there should be any special difficulty here. That certain beliefs are true, and others not, are natural facts within the natural world. And so the question of which methods generally lead to truths can be considered as a straightforward branch of applied empirical science. This branch of science would consider the various possible ways in which both individual humans and social institutions develop scientific theories, and investigate which of those ways generally lead to true theories. (For more on this see my *Reality and Representation*, 1987.)

This kind of realist evaluation of methods is itself a kind of naturalism, and to that extent akin to the Popper–Lakatos strategy for testing methodologies against the history of science. The realist also treats the question of whether certain methods are good for producing truths as an empirical matter, to be decided on the basis of empirical evidence. But precisely because the realist explicitly assumes that truth is what we want of our methods, there is nothing incoherent or circular in the realist's naturalism. The idea isn't that we should somehow distil prescriptions for science from facts about the practice of certain thinkers (while 'bracketing' the question of whether those thinkers' theories are true or not). Rather we simply look to the facts to see which methods are effective means to the unquestionably desirable end of true beliefs.

I know that many philosophers will feel that all this puts the cart before the horse. How can we possibly tell which theories are true, until we've solved the problem of epistemology? But I don't think that this is a real difficulty. It only arises within a traditional Cartesian epistemological perspective, which allows us knowledge of our own minds, and then requires us somehow to build up our knowledge of the rest of the world on that basis alone. I think that we should reject the idea of privileged introspective awareness, and take as our philosophical starting point the fact that we are natural beings within a natural world. And once we do this there is no longer any reason why our epistemology shouldn't be informed by the fact that certain beliefs represent the world correctly and others represent it incorrectly.[4]

Let me now return to Lakatos on the evaluation of methodologies. It is normally assumed that Lakatos, like Popper, is open to the charge of circularity. Thus in his *History of Science and its Rational Reconstructions*, pp. 110–111, Lakatos suggests that an intuitive list of scientifically acceptable historical episodes provides an initial yardstick against which to test a proposed methodology.[5] But later in that article Lakatos shows how the circular appeal to intuitions about proper science can be dispensed with.

Lakatos says that a 'methodological research programme' is entitled to revise the initial intuitive judgements about acceptable historical episodes, *provided* it does so in a 'progressive way' (*op. cit*., pp. 118–119). What Lakatos means, in this context, is that in order for a methodology to overturn the intuition that a certain episode was scientific, its supplementary 'external history' has to come up with a non-rational, causal explanation of why the scientists involved behaved as they did, an explanation in terms of political pressure, or ideological influences, or limitation of resources, or some such.

But this now means that the initial list of acceptable science becomes irrelevant. Clearly, for Lakatos, it is a decidable question whether or not a given episode in the history of thought is the result of non-rational, 'external' influences. So suppose we start with the whole history of human thought, and then simply eliminate those episodes that were due to external influences. The residue then provides the data base against which to test normative methodologies. The right methodology will be the one that succeeds in characterizing all that residue as scientifically rational. On this account, then, there is no need for any intuitive, prior delimitation of 'proper science'. The rational is separated from the irrational simply by excluding from the totality of human thought all episodes which resulted from external causes.[6]

Lakatos thus offers a way of avoiding a circular prior selection of 'scientific' episodes. But, even if this strategy avoids circularity, does it make any sense?

In effect this strategy commits us to the proposition that human beings are always *per se* rational, whenever they are free of corrupting external influences. The acceptability of this presupposition depends on how you think about the notion of rationality.

Recall my original characterization of realism. Realists think of the aim of judgement as correspondence to an independent reality. In consequence, realists will think of a given pattern of thought as being 'rational' just in case it arises from a habit of thought which generally gives rise to true beliefs. From this point of view there is no *a priori* reason whatever why the patterns of thought that human beings engage in when they are not subject to 'external' pressures should be rational. Producing beliefs that correspond to the external world is one thing. Coming naturally to human beings is another.

But now suppose that you reject the idea of truth as correspondence, and opt instead for the anti-realist view that success in judgement is a matter of conformity to rational epistemological standards. The obvious question which then arises is how to identify those rational epistemological standards. The realist way of selecting rational standards – as those which conduce to truth – clearly isn't available to the anti-realist.

But anti-realists can deny that there is a substantial problem here. They can argue that we feel a need to evaluate proposed standards of rationality only because we have in mind that there is some further end to which those standards are oriented – namely, the end of truth. But the existence of this independent end is precisely what anti-realism denies. So there is no need to evaluate standards of rationality. The fact that we human beings find ourselves moved by certain epistemological norms provides its own justification.

Some anti-realist philosophers would be inclined to let the matter rest there (I have in mind particularly the later Wittgenstein and his followers). But there is an obvious difficulty. Different human communities seem to observe different, and indeed often incompatible, epistemological standards. An anti-realist who wants an *objective* notion of success in judgement needs to say something further.

This is where the Lakatosian strategy can come into its own. It's not just any human beings whose intellectual inclinations are self-justifying. Some humans will have their thinking distorted by ideology, or politics, or perhaps by simple shortage of time and resources. So to identify the ideal rational standards for human beings we need to turn away from such cases of external distortion, and look at how humans think when they are free of such handicaps. It is the pattern of thought displayed by the human mind when it is free from external influences that is constitutive of rationality.

Lakatos's proposed strategy for evaluating methodologies makes little sense from a realist viewpoint. But if we reject any concern with truth, and look at things from an explicitly anti-realist perspective, then it seems perfectly cogent.[7]

Let me try to spell out the moral of this section. Because Popper and his followers don't think of methodologies as instruments for getting at the truth, they get into a mess on the question of how to evaluate methodologies: the appeal to the history of science is open to an obvious charge of circularity, and the only good way of avoiding this charge requires us to become whole-heartedly anti-realist about success in judgement. Again, it seems to me that we would do better to part company with Popper at the beginning, and bring the notion of truth explicitly into our epistemology.

NOTES

[1] There is, it is true, a kind of backhanded connection – Popper's epistemology is explicitly designed to reject false theories. But rejection is the easy part of the epistemology of science. On the more substantial question of acceptance, Popper's epistemology still floats quite free of his professed realism. (I here ignore the aberrant 'whiff of inductivism' passage in Popper's (1974, pp. 1192–3). This does argue for a link between methodology and truth – or, more precisely, for a probable link between methodology and high verisimilitude. But it runs counter to the rest of Popper's writings.)

[2] Cf. Laudan's distinction between appraisal for acceptance and appraisal for pursuit (1977, pp. 108–114). However Laudan is still thinking largely in Popperian terms. Although he equates acceptability with beliefworthiness on p. 110, his remarks about truth on pp. 125–127 make it very difficult to see how he can uphold this equation.

[3] See Nola, forthcoming, for a more detailed account of the evolution of Popper's views on these matters.

[4] To those that remain unhappy about taking facts about the natural world for granted in epistemology, let me just observe that contemporary philosophy of science of the kind I am criticizing is perfectly happy to assume facts about what past scientists thought, and indeed about *why* they thought those things, in deciding epistemological questions. These facts are surely far more shaky than, say, facts about the atomic theory of matter. (This tendency to assume historical and psychological facts about past scientific thought, but to 'bracket' physical or chemical facts about the natural world, is, I conjecture, a kind of Cartesian hangover. However, once we realize that the data for contemporary philosophers of science aren't facts about our own minds, but facts about what other people thought, a long time ago, we see that contemporary philosophy of science doesn't make much sense even from a Cartesian standpoint.)

[5] More precisely, he suggests that the intuitions of the scientific community can provide the initial yardstick. But this complication is scarcely much of an improvement on Popper's direct appeal to our own intuitions about good science. For Lakatos still needs some ungrounded intuitions to pick out the 'scientific community' in the first place.

[6] Laudan (1977) similarly takes it that rationality and sociology are mutually exclusive and together divide up the history of science. But for him this is a 'methodological principle, not a metaphysical doctrine' (p. 202), and so he doesn't see it as providing a way of overriding initial intuitions about the scientificity of particular episodes. For some suggestions as to how a realist should see the relation between rationality and sociology, see Papineau, forthcoming.

[7] The interpretation of Lakatos as an anti-realist derives from Hacking (1979). There are plenty of remarks in Lakatos's other writings that express a realist concern with truth. My aim has been to show that, even so, Hacking's interpretation makes perfect sense of Lakatos's approach to the evaluation of methodologies.

REFERENCES

Dummett, M., *Truth and Other Enigmas*, Duckworth, 1978.

Feyerabend, P., 'Consolations for the Specialist', in I. Lakatos and A. Musgrave (eds.) (1970).

Hacking, I., 'Imre Lakatos's Philosophy of Science', *British Journal for the Philosophy of Science*, **30** (1979).

Lakatos, I., 'Falsification and the Methodology of Scientific Research Programmes', in I. Lakatos and A. Musgrave (eds.) (1970).

Lakatos, I., 'History of Science and its Rational Reconstructions', in R. Buck and R. Cohen (eds.), *PSA 1970, Boston Studies in the History and Philosophy of Science VIII*, Reidel, 1970.

Lakatos, I. and Musgrave, A. (eds.), *Criticism and the Growth of Knowledge*, Cambridge University Press, 1970.

Laudan, L., *Progress and its Problems*, University of California Press, 1977.

Nola, R., 'The Status of Popper's Theory of Scientific Method', *British Journal for the Philosophy of Science*, forthcoming, 1987.

Papineau, D., *Reality and Representation*, Blackwell, 1987.

Papineau, D., 'Does the Sociology of Science Discredit Science', in R. Nola (ed.), *Relativism and Realism*, Reidel, forthcoming, 1988.

Popper, K., *Conjectures and Refutations*, Routledge and Kegan Paul, 1963.

Popper, K., *Objective Knowledge*, Clarendon Press, 1972.

Popper, K., 'Replies to My Critics', in P. Schilpp (ed.), *The Philosophy of Karl Popper, Book II*, Open Court, 1974.

Putnam, H., *Reason, Truth, and History*, Cambridge University Press, 1981.

Sellars, W. and Yeatman, R. *1066 and All That*, Methuen, 1930.

University of Cambridge

D.SFENDONI-MENTZOU

POPPER'S PROPENSITIES: AN ONTOLOGICAL INTERPRETATION OF PROBABILITY*

"There is no novel idea in modern and contemporary philosophy which could not ultimately be traced to ancient Greece".

This is not an introductory remark of an enthusiastic Greek philosopher, proud of his ancestors, but a conclusive comment of two prominent non-Greek philosophers, namely Eugene Freeman and Henryk Skolimowski, which goes on as follows:

> To find an anticipation of a novel thought or a doctrine is not difficult. The roots of all important doctrines are as long as our intellectual tradition. To find a *striking* anticipation . . . means that two thinkers quite independently arrived at very similar positions. Such is the case we believe with Peirce and Popper.[1]

Extending this line of thought towards a specific direction, which brings us to the field of probability, we believe we have a *striking* example of that coincidence that brings together two streams of thought, those of Peirce and Popper, revealing at the same time a common source of inspiration: Aristotle. We shall, thus, argue that there exist essential similarities between Popper's *propensity theory* and Peirce's *Tychism*, or his theory of absolute chance. The purpose of this comparison is not so much to suggest an answer to the already discussed, by many significant authors, problems of how much Peirce is contained in Popper's *propensities* or which of the two theories should be judged as more fruitful in an imaginary combat of "Peirce versus Popper."[2] Our main purpose is to shed light upon the *ontological features*[3] of Popper's propensity interpretation of probability, believing that these will be seen even more clearly if they be related both to Peircean Tychism and to the Aristotelian category of *potentiality* or "*δύναμις*".

Seen from this perspective, Popper's propensity theory will be examined mainly as an essential correlate of his indeterminism and his "*metaphysical research programme*".[4] Our interest will thus be focused not on the aspect of propensity revealed in the games of chance or even in Heisenberg's relations in an experimental set-up in the field of Q.M.,[5] but on the aspect related to Popper's *metaphysical* and *cosmological* inter-

K. Gavroglu, Y. Goudaroulis, P. Nicolacopoulos (eds.), Imre Lakatos and Theories of Scientific Change. 441–455.

pretation of the dynamic character of a growing and changing physical reality. In this case, we shall argue, Popper's propensity, interpreted as *potentiality*, bears the air not only of Peircean scientific metaphysics, but also – and primarily – of Aristotelian ontology, although Popper explicitly refuses to identify his propensity with the Aristotelian category of potentiality.[6]

The relative concepts that will serve as the back-bone of our discussion will be those of *probability* and *chance*, *determinism* and *indeterminism*, *possibility* and *potentiality*. Being aware of the fact that there is much confusion concerning the content applied to these terms, I must say that I proceed on the ground of a two-fold classification of their various interpretations: (i) the *epistemic* and (ii) the *ontological*, the first being taken as related to our knowledge, while the second as related to the structure of physical reality. Consequently, the first branch of the above classification is connected with the Humean type of *causality*, according to which there is no necessary connection between cause and effect. As this view attributes an *epistemic* character to causal relations rendering *causality* equivalent to *prediction*, it is inevitable that it leads to an identification of classical *determinism* with the concept of *causality*. All these characterize what Popper calls the "world of clocks".

On the other hand, the second branch is connected with the type of *ontological* causality, which expresses an inner connection between cause and effect. Such a connection is not restricted within the field of human knowledge, but it is embedded in natural processes. This type of causal connection has no need to be identified with *universal* (*Laplacean*) *determinism*, but only with a kind of *determination*, leaving at the same time room for *indeterminacy*. Such an indeterminism which characterizes Popper's "world of clouds" does not entail the abrogation of physical law. It represents a kind of determination which is based on the inner connection between cause and effect and is realized through the "process of actualization" or "the transition from possible to actual". In this context *possibility* does not have a logical or an epistemic meaning, but rather an *ontological* one, being a character of nature in which case possibility is identical with *potentiality*. Such is the case we believe with Popper's propensities and Peirce's *absolute chance*. Both thinkers aim at giving such an interpretation of physical reality that loosens the strictness of mechanical law leaving at the same time room for *freedom* and *spontaneity*.

But let's follow the story very briefly from the beginning. Popper started with a frequency interpretation of the type of Von Mises; he then passed to a reconstruction of his first theory, being led to the formulation of the propensity interpretation, which the author believes is a new contribution to the solution of the difficulties that arose in the field of Q.M. This theory[7] was further developed towards a specific direction in his Postscript.[8]

The essential features of propensity, as was first proposed, are the following Propensity refers to the probability of a *single event*. The single event possesses its probability owing to the fact that it is an event produced or selected, in accordance with the generating conditions. Consequently, a single event may have a probability even though it may occur only once, for it is a property of the *generating conditions* of the experimental arrangement. The probability statement about a die, say, asserts a *propensity*, *tendency* or *disposition* of the die to display certain stable frequencies in repeated trials, but also of the device and the situation. That is, propensity is attributed to the *whole experimental set-up*. "Propensities are properties of neither particles nor photons nor pennies. They are properties of repeatable experimental arrangements".[9] This idea also attributes to probability a character of being *generated* or *created* at the moment of observation. We cannot measure it as a result of our previous observations; it has a sort of *potential* existence which is actualized at the moment of our observation.[10]

Popper repeatedly emphasizes the fact that propensities are physically real. They are physical properties, abstract relational properties of the physical situation.[11] Nevertheless, he refuses to accept that they could be identified with *Aristotelian potentiality*:

> Like all dispositional properties, propensities exhibit a certain similarity to Aristotelian potentialities. But there is an important difference; they cannot, as Aristotle thought, be inherent in the individual *things*. They are not propensities inherent in the die, or in the penny, but in something a little more abstract, even though physically real: they are relational properties of the experimental arrangement.[12]

We shall argue that (i) this in fact is not a real difference but rather a difference in expression and (ii) Popper does not remain completely consistent in his view throughout all his writings. Although he insists on the fact that the environmental conditions are necessary, sometimes he gives more weight to the propensity of the photon or of the particle.

As Bartley remarks in a footnote of [1983] "When Popper first presented his propensity theory . . . Braithwaite compared Popper's idea of propensity, interestingly, with C. S. Peirce's 'would-be' or 'habit'; and ever since then Popper's theory has been attributed to Peirce".[13] And Bartley continues with the following remark:

But there are of course important differences. For example, just as a field of force may be physically present even when there is no (test) body on which it can act, so a propensity may exist for a coin to fall heads even though it falls only once, and on that occasion it shows tails. There may indeed be a propensity without any fall at all. The most important difference from Peirce's theory, stressed by Popper in 1957, is his relational theory of propensity.[14]

Such a comparison between the two theories, we believe, fails to take into account Peirce's Tychism or his theory of absolute chance[15] which it is our aim to consider in the discussion to follow.

In his article, "On the Doctrine of Chances", 1910, Peirce attributed a very peculiar characteristic to probability:

I am, then, to define the meaning of the statement that the *probability*, that if a die be thrown from a dice box it will turn up a number divisible by three, is one-third. The statement means that the die has a certain 'would-be'; and to say that the die has a certain 'would-be' is to say that it has a property quite analogous to any *habit* that a man might have.[16]

How should the concepts of 'would-be' and 'habit' be understood? It is important to notice that habit as a character of probability is defined not as an absolute but as relational property,

... and just as it would be necessary in order to define a man's habit, to describe how it *would lead him to behave and upon what sort of occasion*[17] ... so to define the die's 'would-be' it is necessary to say how it would lead the die to behave on an occasion that *would bring out the full*[18] consequence of the 'would-be'.[19]

As is obvious, Peirce refers to an idea analogous to Popper's experimental conditions. This means that although the 'would-be' is characterized as a property of the die – and not of the die + the conditions – this property comes out unless the necessary conditions appear. And here we have a striking similarity with Popper's characterization of propensity. Referring to the explanatory power of a monistic field theory, he points out that,

... it could explain, at best, only *the physical properties* of particles; or in other words, their *physical behaviour*; or still more precisely, their *disposition* or *propensity* to behave, under circumstances, in certain ways.[20]

It should be pointed out, of course, that Popper time and again emphasizes the fact that there is an equal weight both on the side of the die or the particle and of the experimental set-up, while on the other hand, Peirce insists on the fact that although the conditions are necessary for the actualization or the coming out of the 'would-be', the 'would-be' or disposition or potentiality is mainly a property of the die. Not denying the obvious fact that such points of disagreement exist, nevertheless we suggest that they represent rather a difference in expression and that there is a real ground for the above views to be reconciled, if the 'would-be' were considered in the context of Peirce's *ontological realism*, or his doctrine of the real existence of possibility. To put it in Peirce's words:

> ... the *will be's* the actually *is's* and the *have been's* are not the sum of all reals. They only cover actuality. There are besides *would-be's* and *can be's* that are real.[21]

That is, reality does not cover only *actuality*, but also *possibility* and *potentiality*, just as Popper claims for the reality of his propensities. Peirce's theory of potentiality is based on his doctrine of the three ontological categories, *Firstness* or possibility, *Secondness* or *actua*lity and *Thirdness* or law. It is through Thirdness or law, that the passage from possibility to actuality can be realized. And this can be accomplished owing to the fact that law is an expression of generality, or of a positive kind of generality "which belongs to conditional necessity"[22] and has a sort of *esse in futuro*. Consequently, law is itself an expression of a tendency towards generalization which is carried through a process of habit-taking.

This brings us to the more general meaning of habit, the analogy of human behaviour being extended to the laws of nature and denoting a *flexibility* for some variances in their application. In fact, it is this flexibility, this 'plasticity' in Popperian terms,[23] based on their *tendency to take habits* that gives them a liberty to operate not as *blind*, *rigid* laws, but as *dynamical* ones. This idea is based on Peirce's belief that *mind* and *matter* are not ultimately separated: We must regard matter as mind whose habits have become fixed so as to loose power of forming them or losing them, or in other words, "*matter is effete* mind".[24]

Law then expressed by the category of Thirdness is characterized by a finality, which at the same time leaves room for variation as regards the intermediate stages that lead to the accomplishment of the end. This is accomplished by the tendency towards habit-taking. This is the essential characteristic of human thought,[25] which consists in a "plasticity of habit".[26]

The doctrine of Thirdness is thus an assertion that the freedom of mind has its roots in the habit-taking character of the Universe itself, which is "radically different in its general form from mechanical law, inasmuch as it would at once cease to operate if it were rigidly obeyed".[27] Consequently, growth represents a "positive violation of law".[28] This violation of law, however, does not imply a complete irregularity or disorderliness.[29] Chance is thus used as the element which will break the absolute regularity of mechanical laws, so that a development of laws will be possible through the process of habit-taking, which is characterized as a tendency of nature itself, and the result will be statistical uniformity.[30]

On the ground of the above analysis, habit is intimately related to *chance*, *indeterminacy* and *potentiality*.[31] It is through these elements that reality as a whole, including the laws which constitute a part of reality is actualized through a dynamical process.

In this context, we believe, the problem of the reconciliation of Peirce with Popper regarding the *probability of the single event* can be solved. As far as we know, the efforts that have been made were based on Peirce's early formulation of his theory of probability and, of course, they were led to the conclusion that finally Peirce was not able to solve the problem of the probability of the *single event*.[32] However, we suggest that the probability of the single event is perfectly compatible with Peirce's Tychism.

For Peirce's Tychism, or his theory of Absolute chance, is an attempt to emphasize the dynamic character of nature, that can only be explained if the limits of the general law of causality are broadened so as to include the laws of probability and chance as well. It is through the concept of chance as expressing an element of *spontaneity*[33] or active potentiality in nature, that the phenomena of change, of generating processes, of diversification and of the appearance of novelty can be accounted for.

Such an interpretation, we believe, bears all the essential characteristics to be found in Popper's *cosmological* and *metaphysical* theory of *propensity* and *indeterminism*. But one can go further and trace a common spring of inspiration in the Aristotelian category of dynamis or potentia. It should also be added that both Peirce and Popper build their theories on a *realistic*, *objectivistic* and *anti-positivistic* view.[34]

In the introductory comments of (1982b), Popper emphasizes the fact that although this volume bears on problems of physics, it leads him to what he calls a 'Metaphysical Epilogue'. The reason is that his work "was inspired not so much by microphysical theory . . . but by the problems of physical cosmology."[35] This also appears clearly in his (1982a), where he states that the discussion will be on a cosmological plane; "I shall discuss the character of our world rather than the meaning of words".[36]

It is interesting to notice that Popper introduces a "*metaphysical research programme*" owing to the fact that he wishes to refer to the twofold character of his cosmological theories: their programmatic character, often shaping and determining the course of scientific research and development; and their unstable, and thus metaphysical character. Obviously, this kind of approach reveals a completely different attitude in his thought from the one presented in the Logic of Scientific Discovery. Being completely aware of the fact, he gives his reasons in (1982b):

> I no longer think, as I once did, that there is a difference between science and metaphysics regarding this most important point ... *as long as a metaphysical theory can be rationally criticized*, I should be inclined to take seriously its implicit claim to be considered, tentatively, as true.[37]

The similarity with Peirce's attitude towards metaphysics is striking:

> Find a scientific man, who proposes to get along without any metaphysics ... and you have found one whose doctrines are thoroughly vitiated by the crude and uncriticized metaphysics with which they are packed ... Every man of us has a metaphysics, and has to have one; and it will influence his life greatly. Far better, then, that that metaphysics should be *criticized* and not be allowed to run loose ... In short there is no escape from the need of a critical examination of 'first principles'.[38]

Now, the main target of Popper's attempt to give his own cosmological and metaphysical picture of the world is determinism of Newtonian Mechanics. As an alternative he offers his propensity interpretation. He claims that '*determinism* is completely baseless"[39] while on the other hand "*indeterminism* forces us to adopt the view that there can be no theory which completely predetermines all events ahead".[40] We try to catch the totality of the world with our nets, "But its mesh will always let some small fish escape: there will always be enough play for indeterminism".[41]

Laplacean doctrine (which Popper calls 'scientific determinism') is a position "with which most physicists though not all ... would have agreed at least prior to 1927".[42]

> Among the few dissenters was Charles Peirce ... He did not question Newton's theory: yet as early as 1892 he showed that his theory, even if true, does not give us any valid reason to believe that clouds are perfect clocks ... Thus Peirce conjectured that the world was not only ruled by *strict Newtonian laws*, but that it was also at the same time ruled by *laws of chance*, or of randomness, or of disorder: by laws of statistical *probability* ... I believe that Peirce was right in holding that all clocks are clouds, to some considerable degree ... I am an indeterminist – like Peirce, Compton, and most other contemporary physicists; and I believe, with most of them, that Einstein was mistaken in trying to hold fast to determinism.[43]

One of the main reasons that Popper was led to indeterminism was the fact that he regarded Laplacean determinism,

> as the most solid and serious difficulty in the way of an account of, and a defence of human freedom, creativity, and responsibility.[44]

This was also one of the main reasons that Peirce was led to the rejection of classical determinism or 'necessitarianism':

> ... by supposing the rigid exactitude of causation to yield ... we gain room to insert mind into our scheme.[45]

We now return to Popper's indeterminism as a corollary of his propensity interpretation, that leads him to a new world-picture, according to which,

> ... all propensities ... are dispositional, and the real state of a physical system, at any moment, may be conceived as the sum total of its dispositions – or its potentialities, or possibilities, or propensities.[46]

Such a view goes hand in hand with his ontological realism. For propensities are viewed as objective or real properties of the whole physical world.[47] A propensity is "a somewhat abstract kind of physical property",[48] nevertheless "*it can be kicked, and it can kick back*".[49] Consequently "they are not merely logical possibilities, but physical possibilities".[50] Furthermore, physical possibilities or potentialities are attributed sometimes by Popper not to the whole experimental set-up, but to particles, a fact that is a partial deviation from his thesis expressed in his early articles on propensity interpretation, and which brings him closer not only to Peirce but also to Aristotle.

In his effort to supersede the dualism of matter and field without sacrificing any advantages it may have for the treatment and solution of physical problems, Popper suggests that we consider quantum theory as a particle theory[51] and claims that,

> ... waves ... are mathematical representations of *propensities*, or of dispositional properties, of the physical situation ... interpretable as propensities of the particles to take up certain states.[52]

His theory, thus, is a theory of matter, or of particles, according to which they are considered as realizations of potentialities.[53] Popper gives particular emphasis to the process of *actualization* or of the *transition from possible to actual*. One of the forerunners of the propensity interpretation, according to him, is Heisenberg who pointed out that the concept of probability was closely related to the concept of potentiality of the natural philosophy of Aristotle:

it is to a certain extent, a transformation of the old 'potentia' concept from a qualitative to a quantitative idea.[54]

This idea, Popper argues, does not fit in the context of the Orthodox interpretation, but it is very essential to his own world view, in which *change* "consists in the realization or actualization of some of these potentialities".[55] It is evident here, we believe, that Popper arrives at a view of the world of change and growth very similar both to Peircean Tychism, and to Aristotelian Metaphysics.

In Book IX of his Metaphysics Aristotle starts with an attempt to analyse Being (*ὄν*).

And since the senses of being are analysable not only into substance or quality or quantity, but also in accordance with potentiality and actuality and function, let us also gain a clear understanding about potentiality and actuality . . . the *potentialilties* which conform to the same type are all *principles*, and derive their meaning from one primary sense of *potency, which is the source of change*[56] in some other thing, or in the same thing *qua* other.[57]

But should we admit with Popper that *potentia*, according to Aristotle belongs only to a thing independently of any other relevant conditions? Here is Aristotle's answer:

We must, however, distinguish when a particular thing exists potentially, and when it does not; *for it does not so exist at any and every time*. E.g. is earth potentially a man? No, but rather when it has already become semen, and perhaps not even then . . . And in all cases where the generative principle is contained in the thing itself, one thing is potentially another when, *if nothing external hindres* (*μηθενός τῶν ἔξωθεν ἐμποδίζοντος*)[58] it will of itself become the other. E.g., the semen is not yet potentially a man; for it must further undergo a change in some other medium. But when by its own generative principle, it has already come to have the necessary attributes, in this state it is now potentially a man, whereas in the former state it has need of another principle.[59]

On the ground of the above evidence, we believe that the distinction made by Popper between his propensities and Aristotle's concept of dynamis, does not essentially exist.

The similarity between the two theories can be further established by Popper's explanation of the phenomenon of change: ". . . we may describe the physical world, says Popper, as consisting of *changing propensities for change*".[60]

An analogous reference to change and growth is also made by Peirce; "Everywhere the main fact is growth and increasing complexity".[61] From this Peirce infers that there must be in nature some agency by which the complexity and diversity of things can be increased. This agency is absolute chance or spontaneity[62] by which the *sui generis* can be accounted for. And here is Popper:

> . . . the transition from the possible to the actual takes place whenever a new state of the world emerges; . . . as long as there is any change, it will always consist in the actualization of certain potentialities.[63]

This brings us to the last essential point raised by Popper in support of the reality of the process of actualization expressed by propensities which cannot fit the rigidity of mechanical law, i.e. biological phenomena.[64] Thus, the aim of Popper's "*indeterministic re-interpretation* of Einstein's programme" as well as the "*objectivistic* and *realistic re-interpretation*[65] of quantum theory", "is a picture of a world in which there is room for biological phenomena, for human freedom and human reason".[66] Popper believes that "Biologists have always worked with propensities" and he emphasizes the fact that "The term 'propensity' itself is, of course, a biological or psychological term".[67] Biological organisms, with their character of far-reaching autarky are endowed with "inherent propensities" which although they are relational ". . . yet, they do resemble Aristotle's inherent potentialities of a thing more than other physical or biological propensities".[68]

Finally, in his "Metaphysical Epilogue", he summarizes his metaphysical programme in the following words:

> Everything is a propensity. Or in the terminology of Aristotle we might say: To be is both to be the actualization of a prior propensity to become, and to be a propensity to become.[69]

NOTES

* An earlier draft of the present paper was written in 1977 while I was studying in Oxford for my doctoral thesis on *C. S. Peirce's Philosophy of Probability and Chance*. This earlier draft was read by R. Harré and W. Newton-Smith. I am indebted to both of them for their fruitful critical remarks. I am also grateful to J. Cohen for valuable discussions.

[1] E. Freeman and H. Skolimowski (1974), pp. 508–9. This contribution consists in three parts: (I) 'Charles Peirce and Objectivity in Philosophy', pp. 464–482, written by E. Freeman; (II) 'Karl Popper and the Objectivity of Scientific Knowledge', pp. 483–508, written by H. Skolimowski; (III) 'Peirce and Popper. Similarities and Differences' pp. 508–515, written in common by both authors. The above joint work provides a very interesting and illuminating comparison of the philosophies of Popper and Peirce in the field of Epistemology and Philosophy of Science. For a profound, highly critical discussion of the

dynamics of Scientific Knowledge in Peirce and Popper see also I. Niiniluoto (1984) especially Chapter 3, 'Notes on Popper as Follower of Whewell and Peirce' pp. 18–60 and Chapter 4, 'The Evolution of Knowledge', pp. 61–110.

[2] See for example, A. White (1972), R. Miller (1975), T. Settle (1974) and (1977).

[3] Not following Popper's choice for the term 'metaphysical' we prefer to use the term 'ontological', as the most appropriate characteristic of his propensity interpretation. Nevertheless, we believe that Popper would not have any objection to our preference, if we consider a footnote of (1982a), p. 7 added in 1981, where he confesses: "... I avoided, or tried to avoid, the term 'ontology' in this book and also in my other books; especially because of the fuss made by some philosophers over 'ontology'. Perhaps it would have been better to explain this term and then use it, rather than to avoid it. However that may be, questions of terminology are never important."

[4] Proposed by Popper in (1982b).

[5] In his numerous writings concerning the interpretation of probability in Quantum mechanics, Popper repeatedly emphasizes the fact that the 'propensity interpretation' is the result of his effort to solve the problem of the role of probability in Q.M., "... What I propose is a new physical hypothesis (or perhaps a metaphysical hypothesis) analogous to the hypothesis of Newtonian forces. It is the hypothesis that every experimental arrangement (and therefore every state of a system) generates physical propensities which can be tested by frequencies. This hypothesis is testable, and it is corroborated by certain quantum experiments. The two-slit experiment for example, may be said to be something like a critical experiment between the purely statistical and the propensity interpretation of probability and to decide the issue against the purely statistical interpretation" Popper (1959), p. 38; See also *ibid*., p. 28, (1963), p. 60 and (1982b), p. 151. However, the aim of the present paper is not to discuss the question of Popper's particular contribution to the explanation of Quantum phenomena, or the question of how much light has been shed by the propensity interpretation on the two-slit experiment, or finally which solution should be judged more preferable after all, 'Copenhagenism' or 'Popperism'. All these problems have been fruitfully discussed for example by H. Krips (1984), P. Suppes (1974) and P. Milne (1985). Without ignoring the significance of the above problems, the issue to be discussed here is the essentially metaphysical and ontological character of the propensity interpretation, which is intimately connected – through the category of potentiality – with Peirce's doctrine of absolute chance and, what is more interesting, with Aristotle's ontology. The assumption of the real existence of potentiality in physical phenomena is a very attractive one and we believe that there are many valid arguments to be found not only in the writings of Aristotle, Peirce and Popper, but also in recent developments of science. However, the question as to whether the idea of *potentiality* defended by Popper's propensity theory can find a real place in the field of Q.M. cannot be discussed in this paper.

[6] See for example, Popper (1959).

[7] Presented in Popper (1950, 1957, 1959, 1967).

[8] Popper (1982a, 1982b and 1983).

[9] Popper (1967), p. 38; See also Popper (1957, pp. 67, 68, (1967), pp. 81, 2, 3, 4; (1982b), p. 81.

[10] See for example, "But we do interpret probability measures, or 'weight' attributed to the possibility as measuring its disposition, or tendency or propensity to realize itself." Popper (1959), p. 36; see also *ibid*., p. 38.

[11] "It was this last point, the interpretation of the two-slit experiment, which ultimately led

me to the propensity theory: it convinced me that probabilities must be 'physically real' . . . not only in the sense that they could influence the experimental results, but also in the sense that they could, under certain circumstances (coherence), interfere, i.e. interact with one another". Popper (1959), p. 28; see also (1967) pp. 83–4.

[12] Popper (1959), pp. 37–8. See also (1982a), pp. 95 and 105.

[13] See for example, I. Hacking (1976), D. H. Mellor (1971), A. White (1972), R. Miller (1975) and Gillies (1973).

[14] Popper (1983), p. 282.

[15] A. White (1972), R. Miller (1975) and E. Madden (1964) are some examples among current writers of those who choose not to take into consideration Peirce's ontological theory of chance. We have earlier discussed this problem and given an analysis of Peirce's Tychism in our (1980).

[16] C.S.P. (2.664). All references in this paper follow the standard practice of designating volume and paragraph in the *Collected Papers of Charles S. Peirce* (1931–58).

[17] Our italics.

[18] Our italics.

[19] C.S.P. (2.664). This idea is also expressed in several other occasions, when Peirce is analyzing the concept of probability as frequency *in the long run*. He time and again refers to the genus/species relation: "In order then, that probability should mean anything, it will be requisite to specify to what species of event it refers and to what genus of event it refers" (5.21). And in another attempt to clarify what he means by probability he remarks, "But to speak of the probability of the event B, without naming the conditions, really has no meaning at all" (2.651). All this has been very explicitly analysed by W. Miller (1973).

[20] Popper (1982b), p. 196.

[21] C.S.P. (8.216).

[22] *Ibid*. (I.427).

[23] See for example Popper (1972), p. 248.

[24] C.S.P. (6.25).

[25] See *ibid*. (6.60), (6.268).

[26] *Ibid*. (6.86).

[27] *Ibid*. (6.613).

[28] *Ibid*.

[29] See *ibid*. Compare also (7.136). This very important idea is stressed also by Popper. See for example, Popper (1982b) p. 72, n. 64.

[30] See (6.97); also (7.514).

[31] See also: "This habit is a generalizing tendency, and as such a generalization, and as such a general, and as such a continuum or continuity. It must have its origin in the original continuity which is inherent in potentiality." (6.204).

[32] R. Miller for example, in his (1973) emphasizes the fact that Peirce explicitely claims that probability referring to a single event has no meaning at all. (see 2.652), while on the other hand Popper has always been emphatic in underlying the fact that his propensity theory mainly serves as a theory of the probability of a single event. The closest that Peirce seems to have come, to an application of probability to singular events, Miller notes, is found in a draft of his report to the National Academy of Science in 1901 in which he discussed the logic of drawing inferences about history from ancient documents and monuments. But even in this case Peirce is seen to be defending again the long run type of probability. (see pp. 126–7). So, Miller's conclusion as to the question whether Peirce can be reconciled with Popper on the

problem of the probability of a single event is explicitly negative (p. 131). The same problem is also discussed by Madden (1964). The author makes an effort to reconcile Peirce's frequency theory of probability with the probability of a single event on the ground of a case given by Peirce in 'The Doctrine of Chances', 1878 (2.652) where the frequency theory of probability is connected with the *community view* of truth. But Madden arrives at the conclusion that "on this ground Peirce's view of the single event is unacceptable because it misses the central point of the problem" (pp. 127–8).

[33] See C.S.P. (1.161).

[34] For an illuminating analysis of the relation between Popper's realistic attitude and his theory of indeterminism see J. W. N. Watkins (1974). See also Popper's own claim in his (1982b), p. 175.

[35] Popper (1982b), pp. 30–31.

[36] *Ibid.* (1982a), p. XXI.

[37] *Ibid.* (1982b), p. 199.

[38] C.S.P. (1.129).

[39] Popper (1982a), p. 41.

[40] *Ibid.* (1982b), p. 182. Compare C.S.P. (6.46), (6.48).

[41] Popper (1982a), p. 47.

[42] *Ibid.*, p. XX.

[43] *Ibid.* (1972), pp. 212 ff.

[44] *Ibid.* (1982a), p. XX.

[45] C.S.P. (6.61).

[46] Popper (1982b), p. 159.

[47] *Ibid.*, p. 87.

[48] It should be noticed that in (1982b), p. 71, Popper stresses the fact that although he has always been emphatic in underlying the role of the 'experimental arrangement' or 'set-up', he nevertheless meant "a real objective situation", which is a part of the physical world. As an answer to Feyerabend's article (1968) where it is concluded that his theory is essentially equivalent to Bohr's, Popper remarks: "this is taking as the basis of one's criticism an almost accidental formulation, disregarding the context, and all I have written on propensities. It is clear from my earlier work and also this book ... that propensities, as I call them, are completely objective and dependent *not* on *our* set-up of experiments, but simply on the physical situation which, in certain cases, may be experimentally controlled".

[49] Popper (1982b), p. 72.

[50] *Ibid.* (1982a), p. 105.

[51] *Ibid.* (1982b), p. 196; see also *ibid.* p. 138.

[52] *Ibid.*, p. 126. Compare also (1982b), p. 127; "... the waves describe dispositional *properties* of the particles"; "Thus we assume, with Dirac, that a particle approaching a polarizer has a certain propensity to pass it, and a complementary propensity not to pass it. (It is the whole arrangement which determines these propensities, of course)". (p. 190).

[53] *Ibid.*, p. 160.

[54] See *ibid.*, pp. 132–3; see also pp. 136, 78, 124, 186.

[55] *Ibid.*, p. 159.

[56] Our italics.

[57] Aristotle, Met. IX, 1045b 33ff–1046a 9ff.

[58] Our italics.

[59] Aristotle, Met. IX, 1048b 43–1049a Iff.

[60] Popper (1982b), p. 160.
[61] C.S.P. (6.58).
[62] *Ibid*. (6.59); see also: "I don't know what you can make out of the meaning of spontaneity but newness, freshness, and diversity" (1.160).
[63] Popper (1982b), p. 185; see also (1982a), p. 145.
[64] See *ibid*. (1982b), p. 208; see also p. 90.
[65] Our italics.
[66] Popper (1982b), p. 160.
[67] *Ibid*., 209.
[68] *Ibid*.
[69] *Ibid*., p. 205.

REFERENCES

Aristotle, *Metaphysics*, Loeb edition. Translated by H. Tredennick. Cambridge, Mass: Harvard University Press, 1968.

Freeman, E. and Skolimowski, H., 'The Search for Objectivity in Peirce and Popper', in P. A. Schilpp (ed.): *The Philosophy of Karl R. Popper*. Open Court, 1974, pp. 464–519.

Feyerabend, P. K., 'On a Recent Critique of Complementarity. Part I', *Philosophy of Science* **35** (1968) 309–331.

Freundlich, Y., 'Copenhagenism and Popperism', *British Journal for the Philosophy of Science* **29** (1978) 145–177.

Gillies, D. A., *An Objective Theory of Probability*. Methuen, 1973.

Hacking, I., *The Emergence of Probability*. Cambridge University Press, 1976.

Krips, H., 'Popper, Propensities and Quantum Theory', *British Journal for the Philosophy of Science* **35** (1984) 253–292.

Madden, E. H., 'Peirce on Probability', in E. Moore and R. Robin (eds), *Studies in the Philosophy of C. S. Peirce*. The University of Massachusetts Press, 1964, pp. 122–140.

Miller, R. W., 'Propensity: Popper or Peirce?', *British Journal for the Philosophy of Science* **26** (1975) 123–132.

Milne, P., 'A Note on Popper, Propensities, and the Two-slit Experiment', *British Journal for the Philosophy of Science* **36** (1985), 66–70.

Peirce, C. S., *The Collected Papers of Charles Sanders Peirce*. A. W. Burks, C. Hartshorne, P. Weiss (eds). Belknapp Press of the Harvard University Press, 1931–58.

Niiniluoto, I., *Is Science Progressive?* Reidel, 1984.

Popper, K. R., *The Logic of Scientific Discovery*. Hutchinson, 1934. Revised edition, 1972.

Popper, K. R., 'Indeterminism in Quantum Physics and Classical Physics: Part I, II', *British Journal for the Philosophy of Science* **1** (1950) 117–33 and 173–95.

Popper, K. R., 'The propensity Theory of the Calculus of Probability and the Quantum Theory', in S. Körner (ed.): *Observation and Interpretation*. Butterworth, 1957.

Popper, K. R., 'The Propensity Interpretation of Probability', *British Journal for the Philosophy of Science* **10** (1959), 25–42.

Popper, K. R., 'Quantum Mechanics without "The Observer"', in Mario Bunge (ed): *Quantum Theory and Reality*. Springer-Verlag, 1967, pp. 7–44.

Popper, K. R., *Objective Knowledge; An Evolutionary Approach*. Oxford University Press, 1972.

Popper, K. R., 'Replies to my Critics', in P. A. Schilpp (ed.): *The Philosophy of Karl Popper*. Open Court, 1974, pp. 964–1197.

Popper, K. R., *The Open Universe: An Argument for Indeterminism*. Vol. II of the Postscript to the Logic of Scientific Discovery. Hutchinson, 1982a.

Popper, K. R., *Quantum Theory and the Schism in Physics*. Vol. III of the Postscript to the Logic of Scientific Discovery. Hutchinson, 1982b.

Popper, K. R., *Realism and the Aim of Science*, Vol. I of the Postscript to the Logic of Scientific Discovery. Hutchinson, 1983.

Settle, T., 'Induction and Probability Unfused', in P. A. Schilpp (ed): *The Philosophy of Karl Popper*. Open Court, 1974, pp. 697–749.

Settle, T., 'Popper versus Peirce on the Probability of Single Cases', *British Journal for the Philosophy of Science* **28** (1977) 177–180.

Sfendoni-Mentzou, D., *C. S. Peirce's Theory of Probability and Chance* (Doctoral Thesis, in Greek) University of Thessaloniki, 1980.

Suppes, P., 'Popper's Analysis of Probability in Quantum Mechanics', in P. A. Schilpp (ed.): *The Philosophy of Karl Popper*. Open Court, 1974, pp. 760–774.

Watkins, J. W. N., 'The Unity of Popper's Thought', in P. A. Schilpp (ed.): *The Philosophy of Karl Popper*, Open Court, 1974, pp. 371–412.

White, A. R., 'The propensity Theory of Probability', *British Journal for the Philosophy of Science* **23** (1972), 35–43.

Aristotle University of Thessaloniki

INDEX